Re-envisioning the Anthropocene Ocean

Re-envisioning the Anthropocene Ocean

Edited by
Robin Kundis Craig & Jeffrey Mathes McCarthy

THE UNIVERSITY OF UTAH PRESS
Salt Lake City

The Defiance House Man colophon is a registered trademark of the University of Utah Press. It is based on a four-foot-tall Ancient Puebloan pictograph (late PIII) near Glen Canyon, Utah.

Library of Congress Cataloging-in-Publication Data

Names: Craig, Robin Kundis, editor. | McCarthy, Jeffrey Mathes, editor.
Title: Re-envisioning the Anthropocene Ocean / Robin Kundis Craig, Jeffrey Mathes McCarthy.
Salt Lake City : University of Utah Press, [2023] | Includes bibliographical references and index.
Identifiers: LCCN 2022950988 | ISBN 9781647691004 (hardback) | ISBN 9781647691011 (paperback) | ISBN 9781647691028 (ebk)
LC record available at https://lccn.loc.gov/2022950988

Errata and further information on this and other titles is available online at UofUpress.com.

Cover art by collage artist, Deborah Shapiro.

Printed and bound in the United States of America.

Contents

Dedicated to those who work toward a better future.

CHAPTER 1

Introduction

Why Re-envision the Anthropocene Ocean?

ROBIN KUNDIS CRAIG AND JEFFREY MATHES MCCARTHY

It is time to take care of the oceans as if our lives depended on it, because they do.

—Sylvia Earle, Marine Biologist

THE PREMISE OF THIS VOLUME is simple: All humans depend intimately on the ocean, and we need to be more cognizant of that fact. Likewise, this is a perilous moment for the world's ocean, and so we need to re-envision our relationship to it.

The list of concerns is familiar—the warming climate changes currents and habitats; acidification threatens reefs and shellfish; overfishing damages marine species and marine ecosystems and destabilizes human food security; plastic fills every nook of sea; oil spills contaminate gulfs and estuaries; and agricultural nutrient runoff spurs ocean dead zones that choke all life. These changing baselines impact land as well as sea. The tides are high enough to drench Venice and Boston and Tuvalu. In tropical latitudes, rowdier storms threaten Miami and San Jose as well as Chittagong and Mumbai. Meanwhile, across all longitudes coral reefs and their fish are dying faster than the people they feed can find other food, which leads to more aggressive fishing for fewer fish.

Careful analysis of these challenges has come from the province of science, where weights and measures and calculations lead to stern findings. However, our volume is informed by the humanities and social sciences, too. In the humanities, nuance and story measure the ocean's shining expanse. Herman Melville tells us that humans are natural "water gazers," drawn to sights of the sea: "Circumnavigate the city of a dreamy Sabbath afternoon. . . . Posted like silent sentinels all around the town, stand thousands upon thousands of mortal men fixed in ocean reveries. . . . [A]s everyone knows, meditation and water are wedded forever."[1] There's a joy to encountering the sea, and people are drawn to that view. Melville continues

Figure 0.1. The World's Ocean. NOAA National Geophysical Data Center. 2009: ETOPO1 1 Arc-Minute Global Relief Model. NOAA National Centers for Environmental Information, https://www.ngdc.noaa.gov/mgg/global/.

with a description of the ocean's glorious pull on human viewers: "But look! Here come more crowds, pacing straight for the water, and seemingly bound for a dive. Strange! Nothing will content them but the extremest limit of the land. . . . They must get just as nigh the water as they can without falling in."[2]

In the twenty-first century, we can identify with this craving for the sea, and yet we know we have reason to be troubled. Today we look with unease at an expanse no longer comforting, no more an obedient partner for trade, not the fecund field our nets ploughed. Today the meditator on water looks seaward in time to revalue the ocean. This volume unites diverse thinkers pursuing the worthy goal of looking again at the ocean, appreciating it, recognizing the challenges of a changing climate, and addressing the ways in which Melville's ocean has lately been changed.

* * *

Humans have long viewed the ocean as a place of mystery, otherworldliness, change and transformation, and danger—but *not* as a complex adaptive system that makes life as we know it possible. Consider Shakespeare's shipwreck story, *The Tempest.*

> Full fathom five thy father lies;
> Of his bones are coral made;

> Those are pearls that were his eyes:
> Nothing of him that doth fade,
> But doth suffer a sea-change
> Into something rich and strange.

The verses tell us that the ocean changes those humans brave enough to venture out upon it. Transformation is the essence of ocean experience. Four hundred years along, transformation is an ocean theme preserved in works as diverse as Herman Melville's *Moby Dick* through Disney's recent *Pirates of the Caribbean* movies. In works like Kipling's *Captains Courageous* (1897) or Jack London's *Sea Wolf* (1904), ocean experience improves the immature and strengthens the weak. This romantic transformation dramatizes people enriched by wild, watery nature.

And yet, the very wildness that redeems some literary characters also threatens readers with monsters, menacing abominations from the deep (and the quarterdeck). The ocean is a perpetual source of monsters, both supplying its own terrifying creatures and creating monsters out of the humans it shapes, as in *Moby Dick*, Jules Verne's *Twenty Thousand Leagues Under the Sea*, and Peter Benchley's *Jaws*. In the words of Werner Herzog, "What would an ocean be without a monster lurking in the dark? It would be like sleep without dreams."[3] Not accidentally does Frankenstein's creature escape to the Arctic Ocean at the end of Mary Shelley's novel, one purpose of which is to ask: who is the bigger monster, the scientist Frankenstein or his creation? The literary ocean is *faerie*, and a journey there is often a journey into the subconscious and unconscious, forcing us to face the darker sides of human nature.

* * *

But there *are* pearls in those depths. The economic, political, and legal ocean beckoned the brave and the strong with risky but undeniable wealth—spices, fish, whales, seals, and of course new lands. In Europe's Age of Exploration, "freedom of the seas" became enshrined in international law, a rule of plunder that did not change significantly until the latter half of the twentieth century and which stands poised to change again through the proposed Biodiversity Beyond National Jurisdiction (BBNJ) Convention that the United Nations is proposing. Legal freedom of the seas coupled with an economic and political paradigm of inexhaustibility—the assumption that humans were too small to meaningfully change the ocean—promoted waves of reckless exploitation, from whales and seals to cod and tuna, and hundreds more species besides.

It took a bit of time for the scientific establishment to catch up with the ocean's size and variety—and, indeed, it is still catching up. While early submarines were used in the American Revolution, and Robert Fulton built his submarine *Nautilus* in 1800, modern oceanography is only about 130 years old and dates to the United States, Britain, and European nations sending out a few expeditions to measure and catalog ocean currents, marine life, and coastlines and seafloors at the end of the nineteenth century.[4] According to the Woods Hole Oceanographic Institution website, "The first scientific expedition to explore the world's oceans and seafloor was the *Challenger* Expedition, from 1872 to 1876, on board the British three-masted warship *HMS Challenger*."[5] It was World War II, however, that really pushed the discipline of oceanography forward. As for individual exploration gear, diving bells, diving suits, and diving helmets evolved over centuries, but Émile Gagnan and Jacques-Yves Cousteau invented the self-contained underwater breathing apparatus—SCUBA—in 1942, and patented the first modern demand regulator in 1943.

Today, fifteen million people identify as snorkelers and SCUBA divers. They testify to the tremendous popular interest in the ocean, even as ocean coasts, sea life, and even sea bottoms cope with tremendous pressure. This contrast between expanding ocean use and diminishing ocean health presents both opportunity and crisis for oceanographers and marine biologists.

Nevertheless, for all the swimmers and scientists splashing about, knowledge of the ocean remains limited. NASA oceanographer Gene Feldman noted in 2009 that "even with all the technology that we have today—satellites, buoys, underwater vehicles, and ship tracks—we have better maps of the surface of Mars and the moon than we do the bottom of the ocean. We know very, very little about most of the ocean."[6] The international Census of Marine Life lasted from 2000 to 2010 and involved more than 2,700 marine scientists seeking to uncover the full extent of marine biodiversity. By the end of the decade-long study, these scientists had

> found and formally described more than 1,200 new marine species, with another 5,000 or more in the pipeline awaiting formal description. They discovered areas in the ocean where animals congregate, from white shark cafés in the open ocean to an evening rush hour in the Mid-Atlantic Ridge to a shoal of fish the size of Manhattan off the coast of New Jersey, USA. They unearthed a rare biosphere

> in the microbial world, where scarce species lie in wait to become dominant if change goes their way, and found species believed to reside at both poles.[7]

The scientists found species thought to be extinct, like a Jurassic shrimp, and life forms no one had ever imagined, like giant mats of microbes.[8] However, "[w]hile unlocking many secrets, investigators also documented long-term and widespread declines in marine life as well as resilience of the ocean in areas where recovery was apparent."[9] In short, there's promise and there's peril, so this volume reflects them both in its search for a fuller understanding of the ocean's situation in the twenty-first century.

The ocean demands multiple perspectives, deep dives, broad views, and magnifying lenses. This volume aims for that multiplicity in its collaboration between experts. Indeed, the volume was conceived at an interdisciplinary conference, Revaluing the Ocean, hosted by the University of Utah. Far inland, we gathered thinkers from literature, history, law, politics, Indigenous cultural studies, art, film, oceanography, geology, and more. Our essay collection works from a similar desire to integrate insights from broad horizons.

Re-envisioning the ocean means, at least in part, that the literary, economic, political, and legal narratives of the ocean need to better reflect the emerging scientific narrative about humanity's changing relationship to the ocean. We therefore begin this book with an overview of that scientific story.

* * *

> *For most of history, man has had to fight nature to survive; in this century he is beginning to realize that, in order to survive, he must protect it.*
>
> —Jacques-Yves Cousteau, ocean explorer and documentary filmmaker

Part of re-envisioning the Anthropocene ocean requires the deep psychological and cultural acknowledgments that humanity needs the ocean to survive. We can start with breathing, the simple and reflexive act of drawing air into our lungs that appears on first investigation to have nothing whatsoever to do with the ocean. What we need in breathing (and what all other animals on the planet need, it should be noted), of course, is oxygen, not saltwater. Most of that oxygen, it turns out, comes from marine plants—both the big plants like seaweeds (kelp) and seagrasses and the tiny plants known as phytoplankton. Like all plants, seaweeds, seagrasses, and phytoplankton engage in photosynthesis, turning carbon dioxide and water into sugars and starches. The byproduct of photosynthesis is oxygen.

Calculating exactly how much oxygen marine plants contribute to the atmosphere turns out to be a bit of a scientific problem. Specifically, it's difficult to calculate how many phytoplankton actually live in the ocean, especially because their numbers change, sometimes dramatically, in response to seasonal changes in light and warmth and fluctuating concentrations of nutrients. Nevertheless, these tiny phytoplankton are critically important to humans. As *National Geographic* points out:

> One type of phytoplankton, *Prochlorococcus*, releases countless tons of oxygen into the atmosphere. It is so small that millions can fit in a drop of water. *Prochlorococcus* has achieved fame as perhaps the most abundant photosynthetic organism on the planet. Dr. Sylvia A. Earle, a National Geographic Explorer, has estimated that *Prochlorococcus* provides the oxygen for one in every five breaths we take.[10]

Because of the difficulties in calculating exactly how productive phytoplankton are, the scientific estimates for how much atmospheric oxygen marine plants contribute cover a large range—anywhere from 50 percent to 85 percent of the oxygen in the Earth's atmosphere—with a consensus estimate of about 70 percent. And that's in addition to the oxygen that stays dissolved in seawater, allowing fish and other marine creatures that get their oxygen from water rather than air to survive.

If you want to travel back into geologic time, marine plants become even more important to our current ability to breathe freely. Marine plants have been photosynthesizing—and contributing oxygen to the atmosphere—for *billions* of years. Indeed, "[t]he oldest known fossil is from a marine cyanobacterium, a tiny-blue green photosynthesizer that was releasing oxygen 3.5 billion years ago."[11] In contrast, the first land plants appeared only 470 million years ago, having evolved from their marine counterparts.[12]

Marine photosynthesizers are also an example of how contemporary science is allowing humans to re-envision their relationship to the ocean in the twenty-first century. Scientists discovered *Prochlorococcus* only in 1988, and they are continuing to elucidate its importance more than thirty years later.[13] It turns out that *Prochlorococcus* conducts about 5 percent of total global photosynthesis, influences climate, and probably participated in global evolution.[14] *Prochlorococcus* thus provides an important and concrete example of the fact that humans are still discovering what's important about the ocean, including how it affects our own survival.

Food security is another reason humans need to re-envision their relationship to the ocean. Ocean fisheries and marine aquaculture are important but often overlooked components of world food security, discussions of which tend to focus on land-based crops and livestock. However, one billion people—that is, roughly 2 of every 15 people on the planet—rely on fish as their *primary* source of protein.[15] About 3.3 billion people get at least 20 percent of their protein from fish.[16] In addition, fish is a critical source of protein (30 percent or more of protein consumed) in many specific countries around the world, including several countries in Africa, Japan, Greenland, Taiwan, Indonesia, and several South Pacific island nations.[17] Seafood consumption, moreover, is increasing faster than both population growth and consumption of other animal protein, particularly among the world's developing nations.[18]

Global equity, particularly with regard to small island developing states, Native Alaskans, and Oceania,[19] and human health are thus both intimately tied to our management of marine biodiversity. These are also often some of the human communities most vulnerable to climate change and which are still dealing with a legacy of colonialism. The narratives of their evolving relationships to the ocean can recast both scientific prognostication and postcolonial legacies and thus provide different visions of human intimacy and coexistence with the marine realm and its resources.

Nevertheless, humanity fully uses the sustainable supply of wild seafood and probably more. The United Nations Food & Agriculture Organization (FAO) publishes biannual reports, *The State of World Fisheries and Aquaculture*, that are considered the most reliable source of global marine fisheries information. In a recent report (2018), the FAO notes that wild marine fisheries have been stagnant since the 1980s, despite increased fishing effort, and that aquaculture now accounts for 47 percent of all seafood produced.[20]

Our most basic needs, therefore—oxygen and food—depend on an ocean that is changing, and changing in ways that are likely *not* to benefit humans. But scientific truth cannot lead us to despair. While re-envisioning the ocean does require that we reimagine our dependence upon the blue ecosystems in terms of vulnerability as well as exploitative power, new narratives sampled in this volume also allow us to recast ourselves as protectors of that saltwater realm and effective communicators of our connections to it.

* * *

> *It is a curious situation that the sea, from which life first arose should now be threatened by the activities of one form of that life. But the sea, though changed in a sinister way, will continue to exist; the threat is rather to life itself.*
>
> —Rachel Carson, scientist and activist, *The Sea Around Us*

The ocean is changing. Climate change and its "evil twin," ocean acidification, have numerous ramifications for the ocean, both alone and in combination with other stressors, like marine pollution. Re-envisioning the ocean means seeing it as the ocean of the Anthropocene, with this complex of changes in motion.

In 2014, the Intergovernmental Panel on Climate Change (IPCC) published its Fifth Assessment Report on climate change.[21] The IPCC's reports are generally conservative, particularly where the ocean is concerned. Nevertheless, the Fifth Assessment Report provides a good starting assessment of the changes that have already occurred in the ocean as well as projections for the future. These reports show climate warming to be a transformative force for our ocean and show that the ocean is the critical actor for a warming planet.

The world's ocean has been absorbing most—indeed, almost all—of the extra heat produced as a result of the increasing concentrations of greenhouse gases in the atmosphere, a function of the facts that water has a high heat capacity, the ocean has a lot of volume, and ocean currents can take heat to other places and deeper waters.[22] According to the IPCC, the top 75 meters of the ocean have been warming, on average, by 0.11 C per decade from 1971 to 2010.[23] However, ocean warming has also continued to push deeper, and "[i]t is likely that the ocean warmed from 700 to 2000 m[eters] from 1957 to 2009 and from 3000 m[eters] to the bottom for the period 1992 to 2005."[24]

Importantly, ocean warming will continue for many decades. According to the IPCC, "[t]he global ocean will continue to warm during the 21st century, with the strongest warming projected for the surface in tropical and Northern Hemisphere subtropical regions."[25] However, "[a]t greater depth the warming will be most pronounced in the Southern Ocean (*high confidence*),"[26] meaning that Antarctica might experience the most profound changes. A study published in *Nature* on, appropriately, Halloween 2018, indicated that the ocean is warming even faster than the IPCC suggested.[27] While calculation errors immediately came to light that called into question the most extreme of the authors' estimations, the study's main conclusion—that the ocean is warming faster than the IPCC had indicated—remains valid.[28]

In addition, the ocean is the world's largest carbon sink, and the global ocean has been absorbing most of the anthropogenic carbon dioxide since the Industrial Revolution. As a natural part of the Earth's carbon dioxide (CO_2) cycle, the world's ocean has been absorbing much of the "extra" carbon dioxide that humans have been producing, especially since humans began burning fossil fuels on a large scale as a result of the Industrial Revolution.[29] "Over millennia, the ocean will absorb up to 85 percent of the extra carbon people have put into the atmosphere by burning fossil fuels."[30] Currently, however, winds, currents, and ocean temperatures limit how fast the ocean can take carbon dioxide out of the atmosphere.[31] At the beginning of the twenty-first century, the ocean and land ecosystems (mostly plants) were absorbing about half of the anthropogenic emissions of carbon dioxide[32]—roughly 25 percent by land plants and 25 percent by the ocean.[33] In 2006, oceanographers at the National Oceanic and Atmospheric Administration (NOAA) estimated that "[o]ver the past 200 years the oceans have absorbed 525 billion tons of carbon dioxide from the atmosphere, or nearly half of the fossil fuel carbon emissions over this period."[34] The ocean continues to uptake about 22 million tons of carbon dioxide per day.[35]

The troubling consequence of all this carbon absorption is that carbon dioxide chemically reacts with water to form carbonic acid.[36] This acid-forming reaction is lowering the ocean's pH. Ocean water is naturally basic, with an average pH of about 8.16, and that pH level has been remarkably stable over geological time.[37] However, since the Industrial Revolution, the average ocean surface water pH has dropped by 0.1 unit.[38] While this change may seem small, the pH scale is logarithmic, so that a pH decrease of 0.1 unit means that the oceans have become 26 percent more acidic in the last 250 years.[39]

Likewise, a small change in pH can have big results for the sea creatures we depend on. Indeed, together, climate change and ocean acidification have significant implications for marine biodiversity, marine fishing, and ocean aquaculture. As the FAO explained in 2018:

> The main risks for fisheries and aquaculture are reasonably well understood: A number of marine species, depending on their mobility and habitat connection, are responding to climate impacts by shifting their distributions poleward and to deeper waters. The increased uptake of carbon dioxide by oceans, resulting in higher water acidity, is also of particular concern for calcifying organisms in natural environments (including mariculture facilities), although the

> full ecosystem effects are still inconclusive. Competition for water, changes in the water cycle, increased frequency of storms, and sea level rise are all expected to affect both inland fisheries and aquaculture industries.[40]

Primary production in the ocean—that is, the growth of marine plants and phytoplankton that serve as the basis of marine food webs—is expected to decrease by at least 6 percent on global average by 2100, and by as much as 11 percent in the tropics.[41]

Re-envisioning the Anthropocene ocean, therefore, requires a new level of acknowledgment of Shakespeare's sea change: the ocean itself is becoming something strange, destabilizing from its Holocene norms into a realm no human has ever seen. For example, fish are increasingly migrating away from where they are supposed to be.[42] Recent projections indicate that "changes in marine fisheries production may be just as large as those in crop agriculture, which is often claimed to be the sector most affected by climate change," with 85 percent of coastal countries experiencing declines in *both* fisheries and terrestrial agriculture.[43]

Indeed, the ocean itself may become monster, and we are Frankenstein. The ocean is already experiencing a wide range of biological and ecological impacts as a result of ocean acidification, and these impacts—while admittedly still being studied—are only expected to worsen. The ocean is approaching a pH that is unprecedented in human experience—and it is changing *quickly*. For example, according to NOAA scientists, because of ocean acidification, "[a]t present, ocean chemistry is changing at least 100 times more rapidly than it has changed during the 650,000 years preceding our industrial era."[44] This altered chemical reality is likely to last for a long time—at least from a human and ecological perspective. As reported in *Science*, "[i]t takes the ocean about 1000 years to flush carbon dioxide added to surface waters into the deep sea where sediments can eventually neutralize the added acid."[45] Most frighteningly, changes now occurring in the ocean may herald mass destruction of biodiversity everywhere.[46] "We are entering an unknown territory of marine ecosystem change, and exposing organisms to intolerable evolutionary pressure. The next mass extinction event may have already begun."[47] To escape Frankenstein's fate, we need to learn the lesson that Frankenstein did not: take responsibility and nurture the Anthropocene ocean that we have created.

* * *

The least movement is of importance to all nature. The entire ocean is affected by a pebble.
—Blaise Pascal, mathematician and philosopher

All of the growing understanding about how the ocean is changing and why it is important to global health demonstrates the urgency of revaluing humanity's relationship to the ocean now. One place to start this re-envisioning is the global conversation about implementing the international Sustainable Development Goals. In 2015, the United Nations member countries adopted the 2030 Agenda for Sustainable Development, *Transforming Our World*,[48] at the heart of which were seventeen Sustainable Development Goals, or SDGs.[49] In order of the United Nations' numbering, these goals are: (1) no poverty; (2) zero hunger; (3) good health and well-being; (4) quality education; (5) gender equality; (6) clean water and sanitation; (7) affordable and clean energy; (8) decent work and economic growth; (9) industry, innovation, and infrastructure; (10) reduced inequality; (11) sustainable cities and communities; (12) responsible consumption and production; (13) climate action; (14) life below water; (15) life on land; (16) peace, justice, and strong institutions; and (17) partnerships to meet the goals.[50] SDG 14, "Life Below Water," is the SDG for the ocean.

The United Nations clearly cares about SDG 14. In June 2017, it convened "a high-level Oceans Conference in New York to support the implementation of SDG 14." Moreover, the FAO themed its 2018 *State of World Fisheries and Aquaculture* report around SDG 14.[51]

For pretty much everyone else, however, SDG 14 is the lowest-priority SDG to implement. Officially, the 17 SDGs are unprioritized coequal goals. However, nations and NGOs have limited money and other resources, and, as a result, prioritization is rampant as a practical matter. For example, through the "SDGs in Order" project,[52] a team of researchers from the OECD, Bretton Woods II, and GreenHouse took an economic funding approach to prioritizing the SDGs. When the team asked eighty-five expert economists, social scientists, and political scientists[53] to rank the SDGs in order of how they should be pursued, SDG 14 came in dead last.[54] Funders, similarly, target health and education, not the environmental SDGs,[55] and businesses agree with the "SDGs in Order" researchers that SDG 14 is the least important of the SDGs.[56]

More importantly, the world's nations are effectively following the experts' advice and deprioritizing the environmental SDGs, especially SDG 14. In a survey report published in 2018, SDG 14 again came in dead last among the SDGs in perceived importance among governments.[57] "Only 5.4 percent of the respondents included it in their top six priorities, compared to 65.2 percent for quality education or 60 percent for decent work and economic growth."[58]

The ocean, it would seem, has a public relations problem. Notably, some scientists are working to elevate the importance of SDG 14, "highlight[ing] the importance of the ocean for achieving the Sustainable Development Goals, and suggest[ing] that achieving Oceans targets has important co-benefits through supporting diverse aspects of sustainable development and rarely presents negative trade-offs."[59] While embracing these efforts from the sciences, this volume insists that re-envisioning the ocean needs to be a multidisciplinary and interdisciplinary effort to achieve a cultural change in perception. As such, it brings together not only scientists and social scientists but also academics from the humanities, government officials, lawyers, and ocean explorers to highlight multiple facets of the re-envisioning process.

* * *

> *The ocean stirs the heart, inspires the imagination and brings eternal joy to the soul.*
>
> —Robert Wyland, marine artist

The humanities have an important role to play in ocean conservation. Look, for instance, at Rachel Carson's first success—*The Sea Around Us* (1951)—which mixes a scientist's knowledge of the ocean with an artist's hope for storytelling. Carson says we humans are evolved from the saltwater that still shapes our blood and are destined to return to that same sea. "Mankind," she writes, "could not physically re-enter the oceans as the seals and whales had done. But over the centuries, with all the skill and ingenuity and reasoning powers of his mind, he has sought to . . . re-enter it mentally and imaginatively."[60] Part of this volume's task is to reenter the sea mentally, to swim with care amid its wonders and its warnings. In the story of Western culture more broadly, we have the story of humanity's ambivalent journey towards the sea.

Close attention to the ocean has invigorated humanities and social science scholarship in the past few years. Our collection bridges fields to foster a fuller cultural understanding of ocean policy and ocean perception. We have gathered these diverse voices to make the recent "blue turn" in the academy available to a broad audience. Whether we talk of the "Blue

Humanities," "Blue Cultural Studies," or a "Blue Turn" in scholarship, what's at stake is close attention to Ishmael's "watery part of the world" and, in particular, to the social and ecological relations unique to ocean experience. This set of relations is more important than ever in view of climate change's environmental pressures and harsh social dislocations.

Steve Mentz proposed the concept "Blue Humanities" because it offers "an off-shore trajectory that places cultural history in an oceanic rather than a terrestrial context."[61] Going "off-shore" sets us all in the context of an overwhelmingly grand ocean setting. Consequently, the Blue Humanities broadens human stories into extrahuman contexts that destabilize anthropogenic confidence and replaces human conceits of mastery with the dizzying tumult of the moving ocean. Blue humanists of Mentz's ilk are like swimmers who expect disruption, the better to live in a dynamic, fluctuating world. Under this banner, insights from History, Philosophy, Literary Studies, Gender Studies, Ethnic Studies, and the new materialism flow toward deeper understandings of the ocean's role and vulnerability. The benefit of this blue lens is both what it says about the ocean and how it supplements the scientific findings we've outlined above. The humanities interpret scientific findings in terms of lived experience and the power relations that hierarchize people's access to authority, stability, and dinner.

The ocean experience in art and literature has long divided along an axis of scared dissolution before the chaos of the deep and thrilling Romantic becoming. An early expression of the former suspicion starts alongside the world's creation in Genesis: "In the beginning God created the heavens and the earth. And the earth was without form, and void; and darkness was upon the face of the deep." On the third day the seas are shaped: "And God said, Let the waters under the heavens be gathered together unto one place, and let the dry land appear; and it was so. And God called the dry land Earth; and the gathering together of the waters He called the Seas."[62] The ocean is a space of formless chaos gathered into vast unknowables separated by tracts of earth. God redeemed the land from the void of sea.

What Genesis calls "the deep" is the primordial stuff of chaos also known as *the abyss*. A look into the etymology shows that "the deep" is also Hebrew *tehom*, meaning the primordial waters out of which God brought the ordered world in the Genesis verses quoted above. The development to note is that the word *tehom* came to mean the depths of the sea, the dark abyss, and so the sea itself was a godless element, not unlike the underworld realm of the dead.[63] Indeed, later in Genesis, the waters become an instrument of God's wrath. In Noah's story, the sea wipes the earth back into formless and hopeless chaos, undoing creation and punishing miscreant humans. In this sense, the sea is hell, and those who go out to sea in ships take great risks with both body and soul.

The Old Testament begins with Genesis and the New Testament ends in Revelation, and Revelation offers us a similar warning about the ocean. The redeemed world of the New Jerusalem imagines harmony and to that end promises "no more sea." Revelations 21:1 reads, "And I saw a new heaven and a new earth: for the first heaven and the first earth were passed away; and there was no more sea."[64] This late chapter in this last book of the Bible promises God's grace, and one manifestation of His grace is getting rid of that evil, dangerous ocean.

The Bible thus gives Western civilization reason to hold the ocean at suspicious arm's length. From Genesis to Revelation, the ocean is a chaotic, angry element. Eden, by contrast, is a garden where all is in order, and that terrestrial paradise stands in contrast to the ocean's sinful disorder.

This outlook extends from the Bible into other cultural productions in the West. For example, art from medieval Europe depicts the ocean as a space of bitter, hard labor. The anonymous Old English poem "The Seafarer" is melancholic about ocean life:

> hunger tearing within
> the sea-wearied mind. He does not know this fact
> who dwells most merrily on dry land—
> how I, wretchedly sorrowful, lived a winter
> on the ice-cold sea.[65]

From the Icelandic Sagas to Shakespeare's *Tempest* all the way to Rembrandt's seascapes, the tone is minatory, and the subject more often sinks than floats.

Stepping into the eighteenth century, however, we can see a transition, and a new aesthetic that led writers, painters, and poets to make the ocean something positive. The Romantic era gave us the inspiration of unfettered waves, and the improving influence of wild tests at sea. Byron wrote:

> There is a pleasure in the pathless woods,
> There is a rapture on the lonely shore,
> There is society where none intrudes,
> By the deep Sea, and music in its roar:
> I love not Man the less, but Nature more.[66]

"Both jaws, like enormous shears, bit the craft completely in twain."

—*Page 510.*

Figure 0.2. Melville's ocean can bite back. Illustration for *Moby Dick*, Augustus Burnham Shute. Circa 1892. Public domain.

This appreciation for wild nature is a long way from Genesis, and for Byron and Coleridge and Mary Shelley the ocean became a wilderness from which the brave hero could find insight and improvement.[67] Their male protagonists were transformed by direct contact with the wild waters that Revelation warned us about. Richard Henry Dana spent two years before the mast (in the 1840 book of that name) and Melville's Ishmael likewise runs away to sea seeking thrill and enrichment: "for nowadays, the whale-fishery furnishes an asylum for many romantic, melancholy, and absent minded young men, disgusted with the carking cares of earth, and seeking sentiment in tar and blubber."[68] These heroes were not just laborers but explorers of the soul who found their best selves aboard ship, while sin and temptation brewed on land. Byron's Childe Harold sums up the ocean's Romantic redemption: "Man marks the earth with ruin,—his control stops at the shore."[69]

Literary scholar Margaret Cohen characterizes the ocean's transition from unfriendly threat to uplifting beauty as "the sublimation of the sea."[70] Her study *The Novel and the Sea* (2010) traces a story of ocean revaluation that aestheticized the water and lifted it into popular imagination. Moreover, Cohen attends to "craft"[71]—the physical knowledge of the working ship that complicates the old Romantic cliché of aristocrats afloat. For her, this careful attention with which sailors explored the terraqueous globe makes them modern heroes of technology as well as mind.

Cohen's ideas conclude this long sea journey from a culture's distrust to celebration. Our introduction can only give the merest outline of so grand a subject—the ocean and society—but it is enough to suggest that the Romantic ocean is the one we live with now. The complication, though, is that our labor, our "craft," has so overrun the ocean with plastic and nitrogen and nets that we threaten to make the sea a hell again. So, this overview brings the ocean full circle from the roots of Judeo-Christian culture through the Romantic movement's celebration, to our own time of storm and pollution and extinction.

* * *

I am the Ocean
I am water
I am most of this planet
I shaped it
. . .
Every living thing here needs me
I am the source
I am what they crawled out of
Humans are not different
. . .
But humans, they take more than their share
They poison me, then they expect me to feed them
Well it doesn't work that way
. . .
Me—I could not give a damn, with or without humans
I am the Ocean

I covered this entire planet once, and I can always cover it again
And that's all I have to say.

—Nature Is Speaking, transcript of "Harrison Ford Is the Ocean" video[72]

The global ocean is the primary messenger of a warming climate. From the increase in tropical storm ferocity to sea-level rise, the ocean is where the myriad climate change impacts accumulate, accelerate, and burst into human ken. In this precarious relation, stories from human history supplement the scientific knowledge of ice-cores and parts-per-million. Cultural artifacts persist—*ars longa, vita brevis*—because they sustain truths known more deeply than any one generation's exigencies.

How does culture respond to a climate-changed ocean? Looking back can help us look forward. The stone tablets overlooking Fukushima's seaside are sculptural messages from a deep history, and they warn "Build only above this stone," lest short-term attention or greed forget the long knowledge of tsunami. Likewise, since the Old Testament, works have attended to ocean experience and, in particular, drawn lessons from the plight of ships at sea. This cultural heritage can be a useful tool in our reckoning with climate change. Plato's allegory of the Ship of State is the simple tale of a vessel whose crew would be its captain. The resulting confusion and quarrels make the ship vulnerable to storm and chance. The ocean 2,500 years ago was unforgiving of fools and useful as a literary figure, just as it is today. Jonah's case is well known precisely because an ocean storm is the image of God's wrath, and the crew's response—ejecting the sinner—is the story of a community at risk.

As our own society speeds into a climate future of hurricanes, floods, extinction, and hunger, literary ships at sea float forward as useful models. There on the *Pequod* a crew must unite in the face of maniacal tyranny or perish. Aboard Conrad's ships the crew is the metaphor of a society that with solidarity might survive a storm, but with selfishness will sink. The ocean literature of ships at sea offers extended meditations on a prototypical society responding to environmental hazard. For millennia, Western art has seen in ocean experience the very image of its social contract, and imagined there the awful stages of its dissolution. The impending climate transformation of our ocean is impossible to summarize in its variety and bewilderment, but a ship's crew riding it out is one useful trope from our literature.

More recently, insightful artists attend to the ocean for its mingled force as artistic heritage and deliverer of bad news. Look, for example, at plastic

Figure 0.3. Hokusai's familiar print renders ocean beauty and ocean trouble. *The Great Wave off Kanagawa*, Katsushika Hokusai. Circa 1829. Public domain.

pollution and artistic repurposing of Hokusai's "Great Wave." Katsushika Hokusai's iconic "The Great Wave off Kanagawa" (1829) is a woodblock print showing fishing boats in a heaving sea (with a view of Mount Fuji looking calm in the distance). This is a story of human struggle in a hostile if beautiful environment. Today the recognizable image has been repurposed by artists to make points about ocean health. The photographer Chris Jordan created an eight-foot-by-eleven-foot Great Wave from images of plastic trash in the ocean. Titled "Gyre," it depicts 2.4 million pieces of plastic, equal to the estimated number of pounds of plastic that enters the world's oceans every hour."[73] The message from this ocean imagery is that the modern, industrial economy defiles nature. The Pacific gyre of plastic hangs over the contemporary ocean just as Hokusai's Great Wave hangs over the scrambling fishermen in 1829.

"The Great Wave" is a powerful cultural referent, and Bonnie Monteleone's "What Comes Round Goes Round" is the familiar image overlaid with actual plastic shards gathered from the ocean. Monteleone—a chemist and ocean researcher—founded the Plastic Ocean Project, and views her Great Wave as akin to the AIDS quilt that once raised awareness through an approachable, mobile art form. We are nearly two hundred years from

Hokusai's ocean, and these artists suggest it looms over us as it did those small craft—this time thanks to our own systemic excess.

Meanwhile, in the Caribbean, sculptor Jason deCaires Taylor has expanded the interaction between fine art and the ocean by placing his human sculptures in thirty feet of water. "Vicissitudes" (2007), for instance, is a circle of six-foot-high human figures, facing outwards and holding hands in the depths. Ocean painters like Winslow Homer give us the blue-green waves around floating protagonists, but Taylor sets his works into the sea itself. They are images of islanders facing a collapsing ecosystem and a rising sea. Drawn from the faces of actual locals, they are cast in reef-friendly concrete and set to welcome the steady transformation of sea life growing upon them. They dramatize, thereby, a submerged future for islands at the same time they gesture back towards a drowned past for African slaves. This collaboration between art and sea brings Shakespeare's "Full Fathom Five" full circle because the Caribbean Sea—where Renaissance England exploited a "brave new world"—enacts a literal sea change on its subjects via coral growing on their sculptural bones. Taylor's work makes viewers pause on the people and the ecosystems touched first by ocean-based imperialism and now by climate change's ocean impacts. "Vicissitudes" is important to this volume because it looks backwards but also looks forwards to explore that "something rich and strange" coming next.

Historians know the ocean waves as roads of commerce and tools of empire, so it's a short passage from Taylor's sculpture to cultural histories of slavery and oppression that must inform present efforts to revalue the ocean. For Caribbean and African American readers, the ocean's Romantic interpretations are obscured by troubling visions of slave ships and plantation economies. In Derek Walcott's "The Sea is History," he asks us to meditate on all the ships and all the conflicts that have sailed past his home island to St. Lucia. He also gestures toward African slaves transported across the Atlantic for money and thrown sometimes overboard to carpet that Middle Passage with their bones and their spirits. With Walcott and Taylor, the humanities insist that ocean waves are more than their measurable heights or chemical compositions—they carry a profound variety of cultural meanings.

Paul Gilroy's *The Black Atlantic* (1993) theorizes the ocean as a space of creolization where combination and multiplicity enriched the cultures of the Atlantic circle. Black Atlantic identity is characterized by hybridity, and this African–European–South American–Caribbean ocean mixture is a better image of our world than immutable ethnic differences. Historian

Joshua Reid's *The Sea Is My Country: The Maritime World of the Makahs* (2015) traces the ocean's role in Makah society in the Pacific Northwest. When "the People of the Cape" insisted to early European traders "the sea is my country," they claimed the ocean waters around them as a cultural heritage, and they celebrated millennia of indigenous ocean culture.[74] These words are pertinent here because they upend a European legal practice of deeming unfixable ocean resources "common use" or "common property" (freedom of the seas), in contrast to precisely documented land ownership. The Makah upset the assumption of global *mare liberum*. As writers in this volume explain, the legal status of indigenous conceptions of ocean use are in tension with Western law today, and emerge from an overlooked history that may provide models for resilient ocean policy moving forward.

In *The Great Derangement: Climate Change and the Unthinkable* (2016), Amitav Ghosh argues that novelists are guilty of obscuring the gathering menace of climate change by focusing on other, mundane experiences:

> At exactly the same time when it has become clear that global warming is in every sense a collective predicament, humanity finds itself in the thrall of a dominant culture in which the idea of the collective has been exiled from politics, economics, and literature alike.[75]

The collective brings Ghosh to geographical reordering of climate change. In his work, the climate-changed ocean is an Asian concern. He shows that the biggest populations directly affected by rising seas, changing monsoons, and bigger typhoons are the Indians, the Indonesians, and the Chinese who line the Indian Ocean, the South China Sea, the Bay of Bengal, and the Sea of Japan. Like Walcott's Caribbean, there is a colonial dynamic here as well. Climate change is produced by the great acceleration of Asian economies alongside established Western carbon use. The globe's great capitalist powers rely on these carbon-intensive processes to maintain their hegemony, and so the sea level will continue to rise in Asia. Chinese and Indian stability depends on economic expansion, and Western economies are unlikely to make significant cuts to their carbon budgets because, according to Ghosh, their colonial histories make them unwilling to grant advantages to the Asian people they once ruled. Ghosh's novels *The Hungry Tide* (2004) and *Gun Island* (2019) spotlight working-poor Indians displaced by ocean storms and sea-level rise. These are particular cases of likeable individuals affected directly by the ocean we read about in IPCC reports and quantitative assessments. Ghosh's cultural artifacts are deployed to prod readers toward the global, colonial histories that color the ocean that surrounds them.

* * *

As this introduction suggests, this volume begins the process of re-envisioning of the ocean by re-envisioning the re-envisioners. In the following pages, lawyers, scientists, artists, historians, and literary theorists (and their students) become an interactive community of storytellers rather than isolated disciplinarianists. Many of the chapters reflect personal experience and investment as well as professional endeavor, eliding the lines between research and essay/memoir. The book seeks to foster a kaleidoscopic re-imagining of the Anthropocene ocean. We hope that the reader will hear scientific understanding of change, challenge, and human vulnerability resonating through legal protection efforts and a variety of cultural narratives of connectivity and dependence; will experience coexistent cultural histories both explaining how we got to here and offering multiple visions for the ocean's future; will join in narratives, gaming, and dance that can make the science real and personal and the legal options more politically tangible—in other words, will experience for themselves a sea change in how they imagine the Anthropocene ocean.

Notes

1. Herman Melville, *Moby Dick*, 3rd ed. (New York: W. W. Norton, 2018), 16.
2. Ibid., 17.
3. *Werner Herzog, A Guide for the Perplexed: Conversations with Paul Cronin.* (London: Faber & Faber, 2014), 71.
4. "History of Oceanography," Woods Hole Oceanographic Institution, https://divediscover.whoi.edu/history-of-oceanography/.
5. Ibid.
6. "Oceans: The Great Unknown," National Aeronautics & Space Administration (NASA), October 8, 2009, https://www.nasa.gov/audience/forstudents/5-8/features/oceans-the-great-unknown-58.html/.
7. "About the Census: A Decade of Discovery," Census of Marine Life, 2010, http://www.coml.org/about-census/.
8. Ibid.
9. Ibid.
10. "Save the Plankton, Breathe Freely," *National Geographic*, https://www.nationalgeographic.org/activity/save-the-plankton-breathe-freely/.

11. Kalila Morsink, "With Every Breath You Take, Thank the Ocean," *Smithsonian Ocean*, July 2017, https://ocean.si.edu/ocean-life/plankton/every-breath-you-take-thank-ocean/.
12. Ibid.
13. Elizabeth Pennisi, "Meet the Obscure Microbe that Influences Climate, Ocean Ecosystems, and Perhaps Even Evolution," *Science* Blog, March 9, 2017, https://www.sciencemag.org/news/2017/03/meet-obscure-microbe-influences-climate-ocean-ecosystems-and-perhaps-even-evolution, doi:10.1126/science.aal0873/.
14. Ibid.
15. "The Impact on Communities," Marine Stewardship Council, https://www.msc.org/en-us/what-we-are-doing/oceans-at-risk/the-impact-on-communities/.
16. Ibid.
17. Ibid.
18. United Nations Food & Agriculture Organization, *The State of World Fisheries and Aquaculture 2018: Meeting the Sustainable Development Goals* (Geneva: United Nations, 2018), 2, http://www.fao.org/3/i9540en/i9540en.pdf/ [hereinafter 2018 FAO SOWFA Report].
19. Ibid.
20. Ibid.
21. The IPCC's Fifth Assessment Report consists of four documents: Intergovernmental Panel on Climate Change, *Climate Change 2013: The Physical Science Basis* (2013), https://www.ipcc.ch/report/ar5/wg1/; Intergovernmental Panel on Climate Change, *Climate Change 2014: Impacts, Adaptation, and Vulnerability* (2014), https://www.ipcc.ch/report/ar5/wg2/; Intergovernmental Panel on Climate Change, *Climate Change 2014: Mitigation of Climate Change* (2014), https://www.ipcc.ch/report/ar5/wg3/; and Intergovernmental Panel on Climate Change, *Climate Change 2014: Synthesis Report* (2014), https://www.ipcc.ch/report/ar5/syr/ [hereinafter *2014 IPCC Synthesis Report*].
22. "Ocean Warming," Ocean Scientists for Informed Policy, https://www.oceanscientists.org/index.php/topics/ocean-warming/.
23. *2014 IPCC Synthesis Report*, 4.
24. Ibid., 40.
25. Ibid., 11.
26. Ibid., 60.
27. L. Resplandy et al., "Quantification of Ocean Heat Uptake from Changes in Atmospheric O_2 and CO_2 Composition," *Nature* 563 (2018): 105–7, https://doi.org/10.1038/s41586-018-0651-8/.
28. Christa Marshall, "High-Profile Ocean Warming Paper to Get a Correction," *Science* News, November 14, 2018, https://www.science.org/content/article/high-profile-ocean-warming-paper-get-correction/.

29. Peter M. Cox et al., "Acceleration of Global Warming Due to Carbon-Cycle Feedbacks in a Coupled Climate Model," *Nature* 408 (2000): 184.
30. Holli Riebeek, "The Carbon Cycle," National Aeronautics & Space Administration Earth Observatory, June 16, 2011, https://earthobservatory.nasa.gov/features/CarbonCycle/.
31. Ibid.
32. Cox et al., "Acceleration of Global Warming," 184.
33. "The Ocean Carbon Cycle," *Harvard Magazine*, November–December 2002, https://www.harvardmagazine.com/2002/11/the-ocean-carbon-cycle.html/. Some scientists, however, conclude that the ocean's absorption contribution is even greater: "Over the past two hundred years, the oceans have taken up ~40% of the anthropogenic CO_2 emissions." Richard E. Zeebe et al., "Carbon Emissions and Acidification," *Science* 321 (2008): 52. A 2015 summary report published in *Science* declares that the global ocean has "captured 28% of anthropogenic CO_2 emissions since 1750, leading to ocean acidification." J.-P. Gattuso et al., "Contrasting Futures for Ocean and Society from Different Anthropogenic CO_2 Emissions Scenarios," *Science* 349 (2015): 46.
34. Richard A. Feely et al., *Carbon Dioxide and Our Ocean Legacy* 1 (Washington, DC: NOAA, 2006), https://www.pmel.noaa.gov/pubs/PDF/feel2899/feel2899.pdf/.
35. Ibid.
36. "Ocean Acidification, Explained," *National Geographic,* August 7, 2019, https://www.nationalgeographic.com/environment/article/critical-issues-ocean-acidification/.
37. European Science Foundation, "Ocean Acidification: Another Undesired Side Effect of Fossil Fuel-Burning," *ScienceDaily*, May 24, 2008, http://archives.esf.org/hosting-experts/scientific-review-groups/life-earth-and-environmental-sciences-lee/news/ext-news-singleview/article/ocean-acidification-another-undesired-side-effect-of-fossil-fuel-burning-439.html/.
38. *National Geographic*, "Ocean Acidification, Explained."
39. Ibid.
40. 2018 FAO SOWFA, 131.
41. Ibid.
42. Ibid.
43. Ibid.
44. Richard A. Feely et al., *Carbon Dioxide and Our Ocean Legacy* 2 (Washington, DC: NOAA, 2006); see also Richard A. Kerr, "Ocean Acidification Unprecedented, Unsettling," *Science* 328 (2010): 1500 (emphasizing the speed of current ocean acidification).
45. Kerr, "Ocean Acidification, Unprecedented," 1500–01.
46. Ibid., 1500.
47. Ibid.

48. United Nations General Assembly, *Transforming Our World: The 2030 Agenda for Sustainable Development* (A/Res/70/1) (September 25, 2015), https://sdgs.un.org/2030agenda/.
49. "The 17 Goals," United Nations, https://sdgs.un.org/goals/.
50. Ibid.
51. 2018 FAO SOWFA Report, vi.
52. Jeff Leitner, "SDGs in Order: The First Ever Sequence for Tackling the Most Important Problems in the World: FAQ," New America & OECD (2017), https://www.sdgsinorder.org/faq/.
53. Ibid.
54. Leitner, "SDGs in Order: Goals in Order," https://www.sdgsinorder.org/goals.
55. "Sustainable Development Goals," SDG Funders, https://sdgfunders.org/sdgs/.
56. "Latest Trends and Challenges Regarding Business Integration of the Sustainable Development Goals," World Business Council for Sustainable Development, July 10, 2018, https://www.wbcsd.org/Programs/People/Sustainable-Development-Goals/News/Latest-trends-and-challenges-regarding-business-integration-of-the-SDGs/.
57. Tim McDonnell, "The U.N. Goal that Doesn't Get a Lot of Respect," *NPR Goats and Soda: Stories of Life in a Changing World*, May 31, 2018, https://www.npr.org/sections/goatsandsoda/2018/05/31/614493772/the-u-n-goal-that-doesnt-get-a-lot-of-respect/.
58. Ibid.
59. Gerald G. Singh et al., "A Rapid Assessment of Co-benefits and Trade-offs among Sustainable Development Goals," *Marine Policy* 93 (2018): 223, 228; Mara Ntona & Elisa Morgera, "Connecting SDG 14 with the Other Sustainable Development Goals through Marine Spatial Planning," *Marine Policy* 93 (2018): 214, 215, 220.
60. Rachel Carson, *The Sea Around Us* (New York: Oxford University Press, 2018), 14.
61. Steven Mentz, "Blue Humanities," in *Posthuman Glossary*, ed. Rosi Braidotti (London: Bloomsbury, 2018), 69–72.
62. Gen. 1:8–9 (King James Version).
63. Robert Stoops, "Abyss," in *The Oxford Companion to the Bible*, ed. Bruce M. Metzger and Michael David Coogan (Oxford, UK: Oxford University Press, 1993), 6.
64. Rev. 21:1 (King James Version).
65. Anonymous, "The Seafarer," as translated by Dr. Aaron K. Hostetter, Old English Poetry Project, https://oldenglishpoetry.camden.rutgers.edu/the-seafarer/.
66. From Lord Byron, Childe Harold's Pilgrimage (1866), Canto IV, Stanza 178.

67. Indeed, Shelley's husband, Percy, was drowned when his yacht sank in a squall.
68. Melville, *Moby Dick*, 128–29.
69. Lord Byron, *Childe Harold's Pilgrimage* Canto IV, Stanza 179, (1866).
70. Margaret Cohen, *The Novel and the Sea* (Princeton, NJ: Princeton University Press, 2010), esp. 106–31; quote on p. 27.
71. Ibid., 15–58 ("The Mariner's Craft").
72. Conservation International. "Harrison Ford Is the Ocean," YouTube video, December 1, 2021, https://www.youtube.com/watch?v=rM6txLtoaoc/.
73. "Gyre, 2009," *Running the Numbers II: Portraits of Global Mass Culture*, Chris Jordan Photographic Arts, http://www.chrisjordan.com/gallery/rtn2/#gyre/.
74. Joshua Reid, *The Sea is My Country: The Maritime World of the Makahs* (New Haven: Yale University Press, 2015).
75. Amitav Ghosh, *The Great Derangement* (Chicago: University of Chicago Press, 2017).

PART I

Re-envisioning the Ocean as Connection

Editors' Introduction to Part I

THE OCEAN CONNECTS us all. Climate change connects us all. An unfolding climate catastrophe demands that scholars and lawmakers and artists and scientists look at the ocean in new ways. The four essays in this section use humanities tools to explore that connection's character. A humanist's method is likely to include close reading, critical analysis, speculative logic, and a historical perspective. These essays push readers to reconsider familiar ocean narratives. In particular, the essays probe at cultural preconceptions about migration, about shipwreck, about equilibrium, and about climate change.

Rachel Carson's *The Sea Around Us* argues that we inhabit an ocean planet: "[The sailor's] world is a watery world, a planet dominated by its covering mantle of ocean, in which the continents are but transient intrusions of land above the surface of the all-encircling sea."[1] Seventy years on, Carson's recognition that the ocean is Earth's dominant feature informs contemporary environmentalism. Environmental humanists look to the sea for stories of resilience, for new theories of connection, for replacements to old histories of isolation, for celebrations of creolization and hybridity, and for theories of immersive interconnection. Where environmental scholarship has been "green" in honor of land, it is expanding to include "blue" perspectives that highlight the ocean's importance. The emerging research area of "Blue Humanities" recognizes the defining influence of ocean habitats on human culture and insists on the centrality of ocean imagination to environmental scholarship.

Here in Part I, the four essays revalue the ocean in related ways. First, they emphasize the significance of a changing climate to the ocean and, conversely, the significance of the ocean to understanding a changing climate. In McCarthy and Bassi's telling, the coming storms present turbulent tests to a land population that might learn from mariners and island dwellers long versed in storm disruption. Venice and ships at sea model the cooperation-amidst-distress our Anthropocene era will demand. Mentz and Cunningham show the ocean connecting distant cultures, and they foreground globalization's colonialist roots. Cunningham celebrates Oceania and indigenous wayfaring accomplishments that transform the Pacific from an obstacle into a unifying bridge. Mentz insists ocean-crossing slave ships birthed modernity as an interconnected global network of oppression and production. Polynesian and Caribbean relations to the ocean are inextricably wrapped in the history of colonial exploitation that continues to shape their vulnerability and resilience at the forefront of climate change.

These are hard stories. Still, close study of the ocean's cultural significance also delivers hope because the ocean commons presented first as tragedy can also exemplify possibility. After all, re-envisioning the Anthropocene ocean involves embracing the ocean as a medium of connectedness and hence as an agent of mutual transformation. McCarthy's reading of ships at sea, Cunningham's salute to Polynesian tradition, Bassi's presentation of Venice, and Mentz's theory of ocean origins all place the ocean at the center of constructive responses to climate change. Indeed, all four essays linger on models of alliance springing up in the wake of climate catastrophe. Every culture is the result of human interaction with other cultures *and* with environment. Section I examines cultures transformed by their ocean connections. In this broad-ranging treatment, the common theme is the ocean's centrality to a resilient future.

In conclusion, let's remind ourselves of the ocean's centrality to life on planet Earth. For starters, the ocean covers 70 percent of the planet, and more than three-quarters of all life on Earth lives in or on the ocean. Next, the ongoing climate emergency reminds us that the ocean has captured 90 percent of the warming from surging carbon dioxide emissions, and that it generates half of the oxygen we breathe. Finally, revaluing the ocean is fundamental to the planet's future, both as an ecosystem and as a host to human activity. Interdisciplinary study is a necessary component of such a broad and diverse topic as the ocean, and so this volume's sections approach the ocean from multiple, complementary perspectives.

The humanities scholars in Section I interpret artifacts of ocean experience to understand the past and build the future. The sailing canoe *Hōkūle'a* traveled from Hawaii to Tahiti in 1976 to revive a cultural legacy of ocean exploration and ingenuity. In this sense, the sailing-vessel reconstruction and voyage looked backwards to understand and forwards to reimagine. Humanities scholars work with ocean stories in a similar way. The essays collected here study ocean narratives first to understand historical attitudes and then to imagine a resilient ocean future.

Note

1. Rachel Carson, *The Sea Around Us* (New York: Oxford University Press, 1951), 15.

CHAPTER 2

Literary Oceans

Ship, Crew, Climate

JEFFREY MATHES MCCARTHY

LET'S BEGIN WITH *MOBY DICK* and words from its marvelous narrator, Ishmael: "[H]aving little or no money in my purse, and nothing particular to interest me on shore, I thought I would sail about a little and see the watery part of the world."[1] For Ishmael, what matters most happens at sea. I'm here to echo Ishmael and tell you something you already know: the ocean matters. The ocean—the actual physical thing of tides and waves that Ishmael called "the watery part of the world"—is the primary space in which climate change manifests. Where that watery world overlaps with human lives, we get the fiercest impacts of a hotter planet. Scientists and citizens are recording bigger storms, changing currents, rising seas, melting ice, eroding shores . . . all at the moment when half of humanity lives within one hundred miles of an ocean. Arthur C. Clarke was right when he suggested that the name *Earth* is a misnomer, and that our planet should be named *Ocean*. Yet we are trained by art and habit to experience the ocean as a spectacle and not as the crucial, defining part of our life that it actually is. So, this chapter will study literary oceans where communities collide with overpowering marine forces.

Western culture's aesthetic traditions separate us from the ocean by treating it as something to observe when, in fact, we live with and amidst it. Immersion, participation, entanglement, absorption—this chapter highlights texts that make human beings active participants with the ocean. We will look closely at Joseph Conrad's short novel *Typhoon* and its story of one large storm blasting one ship's diverse human population. A participatory relation is what we should foreground in the era of climate change because the ocean is the vehicle of climate impacts, the space where climate problems compound, the medium through which environmental justice is denied, and the object of artistic creations that comment on human relations to planetary forces.

To this end, I suggest the boat at sea is a useful figure for society beset by a changing, threatening climate. I am particularly interested in how ships' crews respond to upheaval in natural systems. In this, I am much influenced by Steve Mentz's excellent work on shipwreck in early modern literature. In Mentz's work, the radical disruption of shipwreck models the experience of social life we will all face under climate change. The difference, though, is that I focus on the crews that do not sink, and thus take the ocean as a venue for succeeding at reimagined social relations.

Philip Steinberg tells us the ocean is not a place but a construction, a construction wherein "social contradictions are worked through, social change transpires, and future social relations are imagined."[2] We can test this claim by looking to Conrad's novels from a century ago. In the twenty-first century, representations of boats at sea are influenced by culturally produced ways of seeing that are likewise useful for understanding and surviving a changing climate. For instance, an aesthetic tradition like the sublime that finds big weather events thrilling to observe is unhelpful in the Anthropocene. In this chapter, I want to criticize our received aesthetic category of the sublime because this cultural lens translates natural forces into spectacle. You're here watching the iceberg or the flood, and the consequences are happening somewhere else. This model of individual spectatorship is central to Western culture, and I will suggest that, right now, climate change demands we learn to value participatory ways of seeing and accept aesthetic modes that enact participation in one linked fate.

The ocean has been overlooked by humanities scholars, and environmental criticism has been land based. We "greens"—the ecocritics, environmental historians, environmental philosophers—study the pastoral and the forest; we are backpackers, we leave no trace, we think of Eden and garden metaphors.[3] Well, 70 percent of this planet is ocean. Emphasizing the blue alongside the green changes the conversation from solid ground to inconstant liquid, appreciates the moving wave in place of the stable garden, allows mobility alongside rootedness. Despite the ocean's size and influence, many climate commentators, and humanists in particular, overlook it. Margaret Cohen says literary criticism has ignored the ocean and treated "even those novels with oceangoing themes as allegories of processes back on land."[4] The ocean here is only a setting for terrestrial human dramas.

Likewise, the photographer Allan Sekula identifies this repression across the twentieth century in his movie *The Forgotten Space* and calls it, quite simply, "forgetting the sea."[5] The point is, our planet's defining feature, and the one most affected by a changing climate, is the one that most needs renewed attention. This is especially true in the coming era of climate displacement, with refugees and unraveling natural systems challenging terrestrial borders and social hierarchies.

Joseph Conrad's Weather

The author Joseph Conrad made his living on the sea and then built his literature around it. His writing sets human subjects into a nature that's tempestuous and threatening. Joseph Conrad's *Typhoon* shows a community under great strain as their social contract comes unglued during a storm at sea. Back in 1902, *Typhoon* was Conrad's most focused look at environmental disruption's social consequences, and it demonstrates my claim that Anthropocene disruption can be experienced as spectatorship but is, in fact, participation.

Typhoon is the story of a merchant ship and a storm. The steamer *Nan-Shan* is crossing the South China Sea with a cargo that includes two hundred Chinese laborers returning home after seven years of work in the tropics. She encounters a typhoon—a vast low-pressure system spinning counterclockwise across sun-warmed seas, gathering energy from that hot water and expelling it as powerful wind. Conrad's characters contend with natural violence to deliver both the physical ship that carries them and the ethical community that sustains them. Within the *Nan-Shan*, the extreme weather threatens both the European sailors and their Chinese travelers. Looking back from a twenty-first-century context of hurricanes and rising seas, Conrad's fiction spotlights environmental strain threatening moral obligation between human beings, and especially between human beings who speak different languages or carry different color skins. In his "Author's Note" to *Typhoon*, Conrad says the crucial element here is "not the bad weather but the extraordinary complication brought into the ship's life at a moment of exceptional stress *by the human element* below her deck."[6] The novel provides readers an inside look at that "human element" under storm conditions and delivers readers a warning about the fragility of human social structures under climate stress.

The ship appears doomed to sink in the great storm. Nonetheless, her captain rouses his first mate to resolve the chaotic conditions besetting the

Chinese laborers below. You see, they came aboard each with a box holding their wages and rewards from seven years' colonial labor. The violent storm broke those boxes free and shattered them, leading two hundred panicked workers to chase after their small earnings in the tumult. These people are buffeted and tossed by the *Nan-Shan*'s gyrations, and the hold becomes a chaos of the dispossessed—poor people in a global system uncertain whether to cling to their paltry wages or to their own abused bodies. The view by lantern is a metaphor of climate-induced disaster: "[A] mass of writhing bodies piled up to port detached itself from the ship's side and skidding, inert and struggling, shifted to starboard with a dull, brutal thump."[7] Captain MacWhirr shouts orders to his young mate, Jukes, above the storm, "Ought to see . . . what's the matter" and insists he rouse himself from storm survival to aid these luckless others. Jukes's role in this story is to learn about human behavior and nature's storms at the same time. The narrative is structured so that we readers learn along with him. Jukes would merely observe if he could—"indifferent, as if rendered irresponsible by the force of the hurricane"[8]—content to look into the hold and aestheticize the people there, but the captain insists that he and they are participants with linked fates. "Had to do what's fair by them," says the captain. "It may not matter in the end," says Jukes. The captain responds, "Give them the same chance as ourselves. . . . Couldn't let that go on in my ship [even] if I knew she hadn't five minutes to live."[9] Captain MacWhirr prescribes storm ethics to the younger Jukes—you must act instead of look on. The moral center of the story is Jukes's halting recognition of another group's trouble and the captain's decision to address this misery no matter the ship's physical condition.

The *Nan-Shan*'s community in distress before an enraged ocean is a climate change parallel for today. Philosopher Glenn Albrecht calls our climate reality the "new abnormal": "In the Anthropocene, the so-called new normal—or what I prefer to conceptualize as the new abnormal—life is characterized by uncertainty, unpredictability, genuine chaos, and relentless change."[10] Art about the ocean can recognize this "new abnormal" and, I think, dramatize and reimagine those new conditions for an otherwise dizzied audience. Ships' crews enact these relations under fraught circumstances. Put differently, ships on an unpredictable sea give us a blueprint for humanity working with unstable elements. We need such blueprints for the new abnormal.

For Captain MacWhirr, the shipwreck threatening the *Nan-Shan* is also an ethical threat, and the people who share this frail craft must acknowledge one another's common humanity in a collapsing climate. The Europeans are as vulnerable as the Chinese—one officer goes mad and attacks his captain, other crewmembers curse and bicker. *Typhoon* makes its eponymous weather event the vehicle for exploring the ethical imperatives that define a community under stress. In other words, within the *Nan-Shan*, the extreme weather event threatens a human community with social chaos and with the breakdown of moral obligation between people. Conrad attaches social relations to the weather: "This is the disintegrating power of a great wind: it isolates one from one's kind."[11] As a genre, the novel succeeds by spotlighting psychological subtleties. Conrad's other ocean novels put similar ships under strain, but *Typhoon* focuses most clearly on weather's impact on crew. Thus, a novel like *Typhoon* can track the impact of a great storm on a single person's relation to his strained community. Tellingly, the violent storm in *Typhoon* delivers isolation and social disintegration to test sailors. Conrad presents hard weather as the acid test for individual character: "A wrestle with wind and weather has a moral value like the primitive acts of faith on which may be built a doctrine of salvation and a rule of life."[12]

The precarious flux of ocean experience is, I would say, a preview of climate uncertainty and its social consequences. Conrad forces readers to linger on a community's ethical precarity when Jukes at first refuses to assist the poor, dispossessed, subaltern passengers below decks. Steve Mentz's *Shipwreck Modernity* (2015) makes ships under stress a trope of modernity under ecological strain: "[S]hipwreck stories represent the human experience of natural hostility, narrating humankind's failed attempts to navigate an uncertain world."[13] In *Typhoon*, this attempt "to navigate an uncertain world" is as much moral as it is physical. One observer says of the cargo hold: "A regular little hell in there."[14] The superstorm and the struggling ship intensify social friction in ways that resonate with contemporary models of Anthropocene disruption. The IPCC assessment report warns, "Climate change is projected to increase displacement of people. . . . Displacement risk increases when populations that lack the resources for planned migration experience higher exposure to extreme weather events."[15]

How people will deal with challenging natural forces is the subject of much Anthropocene speculation and is, likewise, the subject of sea fiction. The philosopher Peter Sloterdijk writes: "It is characteristic of being human that human beings are presented with tasks that are too difficult for them, without having the option of avoiding them because of their difficulty."[16] This is the stuff of Greek tragedy, with its striving hero fated to die. When fierce hurricanes sweep our coasts, when the sea rises to circle Boston, when

Caribbean islands are crossed by waves, it's then we must join together to engage in tasks too difficult for any one of us. The success of that engagement depends upon a collaborative, practical intelligence that Conrad himself calls "the solidarity of the craft." Margaret Cohen's *The Novel and the Sea* pinpoints "craft" as the embodied knowledge with which professional mariners negotiate the lethal challenges of the unruly sea. The key dynamic of craft is skilled participation, while the worst quality of climate denial now is passive observation. To my mind, climate change emergencies will enact community responses already modeled by crews at sea facing tumultuous natural forces.

The Sublime

A culture's habitual way of seeing can hinder or assist its response to natural threats like hurricanes and droughts, and hinder or assist its response to the sociocultural consequences like displaced people, nativist violence, or national competitions for newly scarce resources. One barrier to communal response is an aesthetic tradition that treats storms and waves as a spectacle mixing delight with terror—the sublime.

Let's put the sublime to the test of climate change. We can start with Edmund Burke, who is certainly not the only theorist of the sublime, but one of the most influential. In his *Philosophical Enquiry into the Origin of Our Ideas of the Sublime and the Beautiful* (1756), Burke presents the sublime as a particular mode of heightened attention: "[W]hatever is fitted in any sort to excite the ideas of pain and danger, whatever is in any sort terrible, or is conversant about terrible objects . . . is a source of the sublime."[17] Burke situates all this "pain and danger" at a crucial remove from the perceiver who makes sense of it all: "When danger or pain press too nearly, they are incapable of giving any delight, and are simply terrible; but at certain distances, and with certain modifications, they may be, and they are, delightful."[18] However, Conrad's Jukes is not allowed to withdraw into that distance that would make the raging sea a splendid spectacle. Instead, he must act and participate in the lives of the dispossessed and work toward his own, connected, salvation.

The twenty-first-century ocean is often presented to Western viewers in the shape of the sublime. A familiar example of treating ocean storms as spectacle is the Weather Channel's coverage of hurricanes each autumn; it emphasizes bringing you close to the fury of hurricanes and plays up the tension of countdowns before impact on vulnerable communities. Weather Channel correspondents enter American living rooms and kitchens battered by storm winds from some pier in Haiti or flooded street in Puerto Rico. Waves crash behind them, a house slides into the water, the lens blurs with raindrops. Next a commercial for soap and we switch over to Sports Center. The point is that storms and sufferings are packaged into a sublime ratings mover—distant titillation occasioning strong feelings. In contrast, Conrad's ocean storm obliges a ship's mate to participate because the storm breaks through the aesthetic distance Burke defines as central to the sublime.

A category-5 hurricane displacing island communities too readily becomes entertainment on my television. The fact that a changing climate will deliver us bigger storms and outsized human consequences means that we need to analyze the terms by which we interpret other people's trouble. The sublime is an influential category of perception leading us toward a stimulating distance from the terrible object—too far from the storm surge or the forest fire and it barely registers, too close and the viewer's delight is displaced by fear. Burke's peer Joseph Addison likewise indicates the necessity of an aesthetic space between the scary force and the viewer who would be lifted by "pleasing astonishment." Addison locates the sublime at sea:

> Of all objects that I have ever seen, there is none which affects my imagination so much as the sea or ocean. I cannot see the heavings of this prodigious bulk of waters, even in a calm, without a very pleasing astonishment; but when it is worked up in a tempest, so that the horizon on every side is nothing but foaming billows and floating mountains, it is impossible to describe the agreeable horror that rises from such a prospect.[19]

In this titillating distance, the grand tradition of the sublime is perhaps helpful to the Romantic poet on holiday, but dangerously inappropriate to a world where superstorms, sea-level rise, and drought afflict the least wealthy the most violently.

Shipwreck

"I shouldn't like to lose her," says Captain MacWhirr of his ship and, by extension, the community in her. Within the *Nan-Shan*, the extreme weather threatens a human community with social chaos and the breakdown of moral obligation between people. If we want to emphasize a climate reading, this shipboard community can be read as the global economy unsettled by extreme weather. The steamer was built in Glasgow, Scotland, purchased by an English conglomerate, flagged for convenience in Siam, crewed by Brits, and carried Chinese laborers from "a few years of work in various tropical colonies" to their port of Fu-Chau.[20] So, when a storm seems likely to sink her, the *Nan-Shan* becomes an international order under duress, and the ocean itself an intermittent homeland for a global economy of contingent locales, not territorial sovereignty. In our own twenty-first-century context of superstorms, rising seas, and intensifying weather, humanity navigates an unfriendly environment, and the poorest humans have the fewest resources—"inert and struggling"—to negotiate that environment. In this, *Typhoon* shows a society under pressure resisting the threat of anarchy from one side and of racist demagoguery from the other.

Chinua Achebe famously called Conrad "a bloody racist" for the very good reason that *Heart of Darkness* obscures the actuality of African lives and turns the continent of Africa into a symbol of one white man's fall.[21] Amidst these failings, we can see that Conrad approaches race and power in ways that are productive for planning into the Anthropocene. For instance, *Typhoon* gives readers an ethical mirror to recognize themselves as part of a similar unraveling social fabric. Jukes is instructed to interact, not sit back and wallow. Conrad is notoriously inadequate in his relation to people of color, and in *Typhoon* he names Chinese laborers "coolies" and is otherwise offensive by today's standards. However, *Typhoon* does oblige readers to engage the ethical questions at the heart of the climate change experience: What do I owe displaced people? How will I respond to natural disaster? What do I have in common with people from other cultures and other races? These questions matter now. The World Bank's recent *Groundswell* study says climate will displace 140 million human beings within Latin America, South Asia, and sub-Saharan Africa by 2050. Thus, while Conrad's treatment of other races is not what today's reader would recognize as progressive or ideal, in *Typhoon*, as in *Heart of Darkness*, the story's racial reckoning brings us to a doorway through which each of us must step to frame the climate world of 2030, 2040, and 2050.

Philosopher Michel Serres insists we must renegotiate a "natural contract" between people and the earth, wherein humans pay attention to this world's messages. For Serres, this natural contract is latent in life aboard ships, and he says sailing life necessitates an instructive commitment between crewmembers:

> On board, social existence never ceases, and no one can retire to his private tent, as the infantry warrior Achilles did long ago. On a boat, there's no refuge on which to pitch a tent, for the collectivity is enclosed by the strict definition of the guardrails: outside the barrier is death by drowning. This total social state . . . holds seagoers to the law of politeness, where "polite" means politic or political. . . . They get the social contract directly from nature.[22]

In other words, everyone aboard ship is a participant. There are no distant spectators on Conrad's *Nan-Shan* and no sublime aestheticizing of the experience. The situation emphasizes the common outcome, Jukes realizes, despite rank or race. Nature has the attention of people in Bangladesh and the Bahamas, but many more see it televised and framed by a habit of spectatorship. To offer an environmental reading of Conrad like this one is to warn us that our unfolding encounter with a disrupted climate is not a spectacle in the distance but an ethical and physical threat to all passengers on planet Earth.

Winslow Homer's 1899 painting *The Gulf Stream* uses the Atlantic Ocean to underline the consequences of stormy nature and, for us, nature's "new abnormal." Homer's *The Gulf Stream* makes the ocean a place where ships are wrecked and offers a subtle difference from Conrad's mariners in *Typhoon*. This surviving mariner has no rudder and thus no steering, has no mast and thus no locomotion, lists to port and thus no stability, feels a squall coming, and sees a bigger ship disappearing into the horizon. Here is Albrecht's "new abnormal" in its worst case. The agency of engines and buoyancy that make *Typhoon* about responding to risk are absent here—there's nothing much he can do with his situation at sea—and the painting is about a dissolution, a consequence of powerful nature unraveling human systems. Art critic Bryson Burroughs observes that the painting "assumes the proportion of a great allegory,"[23] and for me that allegory is about humans dying in nature. When Burroughs says Homer's painting is allegorical, he means it offers some moral insight that we can apply toward the issues of our time. Like these nineteenth-century sailors, our own actions, however contingent, have put us at odds with the natural world we influence. Hence shipwreck becomes an allegory of our present condition. I believe this imminent shipwreck supplements the things we've learned from *Typhoon*.

First, art like *The Gulf Stream* gives us a way to make sense of human experience in a planetary climate so multifarious that the comprehension fails for want of human-sized references. Second, this image of a human body adrift in a harsh environment delivers an Anthropocene epiphany—we grasp our dire situation when it destroys our comfort. Third, viewing *The Gulf Stream* from a climate change perspective makes shipwreck important as the allegory of humanity powerless before a ferocious nature. This is shipwreck at its least helpful—no reinvention, only suffering. It's better to recognize Homer's passive suffering as one end of a spectrum whose other end gives us artifacts where wreckage can be an opportunity for reinvention: Derek Walcott's *Omeros*, Jason deCaires Taylor's Grenada underwater sculpture garden, and Shakespeare's *Tempest*, to name just three.

"Shipwreck is . . . a trope," Steve Mentz claims, "for the conflict between human bodies and nonhuman power."[24] Mentz makes shipwreck a site of conflict, and he argues shipwreck represents our climate condition where the expectation of stability is replaced by duress. But I would go further to say *The Gulf Stream* is an important supplement to our reading of Conrad's *Typhoon* because sometimes shipwreck, like climate change, offers no recovery, only consequences. In Homer's painting the black man is doomed, a few sugarcane stalks his only sustenance, and sharks or drowning the only outcome of the approaching squall. He turns his head away, resolute, but his body is the site where a violent climate expresses itself. Meantime, the developed world sails past unable or unwilling to see the destruction playing out for the less fortunate. In this way, the painting offers us an image of the Anthropocene that the data cannot.

National meteorological services and scientific agencies tell us the ocean is now a space of superheated storms, a place where risk turns into casualties, a place where instead of agency humans have only consequences. But the painting and Conrad's novel show that not all humans are equally affected by climate change. Given this reading, *Typhoon* prefigures the choice wealthier nations face in the next fifty years—the sublime spectacle of hurricane waves and flooded streets can inspire us to get closer and participate alongside others in a disrupted nature, or we can keep our distance, consume the "agreeable terror," and know the guilty pleasure of a spectacle seen from our own security.

One risk of distance is blaming the victims. The IPCC report *Climate Change 2014* warns, "climate change impacts are projected to slow down economic growth, make poverty reduction more difficult, further erode food security and prolong existing and create new poverty traps."[25] Captain MacWhirr appears to have made the right choices, but shipboard authority empowers tyrants like Captain Ahab and Captain Queeg, too. The social tensions consequent to climate disruption can empower authoritarian voices to promise stability and blame vulnerable groups. Writing in *The Atlantic* magazine, Robinson Meyer imagines Donald Trump as "the first demagogue of the anthropocene" because Trump advances racist intolerance and nativist identity to people suffering from economic stagnation. Meyer writes, "His supporters are drawn to him by a sense of global calamity. . . . This xenophobia is grounded in real-life trends . . . moribund economic growth and the mass-migration of non-white people. Both will likely intensify as the planet warms."[26] Simplifying planetary-scale forces into racial enemies and nationalist bromides tempts the ship of Western democracy toward authoritarian strongmen, when what the coming century requires is an inclusive response to shared hardships.

Society itself has long been symbolized as a ship, so the ship remains a useful figure for reckoning with climate change. From Plato's *Republic* to Melville's *Pequod* to Conrad's fiction, the ship of state, the ship in a storm, the ship beset by internal dissension has been a figure of the body politic facing trouble. W. H. Auden suggests that the ship is especially apt to symbolize a community in trouble: "The ship, then, is only used as a metaphor for society in danger from within or without. When society is normal the image is the City or the Garden."[27] Homer's *Gulf Stream* and Conrad's *Typhoon* become our images of advance notice—they capture the human scale of natural risks recognized while echoing the symbolic tradition of the ship of state. This artwork distills the moment action becomes necessary in the face of a changing nature. What else is the Anthropocene?

Conclusions

Here we can pause to theorize on the relation between art and climate in view of the specific artifacts, I've looked back a hundred years to present. Art about human experience at sea insists that we consider scale. Homer's work gives us one representative human struggling with natural forces that become suddenly threatening. A recurring test for scholarship about climate change is communicating the scale of the changes—they are as vast as the tons of ice melting from Greenland, and as tiny as the parts per million of carbon in the atmosphere. Timescales also alienate as studies look back a million years or project a thousand years into the future when readers are hungry now. This scale misses the human; it baffles and alienates its audience at the very moment that audience needs to see climate change as a human issue. Here is a role for the humanities. A painting like *The Gulf Stream* or a character like Jukes in *Typhoon* can translate climate to a human scale by putting their audiences into a setting that dramatizes taxing natural conditions. In this way, artifacts of ocean experience give us these old tools sharpened toward human life in the Anthropocene.

Amitav Ghosh argues something quite different in *The Great Derangement: Climate Change and the Unthinkable.*[28] For Ghosh, the serious novel fails because the novel is individualist, because it is anti-collective. In Ghosh's telling, we need climate change literature that is global in scope, grander in duration. I disagree. We have those big numbers already—from biologists, oceanographers, and chemists. What the novel can do is bring the human scale, what the paintings can do is explore bodily consequences, make climate results intelligible to an audience alienated by geological time and parts per million. So, I agree, this new climate reality demands new modes of description, new ways to understand nature, new narratives for the human future. I suggest we look to representations of individuals whose plight moves us to identification, moves us out of watching and into action.

Creative work can make vast social forces legible—think *Uncle Tom's Cabin, Things Fall Apart, Guernica*—and move people to collective action. My examples from Conrad suggest something more subtle. Conrad's *Typhoon* is recorded via individual characters who combine to create a collective responding well. Remember Serres's point about shipboard life: "On a boat ... the collectivity is enclosed by the strict definition of the guardrails."[29] There is no survival beyond the collective. Conrad's mode of narration is notable for its interlocking letters home, fragmentary testimonies, and multiple views. These formal narrative choices underline the shipboard reality of a shared plight. Thus, on Conrad's ship, we can simultaneously experience Ghosh's collective while also getting to know the individual personalities that compose it. Stories and paintings have long been the instruments by which communities broadcast an ethic or distill an ethos, and this is very different work than the nature-writing tradition of the sublime. Rather than the aesthetic category of uplifted wonder that fills the mind with grand ideas, climate change calls for an ethics of active engagement, and the ocean is the perfect test case for these contrary responses.

In my telling, the ocean is both vehicle of a changing climate and apt image for human crews reckoning with the coming instability. *Typhoon* warns that our unfolding encounter with a disrupted climate is not a spectacle in the distance but an ethical threat to all passengers on Planet Earth. Let's not, I suggest to the wealthy and comfortable, make others' duress an entertaining thrill of "great though terrible" scenes. Representations of ships could instead foster a mode of attention that foregrounds inclusion, that perceives not a spectacle but a shared condition in which to participate actively. Peter Sloterdijk warned us earlier in this chapter that being human means wrestling tasks that are too difficult for us, but wrestling them nonetheless. This the crew of the *Nan-Shan* accomplished, and this the neighbors of a climate-changing ocean must undertake. So, in this volume, including those who study climate change from the sciences, from the law schools, and from the humanities, I add my hope that art from the ocean can focus our culture's attention on the ethical stakes in our climate future and in the difference between observing and participating.

Notes

1. Herman Melville, *Moby Dick*, ed. Hershel Parker (New York: W. W. Norton, 2018), 16.
2. Philip Steinberg, *The Social Construction of the Ocean* (Cambridge: Cambridge University Press, 2001), 209.
3. Happily, there are exceptional scholars focused on ocean research in the humanities, and I am fortunate to paddle along behind Margaret Cohen, Steve Mentz, Dan Brayton, and Stacy Alaimo, to name a few.
4. Margaret Cohen, *The Novel and the Sea* (Princeton: Princeton University Press, 2010), 14.
5. Sekula is quoted in Cohen, *The Novel and the Sea*, 48. Similarly, Dan Brayton's book *Shakespeare's Ocean* (Charlottesville: University of Virginia Press, 2012) argues that the problem is a cultural tradition that treats the ocean as separate from humanity, divorcing the aqueous from human concern. He reads Shakespeare as a corrective, advancing "a deep mutuality between humanity and the marine environment" (6).
6. Joseph Conrad, *Typhoon and Other Tales* (New York: Oxford University Press, 2008), xii.
7. Joseph Conrad, *Typhoon and Other Tales* (New York: Oxford University Press, 2008), 35.
8. Ibid., 39.
9. Ibid., 70.
10. Glenn Albrecht, "Exiting the Anthropocene and Entering the Symbiocene," *Psychoterratica* (blog), December 17, 2015, https://glennaalbrecht.wordpress.com/2015/12/.
11. Conrad, *Typhoon*, 46.
12. Joseph Conrad, letter to William Blackwood, August 1901, in *Collected Letters of Joseph Conrad* vol. 2 (Cambridge: Cambridge University Press, 1986), 354.
13. Steve Mentz, *Shipwreck Modernity* (Minneapolis: University of Minnesota Press, 2015), xxv.

14. Conrad, *Typhoon*, 37.
15. IPCC, Fifth Assessment Report, 2014, 73.
16. Peter Sloterdijk, "*Rules for the Human Zoo*: A response to the *Letter on Humanism*," trans. Mary Varney Rorty, *Environment and Planning D: Society and Space* 27, no. 1 (2009): 24, https://doi.org/10.1068/dst3/.
17. Edmund Burke, *Philosophical Enquiry into the Origin of Our Ideas of the Sublime and the Beautiful* (New York: Oxford University Press, 1998), 99.
18. Burke, *Enquiry*, 101.
19. Joseph Addison, *The Spectator*, no. 489 (1712): 69.
20. Conrad, *Typhoon*, 16.
21. Chinua Achebe, "An Image of Africa," *The Massachusetts Review* 18, no. 4 (1977): 790.
22. Michel Serres, *The Natural Contract*, trans. Elizabeth MacArthur (Ann Arbor: University of Michigan Press, 1995), 40.
23. Bryson Burroughs qtd. in Natalie Spassky, "Winslow Homer at the Metropolitan Museum of Art," *The Metropolitan Museum of Art Bulletin* (Spring 1982): 41.
24. Mentz, *Shipwreck*, xxv.
25. IPCC, 16.
26. Robinson Meyer, "Donald Trump Is the First Demagogue of the Anthropocene," *The Atlantic* (October 2016): 61.
27. W. H. Auden, *The Enchafed Flood: The Iconography of the Sea* (New York: Vintage, 1967), 11.
28. Amitav Ghosh, *The Great Derangement: Climate Change and the Unthinkable* (Chicago: University of Chicago Press, 2016).
29. Michel Serres, *The Natural Contract*, trans. Elizabeth MacArthur (Ann Arbor: University of Michigan Press, 1995), 40.

CHAPTER 3

Creating Ocean

Planetary Immersion and Premodern Globalization

STEVE MENTZ

HOW DID OCEAN arrive on Earth? How was human history shaped by our species' encounter with the great waters? Both the initial immersion of our planet's surface and the first stage of saltwater globalization during the early modern period narrate crucial stages in the long relationship between human culture and the World Ocean. The cultural meanings associated with the Ocean have varied over time, but the encircling seas have always been central to human ideas about ourselves and our environment. To understand Ocean is to make sense of how humans live in and with threatening nature.

I will start at the beginning, by exploring two opposing theories about how water may have arrived on this planet in the first place. Then I will trace a historical process that I call "wet globalization," or the ecological integration of the global world-system after the fifteenth century. These two moments—Ocean's origins, and the human experience of world-changing sea travel—represent key developments in the ongoing story of the human-ocean relationship. The ways these stories get told shape the way we think about the ocean today.[1]

Immersive Origins

Two stories herald Ocean's arrival. In what used to be the most common explanation of the blueness of our planet, Alien water streaked through space on an icy comet that splashed down onto barren rock.[2] In a newer, alternative explanation, Ocean was here all along, its waters stored inside the planetary Core of rocky material gathered together billions of years ago when the planet formed.[3] Water, the element that fills up the Ocean and makes life possible, either dropped from the sky or oozed up from a solid center. It's Alien or Core. These rival possibilities establish Ocean as an ambiguous thing from its prehistoric origins. It is an object with two meanings, two origins, two stories.

The Alien story describes radical intrusion and the imposition of external forces onto a stable if lifeless planetary system. Think of that dry rock in the void, four-and-one-half billion years ago, newly created and circling the sun. Onto its dry, jagged surface splashes the ice comet with its Alien cargo. The newly arrived ice melts, spilling across our planet's surface and reforms our earth as moist and ready for life. So much depends upon this chance encounter in the void, this Alien moisture from above. Space water, while not itself living, creates the conditions without which life cannot begin.

In the Core story, no sky-borne messenger appears from the void. Water instead arrives with the rocks that accrete into the planet on which we walk today. In this version of events, the massive impacts and forces that facilitate planetary creation do not disperse water onto space-traveling comets, but conceal it inside the planet's rocky Core, from which depths, over geological time, it seeps up to fill surface basins. Ocean need not descend from the sky; instead, the great waters emerge, hidden, from the oldest rocks in our planet.

I don't have the expertise to judge the scientific controversy between these opposing stories, but I know what the split between them means. These are stories about the origin of life and the blueness of our planet. Without water, and without a planetary temperature range that includes all three of this substance's phases—solid ice, liquid water, and gaseous vapor—life as we recognize it could not have developed on earth. Ocean origins are birth stories, in the most fundamental sense. Did we come from the sky or from underground? What does it mean that we seem not to be able to choose?

The story of Ocean's fall from the stars reinforces our terrestrial alienation from the soup of life. We fear the ocean, especially in its stormy moods. Sailors and swimmers know that human bodies can't go there to stay. Ocean surrounds our dry homes as a place of risk, vulnerability, and human weakness. We live near the waters, we employ them, and we love them. But they are not our home.

Unless they are. The Core story, championed by astrophysicist Lindy Elkins-Tanton among others, locates water inside our earth from the start. A leader in space exploration and primary investigator for the planned NASA mission to the metal asteroid Psyche in 2022, Elkins-Tanton has studied the chemical composition of water in asteroids and on earth and demonstrated that water can endure the process of planet formation. She describes herself as having become an "evangelist" for all "planets getting their water through their common formation processes, and not by later chance."[4] Perhaps, she speculates, more interstellar planets than just ours have formed with watery Cores that could support life.

Is our ocean Alien or part of our essential Core? The interlaced stories of how humans have imagined and interacted with the ocean over time reveal that both contrasting narratives speak a lasting truth. The ocean is our inhospitable home, but even that oxymoron doesn't quite capture the tension and urgency, dependence and fear, in the human-sea relationship. Humans share some evolutionary characteristics with aquatic mammals, including a layer of subcutaneous fat, relative hairlessness, and bipedalism. The controversial "aquatic ape hypothesis," in which a crucial phase in the evolution of Homo sapiens may have happened in an oceanic or at least terraqueous environment, remains to a large extent unverifiable, but it speaks to an "oceanic feeling" that many of us recognize, even if we don't know where it comes from.[5] Herman Melville wasn't an evolutionary biologist or NASA scientist, but he knew what people love: "They must get just as nigh the water as they possibly can without falling in" (*Moby-Dick*).[6] Poised on the sea's edge, we balance between kinship with and alienation from the watery part of the world. Ocean insinuates its salty fingers into that division and wedges meaning out of both the longing that draws us to the great waters and the fear that drives us away.

I feel both these things. Every day I walk down my street to a crescent moon of gritty sand bearing the unoriginal name "Short Beach." There I look out past a pair of rocky headlands onto Long Island Sound. On clear days, I glimpse the North Shore of Long Island, about twenty-five miles away. Sheltered from northeast-churning hurricanes by the massive glacial moraine of the island, my bay is a relatively calm body of saltwater but, like all inlets, it flows out into the massive encircling currents of the World Ocean. Every day in summer and fall I throw myself into the water's gray-green embrace and think about what it means to put my singular body into the biggest object in the world. It's disorienting and pleasurable and helps me think. According to Diana Nyad, the only person to have swum without a shark cage the hundred miles between Cuba and the United States, swimming creates sensory deprivation and a particular form of physical meditation.[7] Every day I churn sentences through my mind to the rhythm of crawl-stroke arms.

Many of the sentences that I write find me first in those salty waters. The meanings of Alien environment and the waters of our body's Core never strike me more palpably than when I'm swimming, with my head down in the water, ears and nose clogged, minimally aware. I'm a mismatched terrestrial creature partly at ease in the water, relying on repeated movements to keep me moving on the surface. I'm also a fleshy bag of water, matching my fluid center to my aquatic surroundings. Both things, always.

In perhaps the most influential book written in English on the ocean in the twentieth century, *The Sea Around Us* (1951), Rachel Carson places "that great mother of life, the sea" at the Core of the human story.[8] But she also recognizes that when humans return to the great waters, as "in the course of a long ocean voyage," the insights that swim into our imaginations reflect Alienation: the sailor "knows the truth that his world is a water world, a planet dominated by its covering mantle of ocean, in which the continents are but transient intrusions of land above the surface of the all-encircling sea."[9] Few sea writers speak with Carson's particular combination of poetic fervor and scientific exactitude. Her stance of objectivity and clarity may risk monumentalizing or idealizing the sea, but her responsiveness to both experience and emotion remains my polar star. My particular flavor of the blue humanities juxtaposes Carson's precision with Herman Melville's obsession. A theoretical and historical complement to these versions of the sea also appears in the "socially constructed" ocean of geographer Philip Steinberg, who charts the changing political, legal, and economic senses in which humans have understood the ocean in the past half millennium. Steinberg writes a complex social history that emerges from classical Greek geographer Strabo's notion that humans are "amphibious," intimately connected to both land and sea.[10] Bringing together the contrasting strains of Carson's positivism and Steinberg's critical theory makes this chapter intellectually amphibious as well—though I hope in more than a binary sense.

No models of ocean thinking can exhaust the water's undulating surface and unseeable depths. Carson's history and Steinberg's theory map out two entwined planes, but the ocean's verticality plunges down into hidden ways of thinking, including histories of change and human suffering. The Martiniquan poet and philosopher Éduoard Glissant envisions the sea as a site of criminality and retribution. For Glissant, the essential point of origin is neither Alien arrival nor Core secretion but instead the drowned human bodies of the Middle Passage whose remains sediment the Atlantic floor. Following Derek Walcott's poetic insistence that "the sea is history," Glissant imagines "the entire ocean, the entire sea gently collapsing in the end into the pleasures of sand, make one vast beginning, but a beginning whose time is marked by those balls and chains gone green."[11] Sunken bodies and rusting chains present a different order of origin than stories about comets and planetary accretion.

Behind the dizzying plurality of the figures and currents of history and thought represented by Glissant, Walcott, Carson, Steinberg, and others lies the vast bulk of Ocean as object, an immense, moving, vibrant, imagined, and ungraspable collective. Let's dive in.

Wet Globalization and the Premodern Anthropocene

When the supercontinent Pangea fractured and began to drift into distinct continents 750 million years ago, the ecosystems of the newly separated land masses diverged from one another. There was some contact between the massive land mass on which *Homo sapiens* first evolved, Afro-Eurasia, and the distant ecosystems of the Americas, but not much. During the most recent Ice Age, which ended nearly 12,000 years ago, water levels were low enough that many animals, including humans, traveled across the land bridge connecting Siberia to North America. But after the ice melted and the seas rose, the separation of the ecospheres became closer to absolute. Ocean made the continents largely Alien to each other, at least in human terms. Birds crossed the oceans, a small number of Viking ships crossed the North Atlantic around 1000 CE, and the range of maritime exploration of Pacific Island cultures remains difficult to discern precisely. Nevertheless, in broad terms, the living networks and human ecologies of the Americas remained separate from those of the larger connected land mass of Afro-Eurasia from before the dawn of history until the late fifteenth century. During many thousands of years, animal, plant, viral, and human ecologies on each side of the Atlantic and Pacific basins developed in isolation from each other. The overly familiar date of 1492 fingers Columbus as the start of a new era of ecological globalization, but the arrival of the Portuguese fleet led by Vasco da Gama in India in 1499 also marked a key east-facing node in what would soon develop into a global maritime network of trade, violence, colonization, and eventually empire. In the phrase of Earth system scientists Simon Lewis and Mark Maslin, the period humanities scholarship calls "early modern" witnessed the creation of a "New Pangea" that relinked the ecologies and economies of the once-sundered continents.[12]

Many different names have been proposed for the period in which European sailors took to the World Ocean, encircled the globe, and began to connect the global spaces of World History. The environmental historian Alfred Crosby influentially proposed the term "Columbian Exchange" in 1972, inaugurating a tradition that would treat Columbus's voyages to the Americas as the essential first step in global transformation.[13] Crosby's focus, however, was on an ecological process, not on a single human individual like Columbus. Crosby primarily engaged with the physical interweaving of the living and nonliving networks of the sundered parts of Pangea, the Americas, and Afro-Eurasia. Bringing these ecosystems back together, he observed, pushed unlike systems into becoming increasing like each other. "That trend toward biological homogeneity," Crosby writes, "is one of the most important aspects of the history of life on this planet since the retreat of the continental glaciers."[14] Charles C. Mann, whose two works of global ecohistory, *1491: New Revelations of the Americas before Columbus* (2005) and *1493: Uncovering the New World Columbus Created* (2011), have done much to communicate the consequences of Crosby's vision, emphasize the "role of exchange, both ecological and economic" in what was "not . . . the discovery of a New World, but its creation."[15] Marxist eco-historian Jason W. Moore has emphasized how the development of frontier capitalism during the early modern period gave rise to a "world-ecology" of appropriation and exploitation. "Capitalism," Moore argues, became during this period above all "a way of organizing nature."[16] Historians of the period from 1400 to 1800 CE also use categories like "early modernity" and the older Eurocentric term "Renaissance" to describe a process of cultural expansion that can also be accurately named "first globalization," to borrow Geoffrey C. Gunn's description of the worldwide trade network that arose after 1500 CE.[17] An older historical tradition referred in Eurocentric terms to the "discovery of the oceans" after the voyages of the 1480s and 1490s to Asia and the New World.[18] In an almost certainly vain effort to avoid canonizing individuals, I have suggested that we omit Columbus and instead describe this period as "wet globalization," because its crucial technology was the ocean-going ship and its antifundamental environment, the ocean.[19]

In previously describing "wet globalization," I have used the phrase "offshore trajectory" because those words emphasize that the process relies upon global movements on and of salt water. Sea travel connected humans, nations, empires, and religions—not to mention plants, animals, viruses, and ecosystems. Globalization operates now and has always operated through sea routes, from the Spanish silver trade that linked the Pacific coast of the Americas to the Philippines to China in the sixteenth century, through recent efforts in the summer of 2018 by Russian container ships to open the Northwest and

Northeast passages through no-longer-icebound Arctic waters. Even though today most individuals, at least most relatively wealthy individuals, travel the globe by air, the goods that comprise the global economy still travel by ships, in standard-sized containers. Shifting our attention from firm ground to the unstable fluid covering most of our planet's surface emphasizes that many events in the Ages of Discovery and Empire emerged through forces and encounters that were largely beyond the control of individual humans, even those well-known figures who have been canonized as "discoverers" or "explorers." It may be too late at this point to rename the "Columbian Exchange" with less focus on one man, though one of the reasons I prefer the phrase "wet globalization" is the way those words capture the impersonal nature of the forces at work during this period. The World Ocean, with its interwoven patterns of currents and prevailing winds, drove the populations of the separated continents back together. No single mariner, nation, or community fueled those voyages by themselves. The New Pangea floats on the ocean.

For Lewis and Maslin, from their Earth system science perspective, wet globalization in the early modern period marks a new phase in anthropogenic climate change. This period, they argue, fixes the Anthropocene and the origins of the modern world ecosystem. In nominating the year 1610 as the "Orbis Spike," a "Golden Spike" marker inaugurating the Anthropocene, they emphasize the more-than-human consequences of this moment in human history:

> In Earth systems terms it is the last globally cool moment before the long-term warmth of the Anthropocene, and the key moment after which Earth's biota becomes progressively globally homogenized . . . thereby setting Earth on a new evolutionary trajectory.[20]

The scientists choose 1610 as an observed minimum for carbon levels, which would subsequently increase rapidly after industrialization and global population growth. While as a humanities scholar who believes in stories, I remain suspicious of all magic dates, their nomination of 1610 makes a valuable contrast to Columbus's 1492. For Lewis and Maslin, the Age of Man begins as an Age of Death: historical estimates of the death rate of Native American populations indicate that at least 70 percent of the pre-Contact population, and perhaps as high as 90 percent, died within the first one hundred and fifty years after the arrival of Europeans and the viruses and bacteria that sailed with them.[21] Total human casualties during the period of Contact and early colonization ranged from 50 to 76 million. It was on a decimated American continent that early European settlers planted their flags: "The arrival of Europeans in America probably killed about 10 per cent of all humans on the planet over the period 1493 to 1650."[22] The death toll of early modern ecological disruption roughly matches the 50 to 80 million killed globally during the much shorter period of World War II, but in 1945 that number represented only around 2 percent of the global human population. The decimation of American populations in the early modern period holds up a horrific mirror in which we can glimpse worst-case scenarios of large-scale climactic disruption. Lewis and Maslin demonstrate that transoceanic travel during the early modern period marks the key moment when human actions dramatically reshaped our environment on a planetary scale:

> The 1610 Orbis Spike marks the beginning of today's globally interconnected economy and ecology, which set Earth on a new evolutionary trajectory.... In narrative terms, the Anthropocene began with widespread colonialism and slavery: it is a story of how people treat the environment and how people treat each other.[23]

The World Ocean flowed with the blood of Native Americans in the years after Contact, and soon after the drowned bodies jettisoned by the Middle Passage would further bloody the waters.

From the red waters of conquest, slavery, and settlement emerged the new ecological order of globalization. The material and symbolic centrality of two genocides—the disease-driven extermination of Native Americans and the intentional displacement of the Middle Passage—locates early modernity under an oceanic cloud. Glissant suggests that the slave ship and the Caribbean waters into which African bodies were thrown capture the hard birth of something new in the world:

> This boat is your womb, a matrix, and yet it expels you. This boat: pregnant with as many dead as living under sentence of death.[24]

The multiplicity that the poet-theorist names Relation launches itself out from the womb of the Middle Passage. As historian Marcus Rediker has shown in his award-winning *The Slave Ship: A Human History* (2007), the ocean-going ships that transported human cargo from African into the New World created global modernity.[25] The living and dead with whom the ships were "pregnant," in Glissant's term, formed the crucible of the global economy and ecology that would define the modern era. Rediker cites W. E. B. DuBois's well-known observation that the slave trade was the "most magnificent drama in the last thousand years of human history,"[26] and Rediker's expertise as maritime historian shows in detail how that tragic drama relied on the practice of transoceanic navigation. The history of the Middle Passage, Rediker concludes, contains at its heart the inchoate "terror"[27] experienced below decks of the slave ship. Around that ship, managing its buoyancy and its direction, a maritime culture steered the world into a new phase of globalization. "The sea is history," intones the sonorous verse of Caribbean Nobel Laureate Derek Walcott.[28] But also, counters the prose of British-Guyanese writer Fred D'Aguiar, "The sea is slavery."[29] Both the poet and the novelist recognize that Ocean traces the fluid connection between history and slavery, and between the Alien presence of the Ocean and its Core contributions to the shaping of human cultures.

Any consideration of wet globalization guides our attention inevitably toward the slave trade, as if drawn gravitationally by cruelty and world-changing evil. The devastation and upheaval wrought by nonhuman agents during the era of wet globalization, in particular the Afro-Eurasian diseases that devastated Native American humans whose bodies lacked antibodies to counter them, may have killed a larger number than did the slave trade—but the moral blindness of the slaver exposes the fundamental inhumanity to fellow humans that was the harbinger of first globalization. As Lewis and Maslin observe, the choice of any hinge-point in the long arc of ecological history leading to the Anthropocene amounts to a narrative choice. Their choice of 1610 and the wet ecological globalization that brought forth the violent birth of the New Pangea emphasizes human cruelty as well as unintended ecological consequences as drivers of climate change. On a fundamental physical level, the ultimate cause of this global catastrophe was the saltwater substrate on which this cruelty floated, across which those viruses and bacteria traveled to the New World, by means of which the separated ecosystems merged into a single global system.

The devastating consequences of wet globalization for the New World included the political collapse of major Native American polities in Mexico, Peru, and elsewhere. The cultures that later arose in the Americas, first as European colonies and later as independent nations, developed as ocean-centric states. Maritime passages to and from Europe and Asia dominated the trade in goods such as sugar, rum, tobacco, and indigo. The human consequences of living inside this global system include a distinctively American obsession with human liberty. As the historian Edmund Morgan has observed, "the growth of freedom experienced in the American Revolution depended more than we would like to admit on the enslavement of more than 20 percent of us at that time."[30] Morgan emphasizes not just the material contributions of slave labor—the slaves that built the White House—but also the ideological gymnastics required to justify a slave-holding nation dedicated to human liberty. The omnipresence of maritime slavery in the early modern New World generated a particular freedom story known as marronage. The fantasy of escaping from slavery to build a free society became historical fact in innumerable locations across the New World. These maroon communities included such disparate groups as the cimarrones in Panama with whom Sir Francis Drake made an anti-Spanish alliance in the 1580s, to assorted larger groups that mixed with Native Americans in the Caribbean, Suriname, French Guinea, and other locations. As detailed by Richard and Sally Price, the Saramanka maroon communities of Suriname maintain today a complex African and American hybrid social and linguistic culture.[31] Maroon peoples in the Americas are not as evident on the global stage as the descendants of European settlers, but the fantasy of marronage—of taking flight into liberty—represents an essential dream of the New World. This dream of freedom was not exclusively Oceanic—maroons vanished into the high country of Jamaica, in one example—but maritime mobility and alienation from their former surroundings motivated and shaped the forming of maroon societies in the Americas.

To maroon one's way from slavery to freedom represents an individual effort to cast oneself into the sea of history and also to swim by one's own power. As the philosopher Neil Roberts wrote in his book *Freedom as Marronage*, this action "is a multidimensional, constant act of flight."[32] In Roberts's view, the flight-into-marronage captures "modernity's underside."[33] Building on Glissant's writings on freedom and marronage in the Caribbean, Roberts develops a "marronage philosophy [that] runs counter to the idea of fixed, determinate endings."[34] For Roberts, the unfixedness of marronage rejects Kantian philosophical ideas of freedom and autonomy—but in a saltier key, where unfixedness suggests maritime connection. Acts of flight, escape, and radical difference speak to Roberts's critique of Enlightenment political philosophy, but they also speak to the destabilizing process of exchanging solid ground for liquid sea. Glissant describes the birth of global modernity through the contrast between the Mediterranean, "an inner sea surrounded by lands," and the fracturing Caribbean, "a sea that explodes the scattered lands into an arc."[35] In this "sea that diffracts,"[36] human culture assumes maritime multiplicity.

Wet globalization links the flight into radical freedom described by Glissant and Roberts with the resuturing of the global ecology into a "New Pangea" described by Lewis and Maslin. This premodern Anthropocene—the world humans have built, intentionally and not, from the late fifteenth century forward—relies on and is unimaginable without the structural movement and violence of the sea. "We are the Ocean," the claim that Epeli Hau'ofa makes in relation to Pacific Island cultures, can be extended to modern world history and its bloody global conflicts.

Conclusions

Considering the human experience of the ocean during the early modern period and through the stories we tell about the arrival of water on our planet suggests that the great waters are both central and mysterious to our species throughout history. The Alien or Core metaphors that describe the origins of water on our planet also represent the way the sea shapes our history and our present. In one important strain of ocean-thinking, Ocean is Us: there is no human civilization without oceanic connection. In fact, the global expansion we have begun to call the Anthropocene may not be separable from the transoceanic turn of human cultures that surged into view roughly half a millennium ago. On the other hand, however, Ocean remains determinably Alien. The 70 percent of the planet's surface covered by salt water represents an environment in which human beings cannot live, at least not for long. As historical and cultural scholarship in what has become known as the "blue humanities" continues to explore the relationship between humans and oceans, some may seek a possible integration between the Alien and Core perspectives.[37] It may also prove, however, that it is disjuncture and rupture that define the human experience of the place on earth that is necessary to life but in which we cannot live long.

Notes

1. The following material adapts two chapters from my forthcoming book in the Object Lessons series, Steve Mentz, *Ocean* (London: Bloomsbury, 2020).
2. For recent consideration of the nature of water on comets, based on 2018 observations, see https://www.space.com/water-from-comets-46p-study.html/.

3. On the earth's core as a "reservoir of water," see https://www.nature.com/articles/s41561-020-0578-1/.
4. Jeffrey Jerome Cohen and Lindy Elkins-Tanton, *Earth* (London: Bloomsbury, 2017), 19.
5. Elaine Morgan, *The Aquatic Ape Hypothesis*, 2nd ed. (London: Souvenir Press, 2017).
6. Herman Melville, *Moby-Dick*, ed. Hershel Parker and Harrison Hayford (New York: Norton, 2002), 19.
7. Diana Nyad, *Other Shores* (New York: Random House, 1978).
8. Rachel Carson, *The Sea Around Us* (New York: Oxford University Press, 1951), 3.
9. Ibid., 15.
10. Philip Steinberg, *The Social Construction of the Ocean* (Cambridge: Cambridge University Press, 2001).
11. Éduoard Glissant, *Poetics of Relation*, trans. Betsy Wing (Ann Arbor: University of Michigan Press, 1997), 6.
12. Simon L. Lewis and Mark A. Maslin, *The Human Planet: How We Created the Anthropocene* (New Haven: Yale University Press, 2018), 166.
13. Alfred Crosby, *The Columbian Exchange: Biological and Cultural Consequences of 1492*, 30th anniversary ed. (Santa Barbara: Praeger, 2003), 3.
14. Ibid., 3.
15. Charles C. Mann, *1491: New Revelations of the Americas before Columbus* (New York: Vintage, 2005); Charles C. Mann, *1493: Uncovering the New World Columbus Created* (New York: Vintage, 2001), xxiv.
16. Jason W. Moore, *Capitalism in the Web of Life: Ecology and the Accumulation of Capital* (London: Verso, 2015), 3, 78.
17. Geoffrey C. Gunn, *First Globalization: The Eurasian Exchange, 1500–1800* (London: Rowman and Littlefield, 2003).
18. J. H. Parry, *The Discovery of the Sea* (Berkeley: University of California Press, 1974).
19. Steve Mentz, *Shipwreck Modernity: Ecologies of Globalization, 1550–1719* (Minneapolis: University of Minnesota Press, 2015), xxix.
20. Lewis and Maslin, *Human Planet*, 318.
21. Ibid., 156.
22. Ibid., 158.
23. Ibid., 13.
24. Glissant, *Poetics of Relation*, 6.
25. Marcus Rediker, *The Slave Ship: A Human History* (New York: Viking, 2007), 348, 354.
26. Ibid., 348.
27. Ibid., 354.
28. Derek Walcott, "The Sea Is History," in *Selected Poems*, ed. Edward Baugh (New York: FSG, 2007), 137–39.

29. Fred D'Aguiar, *Feeding the Ghosts* (New York: Ecco, 1999), 3.
30. Edmund Morgan, *American Slavery, American Freedom: The Ordeal of Colonial Virginia* (New York: W. W. Norton, 1975).
31. Richard Price, *Travels with Tooy: History, Memory, and the African-American Imagination* (Chicago: University of Chicago Press, 2007).
32. Neil Roberts, *Freedom as Marronage* (Chicago: University of Chicago Press, 2015), 9.
33. Ibid., 23, citing Enrique Dussel.
34. Ibid., 174.
35. Ibid., 33.
36. Ibid.
37. For a rough definition of the "blue humanities," see John R. Gillis, "The Blue Humanities," *Humanities: The Magazine for the National Endowment for the Humanities* 34, no. 3 (2013): https://www.neh.gov/humanities/2013/mayjune/feature/the-blue-humanities/.

CHAPTER 4

Minds Tossing on the Ocean

Venice, the Sea, and the Crisis of Imagination

SHAUL BASSI

She was a maiden City, bright and free;
No guile seduced, no force could violate;
And, when she took unto herself a mate,
She must espouse the everlasting Sea.
—William Wordsworth, "On the Extinction of the Venetian Republic" (1807)

EVERY YEAR VENICE marries the sea. On Ascension Day in the month of May, a ring is tossed into the waters that connect the lagoon with the Adriatic, marking the symbiotic relationship between the city and the sea, but more extensively the whole Mediterranean and all the marine routes that made the city a prosperous and powerful maritime republic. This ritual may have begun as early as the year 1000, after a victorious naval campaign against Slavic pirates, and it was certainly attributed sacred quality in 1177, when Doge Ziani brokered a peace treaty between Pope Alexander III and Emperor Frederick Barbarossa, receiving a golden ring from the former to perform the ceremony "so that posterity knows that the lordship of the sea is yours, held by you as an ancient possession and by right of conquest."[1] Since then the ceremony has acquired new meanings; by the early seventeenth century the Sensa (as the ascension is called in Venice) had already become a visitor attraction, and Venetians extended its duration and launched a new Bucintoro, the majestic state barge ship decorated with a rich array of sea creatures. Represented in realistic detail in various paintings (from Tintoretto to Turner), it is given its most famous allegorical figuration in *Neptune Offering to Venice the Riches of the Sea* (ca. 1745). This painting, commissioned to Giovanni Battista Tiepolo by the Venetian Republic, shows Neptune pouring out the treasures of the sea and the riches of commerce before Venice, depicted as a benevolent queen petting a very sleepy lion.

It has been aptly called "a masterpiece of nostalgia,"[2] because even though Venetians could not predict that the republic would end only a few decades later, the dominion over the sea was by then drastically curtailed by the "wet globalization"[3] of powerful nation-states such as England, Spain, and Portugal, getting the upper hand on the new Atlantic routes. The Sensa then completed its metamorphosis into a "giant spectacle for tourists"[4] and some foreign visitors were quick to capture the widening gap between the pomp and pride of the circumstance and the underlying truth. In 1741 the poet Thomas Gray recorded going to Venice "to see the old Doge wed the Adriatic Whore."[5] After the fall of the republic in 1797, as we read in Wordsworth's famous poem used here as epigraph, the rite became a memory and iconic symbol of a glorious past. In our times it exists as a historical reenactment, with the mayor of the city symbolically renewing the ancient rite mostly to celebrate the traditions of rowing in the inner spaces of the lagoon.

This festive event embodies the entanglements of a physical and geographical location of Venice with its history of symbolic and artistic representations, providing an important case study for the general observation made by Steve Mentz in chapter three of this volume: "Poised on the sea's edge, we balance between kinship with and alienation from the watery part of the world."[6] In this essay I argue that reconfiguring the relationship between Venice and the sea in culture and the arts may provide a specific contribution to a revision of our global cultural and social imaginary of the ocean in times of environmental crisis.

Such revision requires a brief overview of some relevant tropes. In an imaginative inversion of the island's relation to the sea, Venetian writer Tiziano Scarpa literalizes the common metaphor of Venice as a fish suggested by its peculiar shape.

> Venice has always existed as you see it today, more or less. It's been sailing since the dawn of time; it's put in at every port, it's rubbed up against every shore, quay and landing-stage: Middle Eastern pearls, transparent Phoenician sand, Greek seashells, Byzantine seaweed all accreted on its scales. But one day it felt all the weight of those scales, those fragments and splinters that had permanently accumulated on its skin; it felt the weight of the incrustations it was carrying around. Its flippers grew too heavy to slip among the currents. It decided to climb once and for all into one of the most northerly and sheltered inlets of the Mediterranean, and rest there.[7]

Lyrically capturing the long history of transcontinental and intercultural commercial exchange that made the fortunes of the republic, the author finally points to the fateful historical moment of the year 1846, when it became possible to reach Venice via land, imagining the bridge as the fishing line that has finally hooked Venice to the terra firma.

Venice largely depends on its marine dimension, and yet, as Deborah Howard has recently observed, its visual culture (which includes one of the most celebrated schools in Western painting) has been very reluctant to represent the sea.[8] This seeming paradox is epitomized by the grandiose tessellated floor of Saint Mark's Basilica, where a large rectangle has been known as "the sea" since the sixteenth century: the sea contains Venice but it is Venice that tries to contain the sea, or at least to tame it, privileging the quiet waters of the lagoon to the stormy waves of the ocean.

If we shift from the collective to the individual level, the same ambivalence was captured by a writer who never set foot in Venice and yet experienced its fascination at a distance. Life on land and life on water reflect on each other in Shakespeare's *The Merchant of Venice*:

> Your mind is tossing on the ocean,
> There where your argosies with portly sail,
> Like signors and rich burghers on the flood,
> Or as it were the pageants of the sea
> Do overpeer the petty traffickers
> That curtsy to them, do them reverence,
> As they fly by them with their woven wings. (1.1.8–14)[9]

"We fear the ocean, especially in its stormy moods," writes Mentz.[10] Venice aristocracy was proudly composed of travelling traders, but by Shakespeare's time the typical merchant was someone whose property was at sea, while his body and soul were left to wait, hope, pray, and worry on solid ground.

> Should I go to church
> And see the holy edifice of stone
> And not bethink me straight of dangerous rocks,
> Which touching but my gentle vessel's side
> Would scatter all her spices on the stream,
> Enrobe the roaring waters with my silks,

> And in a word, but even now worth this,
> And now worth nothing? (1.1.30–37)

The more prosaic Shylock is even more straightforward with his admonition, as he conjures up the human threats that add to the meteorological hazards.

> But ships are but boards, sailors but men: there be land-rats and water-rats, water-thieves and land-thieves, I mean pirates, and then there is the peril of waters, winds and rocks. (1.3.19–22)

As a modern resident of Venice, I live surrounded by stones and quiet canals, but somehow my mind is also tossing on the ocean. Our family's outdoor life revolves around Santa Maria Formosa, one of the most diverse *campi* (squares) in Venice, with museums and hotels, bars and restaurants, shops and stalls, businesses for residents and tourists. As an exercise, I started looking around me, self-consciously investigating all the historical references to the sea inscribed in the urban text. On the northern facade of the church, instead of the traditional saint, is the statue of Vincenzo Cappello, statesman and admiral of the Venetian navy. The unassuming palazzo on the opposite side bears a plaque dedicated to Sebastiano Venier, the *doge* who fought successfully in the historic battle of Lepanto. The Querini Stampalia foundation, where classic and contemporary art join hands, owes its name to the very ancient Querini family acquiring the Greek island of Astypalaia (Stampalia in Italian) in the fourteenth century. The street that departs on the left is called Ruga Giuffa, a toponym that refers to Armenian merchants coming to Venice from Old Julfa (today in Azerbaijan), while the *calle* on the right side of the church is called Mondo Novo, after an early modern store that attracted its patrons by alluding to the recently discovered "New World." War and peace, battles and trades—the exercise could continue at different levels, and the most unexpected and recent reference is in the recent blockbuster movie *Spider Man: Far From Home* (2019), where the same *campo* is devastated by a gigantic aqueous villain called Water Elemental. Standing on solid ground, we can be constantly reminded of the sea that has made us and that today threatens to unmake us.

* * *

Shifting from fantasy to reality, nowadays water wreaks havoc in the streets in more tangible form. Today the sea generates a different kind of anxiety from the one experienced by Shakespeare's characters and early modern Venetians, and the anticipation of the wedding is supplanted by the fear of obliteration. As the title of a book on the challenge of sea-level rise in the new millennium puts it, the general outlook is "Venice against the sea."[11] *La Serenissima*, the most serene republic, has become the *turbatissima*, the most troubled republic, in Amitav Ghosh's ingenious wordplay.[12] On November 12, 2019, Venice suffered the worst flooding in half a century. The streets routinely flood every November, and we are equipped with multiple warning systems—text messages and loud sirens—that tell us how high the water is going to be: our rubber boots are duly aligned alongside shoes next to the door. The sirens start at 110 cm (3 feet 6 inches) above sea level, which means a few inches above street level and enough to walk ankle-deep in water on many pathways. Up to 140 cm (4 feet 5 inches), signaled by four different distinct tones, it is bad but manageable: shopkeepers know they have to raise their merchandise and electrical appliances, parents have to carry their children. That night it was forecast it could go as high as 150 cm, but many people were still going about their business undeterred. The sirens were heard two hours before the peak. But then they were strangely repeated. And then the last tone was eerily prolonged, and we knew that something really bad was happening. Two powerful winds—the warm, southern sirocco and the cold, northern bora—started howling together, and with the full moon, the rising tide, and the torrential rain, the perfect storm—literally—materialized. New messages announced 160 cm (5 feet 2 inches)—a calamity—then 170 (5 feet 6 inches) . . . 180 (5 feet 9 inches). We all had one fateful measure in mind: the record-breaking 194 (6 feet 3 inches) that overwhelmed the city in 1966, causing the irreversible exodus of tens of thousands of people. This one eventually stopped at 187 (6 feet 1 inch). Minutes into the event, social media started bursting with alarming reports and even fake news of drowned people. We saw clips of streets turned into torrents; the sturdy ferries rocking like paper boats; waterbuses that were sunk and others tossed on land; gondolas piled on one another like a game of pick-up sticks. Outer walls collapsed; trees were uprooted, mangling open-air artworks. A person died on the island of Pellestrina, which divides sea from lagoon. The next morning, we learned of the extensive damage to houses and shops; of the books turned into pulp at bookstores and museums penetrated by the muddy liquids; of the newsstand I pass by every morning washed away with all its merchandise. The floor of St. Mark's Basilica, a millenary religious and civic symbol, was inundated with the worst damage caused by the salt left when the water receded. The only comfort came from the hundreds of people, young and old, who took to the streets to volunteer their help. That day there was also rage and resentment. Everybody was talking about the MOSE, the €5 billion flood-protection system sitting idly on the bottom of the sea, awaiting its technical test three decades after the project approval, after sending many corrupt public servants and businessmen to prison.

And then water came back for a full week, never as high but definitely as steady. Our parameters had suddenly changed. Now 140 cm (4 feet 5 inches), the highest measure for the alarm system, had become almost reassuring. The main change was psychological; *acqua alta* was no longer an annoying occasional occurrence, but a daily regular presence. As a reader of literature, I used to think of the end of Venice as a centuries-old trope. For the first time I felt it was an actual prospect, and I admitted to myself that I might not be able to end my days here. Ultimately the city was left in a schizophrenic situation. On the one hand, it was imperative to assess and report the exceptional damage; on the other hand, precisely those who had suffered the worst devastation advocated a fast return to normalcy in order to bring tourists back and restart the economy.[13]

* * *

Why dwell on Venice? The magnitude of the damage and human loss that I have described pales in comparison to that of natural disasters such as tsunamis in South Asia, or Katrina or Sandy in the United States, and many other nameless catastrophes striking regularly in the global South. Why is this city more noteworthy and media-attractive than other cities?

The conventional answer would be that Venice has an unparalleled concentration of beauty. A city that in 2021 has celebrated 1,600 years since its legendary foundation; a metropolis where wealth was diffused and so was aesthetic competition. Every monastic order desired the most elegant monastery, every parish the most splendid church, every noble the most lavish palazzo, all paid by lucrative mercantile enterprises. And at key historical moments of political and economic crisis, a state self-consciously investing in its own cultural and symbolic capital, magnifying its grandeur as its actual power was declining. A further hypothesis can be borrowed from Elizabeth M. DeLoughrey's illuminating study *Allegories of the Anthropocene*, where she argues that

> allegory has long relied on the figure of the island to engage the scalar telescoping between local and global, island and Earth. The island's simultaneous boundedness and its permeability to travelers—and

therefore its susceptibility to radical change—have made it a useful analogue for the globe as a whole. Of course, the island also represents finitude, a cautionary concept for the Anthropocene epoch of planetary boundaries that include threats to biodiversity and mass extinctions.[14]

Even though DeLoughrey talks about far more vulnerable Pacific islands subject to colonialism and neocolonialism, the pertinence for Venice is confirmed by Serenella Iovino:

> Our global ecological crisis confirms how deeply unstable and delicate is the equilibrium of natural-cultural substances and forces. As the perfect epitome of this fact, Venice represents the discordant harmony of elements upon which human civilization lies. Even more so, it challenges the very possibility of such a harmony. To create a city suspended on a lagoon ... is an exercise in hybridity, not only because it mixes water and land into a new elemental combination, but above all because it is an act of hubris, a violation of ontological pacts. Certainly, hubris may have a creative function, and Venice stays as the luminous splendor of this assumption.[15]

A city ubiquitously represented (and replicated!) worldwide, Venice has been frequently used in the Western imagination and beyond as utopia (a quasi-mythical place outside of time and place), as dystopia (a cautionary trope that suggests ominous apocalyptic futures), and as heterotopia (a site of romance and escapist fantasies). Today Salvatore Settis argues for a more historically nuanced appreciation of the city's global relevance:

> A city with a long history of cosmopolitanism, Venice can still be a testing ground for an inclusive notion of citizenship relevant to our times.... Venice is the paradigm of the historical city, but also of the modern city like Manhattan. It's a thinking machine that allows us to ponder the very idea of the city, citizenship practices, urban life as sediments of history, as the experience of the here and now, as well as a project for a possible future. Its problems are complex in an unparalleled way.[16]

If we agree with Amitav Ghosh that "[t]he climate crisis is also a crisis of culture, and thus of the imagination,"[17] Venice becomes an important vantage point to address this unprecedented challenge, and reimagining its past and future oceanic entanglements is a necessary gesture. This requires dismantling certain recurring rhetorical strategies. In the aftermath of the November flooding, Venice was personified on the cover of *Vogue* magazine as a pale young woman dressed in funeral black and holding a slab of rough-hewn stone carrying the bank account number of the municipality. A far cry from the complacent queen that jadedly accepts Neptune's cash flow in Tiepolo's painting, this modern figuration casts Venice as a passive, sickly victim of disaster whose survival depends on being saved and protected by the viewer's donations. Riches pour into Venice to keep Neptune out, mostly feeding the exorbitant costs of the MOSE. The underlying notion of this combination of necrophilia and technophilia is that of Venice as a crystallized entity in need of being preserved intact and unchanging. As archeologist Diego Calaon remarks: "The world wants Venice saved, so we preserve every single stone, aiming to preserve its material authenticity."[18] What is neglected is that the "natural catastrophe" of the flooding is not natural, deriving from a lethal concurrence of factors. The ecosystem nurtured by the Venice Republic for centuries has been indiscriminately tampered with over the last hundred years; cruise ships higher than the highest building in Venice have been allowed to come into the heart of the city, making the car-free city air dangerously toxic, contributing pollution to global warming, and threatening new intrusive excavations of the lagoon—two major causes of sea-level rise; tourism has been deregulated and left to escalate to overtourism, causing the extreme commodification of every aspect of Venetian life made available for rapid consumption and a rapacious extractivist economy.[19] In Calaon's archaeological and anthropological perspective, past Venetians had a completely different approach to the rising waters, one of constant care and gradual adjustment. The soil was naturally sinking, although the urban fabric was adapting the walking levels generation by generation. Great scientists such as Cristoforo Sabbadino and Benedetto Castelli made crucial contributions to the study of the management of waters. Against the nostalgic preservationists, the technocratic futurists, the disillusioned apocalyptics, and the naïve romantics who place Venice in an idealized sphere outside of history, the role of Venice as a city of innovation with its constant ability to reinvent itself needs to be reclaimed.

What is the role of literature in this context? As a possible answer, I would like to compare the last two important literary texts that have foregrounded the relationship between Venice and the sea, Joseph Brodsky's long essay *Watermark* (1989) and Amitav Ghosh's novel *Gun Island* (2019).

A childhood survivor of the siege of Leningrad and of Soviet persecution, the dissident-poet and Nobel Laureate Brodsky found political asylum in the United States and another home in Venice, to which he dedicated a beautiful lyrical autobiographical meditation.

This city takes one's breath away in every weather, the variety of which, at any rate, is somewhat limited. And if we are indeed partly synonymous with water, which is fully synonymous with time, then one's sentiment toward this place improves the future, contributes to that Adriatic or Atlantic of time which stores our reflections for when we are long gone. Out of them, as out of frayed sepia pictures, time will perhaps be able to fashion, in a collage-like manner, a version of the future better than it would be without them. This way one is a Venetian by definition, because out there, in its equivalent of the Adriatic or Atlantic or Baltic, time—alias water—crochets or weaves our reflections—alias love for this place—into unrepeatable patterns, much like the withered old women dressed in black all over this littoral's islands, forever absorbed in their eye-wrecking lacework. Admittedly, they go blind or mad before they reach the age of fifty, but then they get replaced by their daughters and nieces. Among fishermen's wives, the Parcae never have to advertise for an opening.[20]

Brodsky's sea is both a physical and a metaphysical presence in a city where he found refuge from the political storms he endured in his lifetime and where he was laid to rest. Rather than basking in hackneyed images of the past, Brodsky envisions a peculiar temporality in which Venice, because of its aqueous element, projects humanity to the future. But when his high poetic register gives way to semi-ironic sociological vignettes, the contradictions of the text emerge. The daughters and nieces of the lacemakers can no longer afford to live in Venice, or if they do, they are likely to find it more profitable to rent their deceased mothers' apartments on Airbnb than to inherit their art and craft. And there is an even more consequential ironic dimension to the text, when read thirty years after its publication and twenty-five after its author's premature death, an irony unwittingly captured by Sanna Turoma: "The writing of Brodsky's essay on Venice was initiated by a request from the *Venice Water Authority—Consorzio Venezia Nuova*, the local authority fighting the ecological crisis threatening the environment of the Veneto region and the foundations of Venice."[21] But not only was Consorzio Venezia Nuova *not* the Venice Water Authority, it was also emphatically *not* champion of ecological battles. Brodsky did not live long enough to see this state behemoth, created to protect Venice from the rising tides, being itself engulfed by the largest corruption scandal in the history of the city, one that led to the closing of the Venice Water Authority (whose former director was one of the many beneficiaries of the Consorzio's generous kickbacks), a magistrate that had existed in Venice for centuries and was proudly associated with a sixteenth-century Latin motto engraved in stone:

> By Divine Providence, the city of the Veneti is founded on water, surrounded by water and protected by water instead of by a wall. Hence, whoever dares in any way to damage the public waterways, let him be condemned as an enemy of the Fatherland and punished no less gravely than someone who has undermined a city's walls. Let this Edict remain in force immutable and perpetual.[22]

It would be silly and anachronistic to consider Brodsky complicit with a criminal system that degenerated decades after *Watermark*. The successful activation of the barriers in October 2020, hailed by many skeptics as a necessary move to fend off the threats of the sea and as a lesser evil after the exceptional flooding of the previous year, cannot erase the economic and moral loss inflicted on the city by a technological juggernaut under whose auspices the last great literary text that equals Venice with eternity and timelessness was composed.

Some thirty years after Brodsky, another residency program inspired Amitav Ghosh's *Gun Island* (2019), a novel based on a full immersion (pun intended) into the social fabric of Venice. In *The Great Derangement* he had asked: "Can anyone write about Venice any more without mentioning the *aqua alta*, when the waters of the lagoon swamp the city's streets and courtyards?"[23] Indeed, many continue to describe Venice without the *aqua alta* or dismiss the flooding as an innocuous picturesque feature, in the spirit of the foolish tourist who is invariably caught on camera swimming in Piazza San Marco while people are struggling to salvage their property. Ghosh also added: "Nor can they ignore the relationship that this has with the fact that one of the languages most frequently heard in Venice is Bengali: the men who run the quaint little vegetable stalls and bake the pizzas and even play the accordion are largely Bangladeshi, many of them displaced by the same phenomenon that now threatens their adopted city—sea-level rise."[24] The day after November 12, 2019, one could see many Bangladeshi workers mopping the floors of Venetian shops and selling disposable boots to the flabbergasted tourists. These twin observations coalesced into the plot of *Gun Island*: "Of all the gifts that Bangla had given me, this was by far the most unexpected: that it would help me find a context for myself in this unlikeliest of cities—Banadig, Bundook, Venice."[25]

Against the traditional literary representations of Venice as escapist fantasy, sentimental utopia, or melancholic dying place, Ghosh places the city at the center of a turbulent whirlwind of global phenomena, connecting it, as it were, to two marine ambits: the troubled sea of migration and the rising sea of climate change. Not the first postcolonial writer to challenge the eurocentrism of the Venice literary canon, Ghosh becomes the first to connect the twin forces of capitalism and imperialism to examine the impact of sea-level rise and mass migration on the city. *Gun Island* is a fast-paced novel that has been criticized for being full of improbable events, including a violent storm that was considered an implausible event until November 12. However, the simultaneity and interconnectedness of various dramatic and improbable events is self-consciously thematized in the novel by the narrator and protagonist Dinesh Datta:

> The word "chance" hit me with such force that I lost track of what Piya was saying. Shutting my eyes I silently embraced the word, clinging to it as though it were my last connection with reality.
>
> Yes, of course, it was all chance these unlikely encounters, these improbable intersections between the past and present; that almost fatal accident that had brought me face to face with Rafi in the Ghetto: all of this was pure coincidence, of course it was. To lose sight of that was to risk becoming untethered from reality; chance was the very foundation of reality, of normalcy.[26]

In the vast literary and philosophical exploration of chance, Aby Warburg's examination of "the moody goddess of sea trade—Fortuna" appears particularly relevant to analyze the improbabilities of *Gun Island*. As Franco Moretti reminds us, Warburg showed that, while long associated with "'chance,' 'wealth,' and 'storm wind' (the Italian *fortunale*)," the sea in early Renaissance images was "progressively losing its demonic traits; most memorably, in Giovanni Rucellai's coat of arms she was 'standing in a ship and acting as its mast, gripping the yard in her left hand and the lower end of a swelling sail in her right.'"[27] According to Warburg, this iconographic shift marked the onset of a new era in which the sea was gradually coming under control: "Fortune had become 'calculable and subject to laws,' and, as a result, the old 'merchant venturer' had himself turned into the more rational figure of the 'merchant explorer.'"[28] Moretti also quotes Margaret Cohen's *The Novel and the Sea*, where the adventure of Robinson Crusoe becomes less "a cautionary tale against 'high-risk activities,' and becomes instead a reflection on 'how to undertake them with the best chance of success.'"[29] Against this literary tradition of increasingly dominant seamen (even in their antiheroic Conrad variant), and right after completing his epic marine Ibis trilogy, Ghosh depicts a new type of modern-day traveler who inhabits a very different world. Dinesh Datta, the protagonist and first-person narrator of *Gun Island*, is mostly a character who is done by rather than doing, especially vis-à-vis the global oceanic dynamics. He is the victim of uncanny natural phenomena in India and Venice. He admits he is "only dimly aware of the phenomenon" when he is told that Italy has become a gateway for migrants—"thousands of *rifugiati* are coming across the sea, in boats from Libya and Egypt"[30]—and is subsequently hired as a translator to work on a documentary on the phenomenon of Bangladeshi migrants in Venice. He is a shy observer who is drawn to most of the events of the book by strong-willed, sophisticated, and high-principled women (Cinta, Piya, Gisa) who represent different forms of agency: Cinta is the humanist, Piya (a character from Ghosh's novel *The Hungry Tide*) the marine scientist who studies dolphins, Gisa the activist who documents the migrant crisis. In its praise of intellectual work (and specifically of the invaluable work of translation, which here is not only translation across languages but also across cultural and social codes), the novel also foregrounds the necessary alliance between the humanities, sciences, and political activism.

Crucially, the Venetian sea cartographed by the novel also has historical depth, connecting the seventeenth-century Gun Merchant and the Ghetto inhabitants to the twenty-first-century migrants. Countering the post-Romantic canon that gradually placed Venice in splendid isolation, Ghosh fully embraces the transcontinental commercial and cultural maritime networks that made the fortunes and beauty of the metropolis, a fact still not properly acknowledged in the current popular narratives.

I fell silent, overtaken by an overwhelming feeling of gratitude—towards the Gun Merchant, to his story, to Manasa Devi, and even to that king cobra: it was as if they had broken a spell of bewitchment and set me free.

My eyes wandered to the moonlit sea and I was reminded of a phrase that recurs often in the Merchant legends of Bengal: *sasagara basumati*—"the ocean'd earth." At that moment I felt that I was surrounded by all that was best about our world—the wide open sea, the horizon, the bright moonlight, leaping dolphins, and also the outpouring of hope, goodness, love, charity and generosity that I could feel surging around me.[31]

The distance could not be further from the lyrical irony of Brodsky, for whom the ineffable beauty of Venice was also the opposite of the horrors he had experienced in the first part of his life, a sanctuary from history. Thirty years later Ghosh throws Venice back in the whirlwind of history, which may be the only way to contribute to its survival. Even as he revisits well-honed tropes linked to the city as a place of mystery, suspense, and uncanny encounters, his Venice is emphatically not that of modern fantasies of escapism. It is the city where the characters are forced to experience and face the violence of natural disaster connected to sea-level rise and climate change; where they confront the migration phenomena that displace masses of people and force them in vicious cycles of human trafficking. Hypnotized by its beauty, we tend to forget that Venice has been for centuries an exceptional technological and ecological achievement. In times of crisis Venice has often turned to the arts to reinvent itself and its economy. Under new conditions, we need to follow the same pattern, starting from the indispensable premise that Venice cannot just be a place of culture consumption but of tangible and intangible cultural production. In the worst-case scenario this investment should be directed at imagining a world without Venice, one where the city, submerged by rising seas, will become Angkor Wat or Petra or Chichen Itza, or, even worse, like Atlantis, surviving only in cultural and digital memory. In the best-case scenario, what can be imagined, elaborated, or created here may become relevant not only for itself but also for other coastal cities facing the same challenges.[32]

To try to protect Venice from the rising sea would ultimately not be enough: in their different ways, Brodsky and Ghosh envision a city that could instead become an international laboratory for scientists, scholars, and artists, an ideal place to tackle the environmental crisis and formulate solutions that apply to all coastal cities in the world. Today, encouraging new public and private initiatives are pointing in this direction, forging new alliances between arts and science and placing the relationship between Venice and the sea squarely back at the center of the city and of its imaginary. These include the first Environmental Humanities MA degree in Italy at Ca' Foscari University;[33] the political bid to make Venice the "*world capital of sustainability*"; the collective participation of the European "Bauhaus of the Seas" network; and the opening, in an abandoned church that once held the remains of Marco Polo, of Ocean Space, "a planetary center for . . . catalyzing ocean literacy and advocacy through the arts."[34] Like Ghosh's novel, these are both concrete and symbolic events that promise an alternative future for Venice and challenge the image of the city as a moribund patient begging for international aid. In this light we leave the final word to Brodsky's vision: "turning this place into a capital of scientific research would be a palatable option, especially taking into account the likely advantages of the local phosphorus-rich diet for any mental endeavor."[35]

Notes

I thank Carmen Concilio, Anna Nadotti, Mitchell Duneier, Jeffrey McCarthy, Diego Calaon, Elena Grandi, Daniela Zyman.

1. Thomas F. Madden, *Venice: A New History* (New York: Penguin, 2012), 104.
2. Madden, *Venice*, 316.
3. Steve Mentz, "Creating Ocean: Planetary Immersion and Pre-Modern Globalization" (chapter 3 this volume).
4. Madden, *Venice*, 348.
5. Ibid.
6. Mentz, "Creating Ocean."
7. Tiziano Scarpa, *Venice Is a Fish: A Cultural Guide* (London: Serpent's Tail, 2009), 2.
8. Deborah Howard, "Venice and the Sea," Lecture, Venice, Circolo Italo Britannico, February 10, 2020.
9. This and all subsequent quotes from the play are from William Shakespeare, *The Merchant of Venice* (Arden 3, London: Bloomsbury 2010).
10. Mentz, "Creating Ocean."
11. John Keahey, *Venice Against the Sea: A City Besieged* (New York: Thomas Dunne Books, 2002).
12. At a book launch in Venice in 2019.
13. A situation that was sadly repeated during the COVID-19 pandemic.
14. Elizabeth DeLoughrey, *Allegories of the Anthropocene (*Durham: Duke University Press, 2019), 6.
15. Serenella Iovino, *Ecocriticism and Italy: Ecology, Resistance, and Liberation* (London: Bloomsbury, 2016), 49.
16. Salvatore Settis, *If Venice Dies* (New York: New Vessel Press, 2016),170)
17. Amitav Ghosh, *The Great Derangement: Climate Change and the Unthinkable* (Chicago: University of Chicago Press, 2016), 9.
18. Diego Calaon, "Sinking Venice, Thinking Venice: Cultural Heritage Discourses between Environment, Materiality, Politics and Collective Perceptions," unpublished paper, 2021.
19. Giacomo Salerno, *Per una critica dell'economia turistica. Venezia tra museificazione e mercificazione* (Macerata: Quodlibet, 2020).
20. Joseph Brodsky, *Watermark* (New York: Farrar, Strauss & Giroux, 1992), 124.
21. Sanna Turoma, *Brodsky Abroad: Empire, Tourism, Nostalgia* (Madison: University of Wisconsin Press, 2010), 153.
22. Museo Correr, "Layouts and Collections," https://correr.visitmuve.it/en/il-museo/layout-and-collections/venetian-culture/.

23. Ghosh, *Great Derangement*, 63.
24. Ibid.
25. Amitav Ghosh, *Gun Island* (London: John Murray, 2019), 178.
26. Ibid., 201.
27. Franco Moretti, *The Bourgeois: Between History and Literature* (London: Verso Books, 2013), 26.
28. Ibid., 27.
29. Ibid. The reference is to Margaret Cohen, *The Novel and the Sea* (Princeton: Princeton University Press, 2010).
30. Ghosh, *Gun Island*, 160.
31. Ibid., 295.
32. Even more so after another, unexpected and unprecedented crisis (the COVID-19 pandemic) temporarily emptied out the city and showed the volatility of an economy overwhelmingly dependent on tourism.
33. https://www.unive.it/pag/39007/.
34. "About Ocean Space," Ocean Space, https://www.ocean-space.org/about/.
35. Brodsky, *Watermark*, 115.

CHAPTER 5

Mobilizing Vessels and Voices

"A Climate Movement in the Pacific, for the Pacific, and with the Pacific"

TAYLOR CUNNINGHAM

The Voyaging Revival

On May 1, 1976, a crew of Native Hawai'ians and *haole* (foreign, namely white) researchers set sail on a traditional, double-hulled Polynesian voyaging canoe heading for Tahiti.[1] Named *Hōkūle'a*, or "Star of Gladness"—after Arcturus, the bright zenith star of Hawai'i—she was the culmination of over a decade of collaborative efforts among artists, engineers, sailors, and anthropologists to reconstruct an ancient deep-sea voyaging vessel and train a crew capable of sailing her by traditional navigational methods. The project's goals were twofold: first, to demonstrate to a skeptical scientific community that ancient Polynesians had settled the far-flung islands of the Pacific intentionally and not simply "by accident," as anthropologist Andrew Sharp and others argued; and second, to revive a cultural legacy of voyaging canoes that was nearly "extinct." The Polynesian Voyaging Society, founded in 1973 to support the construction of *Hōkūle'a*, understood the project as a "Voyage of Rediscovery."[2]

Following her May departure, *Hōkūle'a* sailed over 2,500 miles of ocean, tracing a legendary course of cultural and material exchange between Hawai'i and Tahiti. Mau Piailug, a respected navigator from the Micronesian island of Satawal, guided the vessel's way, showing the Hawai'ian crew on board how to sail "in the wake of [the] ancestors."[3] One of the last keepers of the ancient art of wayfinding, Piailug charted a course using only traditional methods: close observations of stars, swells, and marine life. He said of the practice in a 1983 documentary, "I don't need a map or a sextant. I just use my head. I observe the ocean and the sky, and I remember the words of my teachers."[4] While only three decades prior, Norwegian scholar Thor Heyerdahl had famously concluded that ancient Polynesians lacked the technology and skill necessary to settle the Pacific from mainland Asia, Piailug demonstrated that such exceptional sailing feats by traditional means were possible for those who had mastered the necessary knowledge. Contrary to dominant narratives that dismissed indigenous oral migration histories as "myths," early Pacific peoples were intelligent seafarers and deliberate settlers.[5]

When *Hōkūle'a* made landfall in Pape'ete Harbor after thirty-four days at sea, over seventeen thousand Tahitians were waiting to celebrate her arrival. The historic passage represented an important cultural victory for Polynesian peoples. It revitalized ancestral seafaring traditions, strengthened ties among islands throughout the Pacific, and rewrote history on Islanders' own terms. Nainoa Thompson, the first Native Hawai'ian to learn the art of wayfinding (under Piailug's teaching) in over two hundred years, recalls the Tahitian response that day in June 1976:

> None of us were prepared for that kind of cultural response—something very important was happening. [Tahitians] have great traditions and they have great genealogies of canoes and great navigators. What they didn't have was a canoe. And when Hōkūle'a arrived at the beach, there was a spontaneous renewal, I think, of both the affirmation of what a great heritage we come from, but also a renewal of the spirit of who we are as a people today.[6]

Thompson's account suggests that *Hōkūle'a* was only the beginning of a much larger movement to empower Pacific peoples after centuries of territorial and cultural dispossession. Though *Hōkūle'a* was originally conceived as a scientific experiment for a largely Western audience and Hawai'ian contribution to Bicentennial celebrations in the United States, her significance was personal rather than performative.[7] *Hōkūle'a* represented a cultural victory, because she validated Polynesian traditions to Polynesians. Only peripherally did she speak to the interests of colonizers. As a result, *Hōkūle'a* not only delegitimized colonial narratives of the Pacific, but challenged the power relations that continue to support them.

Since her maiden voyage more than forty years ago, *Hōkūle'a* has sailed over 140,000 miles of ocean, drawing physical and emotional connections between islands and coastal communities around the world. More recently, voyages have taken up pressing issues related to global pollution and climate change, placing the concerns of indigenous islanders at the center of international environmental politics. In 2013, *Hōkūleʻa* embarked on a four-year journey to circumnavigate the globe, spreading a message of *mālama honua* (caring for our earth) while collecting stories of hope in ports around the world.[8] Elsewhere in the Pacific, traditional voyaging canoes have become central to grassroots environmental movements like the Pacific Climate Warriors and their well-known "Canoes vs. Coal" action in Australia, as well as in modeling sustainability through projects like Fiji's carbon-neutral vessel, *Uto Ni Yalo* (Heart of the Spirit), which gathers ocean trash around islands.[9] Canoe-building initiatives at universities and community centers throughout the region are also often used as a vehicle for cultural healing among groups grappling with the traumatic historical legacies of European colonialism.[10]

Today, traditional voyaging canoes function as vessels of strength, skill, and cultural continuity for islanders facing sea-level rise, ocean acidification, and storm surges in a climate change present and future. While popular Western media often play into lingering colonial narratives in portraying island communities as powerless populations—the world's first "climate refugees," living on "drowning" or "disappearing" islands—the Pacific voyaging canoe tells a story of cultural ingenuity and resilience. As sensational news stories narrate debates among global powers like Australia, New Zealand, and the United States about how best to rescue poor island nations from their landless fates, Pacific seafarers insist on navigating their own way forward through a world in flux—much as their ancestors did before. Thus, the voyaging canoe simultaneously enacts history and creates it, asserting the agency of its master builders, wayfinders, and crew members to collectively forge their own futures.

The Vaka in Oceania

A child plays alone in shin-deep water. A piece of ocean plastic—what could be mistaken for a small, inflatable raft—floats toward him. The boy plods through the water, crossing in front of a CBSN logo casting its reflection on the waves. The scene cuts to a frame filled with buzz words and phrases: MAN VS. NATURE, KING TIDES, EROSION, EXTINCTION, CARBON EMISSIONS, AND CLIMATE REFUGEES. They move laterally across the screen—like the plastic waste, like the boy. Behind the words is a narrow stretch of paradise where clear, blue waters meet sandy beaches laden with palm trees. Then the cacophony of words exits the screen, leaving an urgent announcement in their wake: Oceans are rising and nations are sinking.[11]

In a 2017 segment for CBS News, a foreign correspondent walks along sea walls, past abandoned homes and flooded fields on a nondescript island in Kiribati. He interviews locals "living on the brink" on landmasses "as narrow as a basketball court." He asks a mother of two what her plans are for the future. "To migrate," she admits, "to leave this country . . . because of climate change." The mother says she does not have a choice; there is no future for her children on the island. In the coming decades, sea-level rise will make life untenable in Kiribati.

Some I-Kiribati have already lost their homes. Another woman points the journalist to the site of her former home, a stretch of beach now under water. "Who do you blame?" he asks her. "I know you're going to hate me—America, the United States," she says, referring to the nation's outsized carbon footprint. The woman's words are later emphasized when President Donald Trump's June 2017 announcement to pull the United States from the Paris Climate Accord airs in Kiribati. "Where is the justice?" asks former Kiribati president Anote Tong.

Highlighting Tong's insistence that I-Kiribati retain their "dignity," the news story later describes a plan for threatened islanders to move to land they purchased in Fiji. The CBS segment, however, strangely dismisses this prospect (among mitigation measures, like rainwater harvesting and mangrove planting) by continually insisting that I-Kiribati have "nowhere to run." Essentially, the story asserts that while coastal communities facing sea-level rise in countries like the United States have safety nets—higher ground, government relief—the people of Kiribati must simply wait to become refugees.

Portraits of islander plight speak directly to legacies of European colonialism that continue to extend their authority over Pacific space and story. Carol Farbotko and Heather Lazarus contest such climate narratives, explaining that representations of helpless islanders on drowning islands are often used as visible proof of an otherwise nebulous, ostensibly invisible, environmental crisis. They depict Pacific Islanders as signifiers of climate change, alongside polar bears and melting ice caps, in sensational—and thereby marketable—stories that exploit them as "victim-commodit[ies]."[12] These narratives, they argue, effectively use the Pacific as a "laboratory" for crisis—a practice that bears the colonial echoes of nuclear Pacific Proving Grounds and waste dumps.[13]

Stories can scar the imagination. Tongan scholar Epeli Hau'ofa writes that colonizer narratives, terms, and maps placed islanders on "mental reservations."[14] "Continental men," as Hau'ofa calls Europeans, came to the Pacific with a landlocked sense of place and drew lines around islands. They told islanders to think of themselves as limited to small, isolated, and peripheral places, while enforcing policies to restrict autonomous movement between them. After many years under colonial rule, says Hau'ofa, some island nations came to believe this story.

Hau'ofa insists, however, that "smallness is [only] a state of mind," and to liberate their thinking, islanders might begin by changing the names of places. Hau'ofa thus calls for Pacific Islanders to repopulate the vast region of Oceania. He reasons that the emphasis on "islands" in "Pacific Islands" restricts place to land. The name, in other words, points to a series of unconnected fragments of earth: "islands in a far sea." They are, as imperial powers envisioned, more readily colonized. Yet, to call the same region "Oceania" is to include ocean space—effectively, to recognize the more expansive, interconnected whole that Pacific peoples actually inhabit: "a sea of islands." This place is vast and complex and growing; it evades a colonial imaginary.[15]

Early Pacific people calculated the size of their territories differently than eighteenth- and nineteenth-century Europeans. They measured places in terms of the land and sea, the heavens above, and the submerged world of fire below. The voyaging canoe not only traced the contours of this world—negotiating reefs, stars, waves, and winds—but also pushed at its edges. One's world, observes Hau'ofa, stretched "as far as one could traverse."[16]

In Oceania today, the voyaging revival is a liberating reminder of more expansive places that transcend colonial maps and stories. Significantly, the power of the *vaka*, or voyaging canoe, emerges from connectivity—relationships maintained through time and across space. The voyaging canoe is of the land, but built for the sea. It embodies a cultural heritage that is continually venturing out. In Polynesian traditions, the master canoe builder shapes the voyaging vessel out of a koa tree, which represents cultural "rootedness." Once seaworthy, the canoe takes to the ocean, carrying with it a people's land and history. Guided by ancestral knowledge, the voyaging canoe has the capacity to transport people and materials through a fluid world. The *vaka* lies at the nexus of earth and ocean, past and present, home and away. It is, in other words, the vehicle of Polynesian history.[17]

Pacific arts scholars Tammy Baker, Sharon Mazer, and Diana Looser observe that the voyaging canoe lies at the heart of many Pacific identities, "as precursor, metaphor, lived reality, and aspirational ideal."[18] It carries cultural heritage, a deep sense of pride, and knowledge, while facilitating multiple belongings across Oceania. Mazer observes in her work with the Maori, for instance, that the voyaging canoe is a dwelling space for culture. She cites a master carver who describes the *waka* as a literal and metaphorical *marae* (house), which can be "likened to an upturned canoe, and the parts of it to the parts of the human body."[19] The *waka* holds and recalls the bodies of people and their ancestors.[20] It is the "idea(l)" of a people—a wellspring of strength from which islanders may draw when moving through new and transitional spaces.[21] Thus, while one may leave one's island in a voyaging canoe, one never leaves one's home behind altogether.

Though we should acknowledge the emotional and physical traumas associated with climate displacement, it is equally important to recognize the cultural resilience of Pacific peoples and their canoe cultures. Anthropologists Wolfgang Kempf and Elfriede Herman point out that a cultural heritage of mobility and trans-Pacific connectivity provide deep-seated "resources" for islanders responding to sea-level rise, ocean acidification, and more aggressive storm systems.[22] While Westerners often assume migration is a cultural crisis, involving the complete dissolution of community identity into a wider global sphere, the people of Oceania have historically understood moving differently. In the "semi-rooted, semi-moving ways of . . . small islands," migration is common throughout the Pacific—for school and work, to visit family, and to exchange goods and services. As a result, islanders have strong religious, economic, and kinship networks around the world—networks that do not all lead to the Global North, where calls to "save the climate refugees" typically emerge.[23] Thus, Kempf and Herman identify among islanders a unique capacity for "place-making and identification that find expression in multiple belongings."[24] Moving, in other words, does not necessarily signal the collapse of cultural identity.

The CBS news segment—what Carol Farbotko might call a "wishful sinking" narrative—assumes island people are limited, rather than enlarged, by ocean space.[25] As a result, it imagines islanders must be stuck—"on the brink"—waiting for an inevitable end: a disappearing people on a disappearing island. As Hau'ofa and others demonstrate, however, islanders are far from stuck.[26] Grounded in their voyaging histories, Pacific peoples know many pathways through the ocean—and the I-Kiribati are no exception to this tradition.

Kiribati is a nation of thirty-three islands and atolls spread between three archipelagos. The aggregate landmass of the islands in the former British colony totals only 726 square kilometers. However, including ocean space—the

200-mile exclusive economic zones that surround the country—Kiribati is actually one of the largest nations in the world, extending over 3.5 million square kilometers. While roughly ninety thousand I-Kiribati live within their nation's borders, many others live in the Solomon Islands, Fiji, and New Zealand, or work abroad on European and Asian ocean-going vessels. Teresia Teaiwa and Juliette Launiuvao observe that Kiribati culture, with its "unique balance of mobility and grounded-ness," skillfully navigates these vast spatial and social geographies. That culture, they argue, is embodied in *te wa*—the Kiribati outrigger canoe.[27]

Teaiwa and Launiuvao place *te wa* firmly within Polynesian voyaging traditions, referring to the vessel as "the Hokule'a of Kiribati."[28] Like the *vaka*, *te wa* is an extension of self—one's body, family, and heritage. In Kiribati, one would not say "my canoe," but rather "canoe of me."[29] I-Kiribati writer Marita Davies likewise observes that "*te wa* symbolizes everything the Kiribati people have achieved." She recalls the millennia over which Pacific peoples explored more than sixteen million square miles of ocean to settle every habitable island in Oceania. This history is clear evidence of her people's independence, resourcefulness, and deep respect for the ocean. That heritage, says Davies, is apparent even in the traditional construction of *te wa*: "Days before the spirit level, the maker would position his workplace on land and face out to sea . . . [using the horizon as] his level. Even before the canoe is introduced into the saltwater, its destiny to sail towards the horizon is already decided."[30]

Davies believes *te wa* will protect and guide I-Kiribati through an uncertain climate future. She points to the mini-sailboat races that Kiribati children often host in the puddles left behind by rainstorms and high tides. The boats the children create for these competitions are highly original—made from discarded plastics, cans, and pandana leaves—and remarkably graceful in the water. These vessels, and their makers, give Davies hope. Despite the realities of sea-level rise, which threaten I-Kiribati lifeways, she is confident her people's traditions will persist in the coming decades: "With each saltwater puddle left behind, there will be a child building their own *te wa* out of an empty soft drink can. They will laugh and cheer as they watch it float upon a splash of water and Kiribati and its culture will hold strong."[31]

Not Drowning—Fighting

On October 17, 2014, thirty indigenous activists from twelve island nations led a flotilla of kayaks, surfboards, and canoes into the world's largest coal export facility—Australia's Newcastle Port.[32] There, hundreds of demonstrators from across Oceania punctuated their calls for an end to fossil fuels with a blockade large enough to disrupt the day's shipping schedule and prevent eight bulk carriers from passing through the port. The protest's indigenous organizers, who form a collective called the Pacific Climate Warriors, came with a clear message: "We are not drowning. We are fighting!"[33]

The "Canoes vs. Coal" action came on the heels of four new coal ventures in Australia, which Prime Minister Tony Abbott approved in December 2013. Among those projects was a thermal coal mine in Queensland that would be permitted to dump its dredge soil fifteen kilometers from the Great Barrier Reef, an ecosystem already beleaguered by several mass coral bleaching events.[34] Meanwhile, only months after Prime Minister Abbott signed off on increased fossil fuel production in Australia, the Marshall Islands evacuated nearly one thousand people when severe king tides inundated its low-lying atolls. Such flooding, which has become common in recent years, represents an ongoing threat to Marshallese homes and livelihoods.[35]

While climate change represents an existential crisis for every community on Earth, historical and geographic circumstances make some communities more vulnerable than others. Thus, environmental changes, and their associated risks, have been distributed unequally among front-line communities around the world—many of whom have placed minimal concentrations of carbon into the atmosphere. Such is the case of low-lying atolls and islands in the Pacific—Kiribati, Tuvalu, the Maldives, and the Marshall Islands being the most widely publicized—that are now grappling with the consequences of global carbon emissions: essentially, paying the costs of industrialization and wealth accumulation elsewhere.[36] For politically and economically marginalized native islanders, this explicit environmental injustice is a direct legacy of colonialism, which privileges the needs and concerns of colonizers over those of the historically colonized.

Since the Rio Earth Summit in 1992, when the United Nations formally recognized the threat of climate change, concerted global action has been slow-moving, largely ineffective, and far from revolutionary. Significantly, climate policy researchers Harald Winkler and Joanna Depledge point out that island states with small economies have been historically underrepresented in international climate talks "marked by the usual positioning and bickering between developed and developing countries." These negotiations typically value communities according to their economic output and understand climate change as an economic crisis that requires market-based solutions. As a result, global climate governance has rarely done more than

"pay lip-service" to marginalized perspectives and rights.[37] Maria Tiimon, a prominent I-Kiribati activist who works to foreground Pacific voices in climate debates, likewise observes that the logic of capitalism presents a significant challenge to her work:

> My climate action journey has been very challenging. To try and convince rich and developed countries that seem to care more about their economy [*sic*] is not an easy task. I have been challenged by many people who are very skeptical on climate issues. At times, it feels as if there is no light at all in the tunnel and I fear for the future of my country Kiribati and, of course, the rest of the world.[38]

The tenor of Tiimon's words is striking. She is clearly disheartened by and impatient with a value system that privileges the economic interests of global powers like Australia over communities facing the permanent loss of their homes. Understandably, she entertains a sense of hopelessness.

Fijian climate activist Fenton Lutunatabua similarly acknowledges the crisis climate change poses for his people: "If you look at the existing [climate] narrative . . . without pessimism, you fail to understand our realities." Lutunatabua is quick to point out, however, that portraits of indigenous suffering and disempowerment offer only a "half-truth," because they fail to account for the enduring strength and spirit of island communities. Rather, he argues, Pacific peoples demonstrate their power in courageous acts of resistance and resilience.

> Our courage as a people is the type of courage that thrives whilst existing in a perpetual state of rehabilitation. Our courage as a people is the type of courage that will build a movement of Pacific Islanders demanding climate justice. Our courage as a people is the type of courage that will stop giant coal ships with handmade canoes.[39]

Recognition of the historical and ongoing injustices that heighten risk for communities and reduce their capacity to adapt to changing circumstances is necessary and productive; however, these stories are not the only ones worth telling. Simultaneously, many Pacific climate activists tell another version of the climate change story—one in which they are warriors fighting for liberation from colonialism, environmental destruction, and fossil fuels. During the 2014 Canoes vs. Coal action in Australia, indigenous islanders organized to carry out this fight over ocean space—blocking coal transportation over the same highways that once facilitated exploration and cultural and material exchange among ancestral Pacific people. The Climate Warriors chose the traditional voyaging canoe as the primary vehicle through which they would wage this war.

Over the months prior to the Newcastle blockade, young indigenous climate activists across the Pacific began building traditional canoes on their home islands. They turned to their forests for materials and to their elders for guidance. While felling trees, giving shape to hulls, rolling coconut fibers into rope, and transporting finished crafts to the ocean, they performed ceremonies to foster strong connections with their ancestors, reinforce ties among community members, and prepare themselves for the journey ahead.[40]

On the day of action, the Pacific Climate Warriors and their Australian allies must have presented a stark contrast to the massive, coal-bearing bulk carriers in Newcastle Port: a flotilla of small, carbon-neutral vessels of varying shapes, colors, materials, and traditions confronting the faceless steel of a fossil-fueled machine. Though the flotilla managed to block eight of the twelve ships scheduled to pass through the port that day, for Australia's coal industry the demonstration was, in all likelihood, only a minor inconvenience. In many ways, a Canoes vs. Coal matchup plainly exposed their uneven power dynamic.

For their part, however, the Pacific Climate Warriors declared victory, and perhaps would have no matter how many ships they managed to block that day, because the blockade was not an end in itself. Rather, the goal of the demonstration was to mobilize vessels and voices. In an interview after the Newcastle action, Climate Warrior Milañ Loeak reflected on its significance:

> You know . . . the flotilla might not have made even the slightest bit of difference to the mining companies, but I think we really did have an effect on a lot of people that were there. You know, there were moments when everybody got emotional, and then the next moment you're all singing and dancing, and I felt that the culture that we brought was beautifully displayed that day, and I think we were effective in that we reached out to the local people [in Newcastle], and we touched lives by bringing what we believed in [to the action].[41]

Climate Warrior George Nacewa likewise spoke to the emotional and spiritual connections the flotilla manifested: "Even though we speak different languages and come from different cultures, we are connected to the land

and we are connected by the ocean."[42] Loeak and Nacewa's reflections point to both local and global entanglements and indigenous and nonindigenous identities. Significantly, the Pacific Climate Warriors formed in partnership with 350.org, a network that connects grassroots climate movements around the world. They also chose not to act alone and called upon Australians to join the demonstration, providing them with kayaks to do so. As a result of this outreach, a diverse gathering of vessels and voices confronted the fossil fuel industry in Australia to demand change.[43]

A resilient movement is, in large part, the result of such extensive social networks. Robin Craig and Melinda Benson identify resilient systems as those which have the capacity to "self-organize," to continually learn and adapt to changing circumstances. Resilient ecosystems are typically biologically diverse and supported by a number of "nonlinear" agential forces.[44] While resilience theory is often applied to our understanding of natural ecosystems, it is also useful for thinking about social ecosystems. Like a resilient ecosystem, a grassroots movement self-organizes from the bottom up along extensive social networks. It is characterized by diversity, a number of perspectives working in tandem to inform, challenge, and ultimately strengthen one another. Though more authoritarian, top-down schemes may respond unilaterally, and thereby more quickly, to emergent problems, consensus-driven bodies with shared values have the potential to respond more intelligently and robustly to changing circumstances. As one Pacific Climate Warrior reflected, "one person can go faster, but togetherness can take us farther."[45]

Following a Warrior training in Kiribati in March 2014, Lutunatabua reflected on what it would take to "stir up the warrior energy" in the Pacific and build a resilient movement:

> A large part of what shapes whether we stand a fighting chance in this battle against the giant fossil fuel industry, depends on how we build and wield the people power base we have as Pacific Islanders and as nations.
>
> We need to build a generation of Climate Warriors trained to stand up for the Pacific and build on our authentic and inherent truths, strengths, and aspirations.
>
> We need to inspire innovativeness and creativity in our young climate warriors, we need to take advantage of new media, and place an emphasis—like never before—on the unity of our region connected by passion and the very ocean that poses a threat to our identity.[46]

Lutunatabua emphasizes the importance of solidarity in building a resilient climate justice movement "in the Pacific, for the Pacific, and with the Pacific."[47] While collective action is a significant means of building social power, without shared values that power may be wielded ineffectively, and without passion it is unlikely to endure. Traditional voyaging canoes play an important role in Pacific movement-building, because they embody a shared cultural heritage around which Climate Warriors and other Pacific activists can gather, celebrate their collective strength, and express their right to participate in global climate governance. As a united front, the Pacific Climate Warriors could thus approach the largest coal port in the world with "people power"—the will of collective aspirations shaped in a strong network of relationships and anchored in the canoe cultures of Oceania.[48]

Crafting Futures

As a young boy, Mau Piailug learned the art of wayfinding from his elders. He memorized the paths of over 150 stars, learned to feel the direction of swells from the rise and fall of a canoe, studied the movements of marine life for signs of land, and examined the shapes and colors of the sky for hints of weather. Guided by the wisdom of his Satawal ancestors, Papa Mau, as he came to be known by his Hawai'ian students, always knew where he was going, because he trusted the significance of what he saw and felt. "This is what we call courage," he said. "With this courage you can travel anywhere in the world and never be lost."[49]

Over the course of his life, Papa Mau had been on many voyages near his home in Micronesia, but none were as long as the course he navigated from Hawai'i to Tahiti in the summer of 1976. Far from Satawal and under unfamiliar skies, Papa Mau guided a Hawai'ian canoe across 2,500 miles of ocean using only traditional methods. Despite the many who believed it could not be done, the month-long voyage was successful. For Mau, however, the journey proved to be frustrating: he was disappointed by his Hawai'ian students' lack of discipline. They needed a teacher to show them how to navigate, but they were not yet ready—in spirit—to learn "the old ways." So, instead of returning to Hawai'i with *Hōkūle'a*, Mau went home to Satawal, leaving one tape-recorded message in his wake: "Do not come look for me; you will not find me." Without a master navigator, the crew had no choice but to sail home to Hawai'i using modern instruments.[50]

In Western traditions, we might call Mau Piailug a compleat, or perfect, mariner: a master of craft. In *The Novel and the Sea*, Margaret Cohen studies this character through the protagonists of early modern sea adventure literature: real and imagined men like James Cook, Robinson Crusoe, and Joseph Conrad who "excel[led] in the art of action."[51] Within this tradition, the compleat mariner is a hero on the high seas, "an icon of effective practice and human ingenuity," navigating a world of "unruly forces that can be negotiated but not controlled."[52] At the mariner's disposal are the many tools of craft: discipline, intuition, physical strength, resourcefulness, patience, and courage. A deviation from the "bourgeois abstraction[s]" and interiorizations typical of modern reason, Cohen defines craft as a form of practical reason that demonstrates the adaptive capacity of "embodied intelligence."[53] While a mariner's craft may make use of theories and ideologies, the practice ultimately gives primacy to "the senses, intuitions, feelings, and the body."[54] This attentiveness to an immediate material world similarly characterizes the craft of a master wayfinder. To chart a course, Piailug uses his instincts and observations. He does not need Global Positioning Systems to show him the way; he can feel it for himself. Like a compleat mariner, Piailug is a powerful expression of human agency in a world of flux.

Of course, there are important differences between a master wayfinder and a master mariner. While eighteenth-century European seafarers needed a nautical compass and sextant to navigate across oceans, Piailug and his ancestors practiced their craft unmediated by instruments. In awe, Nainoa Thompson once said of his teacher: "Mau can unlock the signs of the ocean world and can feel his way through the ocean. Mau is so powerful. . . . It's like magic; Mau knows where something is without seeing it."[55] Piailug's power comes from his keen sensitivity to subtle environmental shifts and signs in an ocean world he knows well. His lived experience of the ocean is thus qualitatively different from that of European masters of craft, who braved the dark waters of a chaotic Other. Moreover, as the master navigator for his village, Piailug contributed vital ocean resources to an atoll community living on less than 350 acres of land. By contrast, the compleat mariner worked in the service of empire: "[their] heroism . . . was pressed into ideological and cultural work for nationalism and capitalism back on land."[56] Whether the compleat mariner understood their seafaring feats in such a light or not, they were effective agents of imperialism. So, as European masters of craft pushed at the edges of the known world to amass an "empire of knowledge," they simultaneously furthered a project of social and environmental conquest, of empowerment through subjugation.[57] The craft of a compleat mariner was thus instrumental in driving others' crafts—that of a Pacific wayfinder, for instance—from the world's oceans.

Nonetheless, Cohen maintains that the mariner's craft "called on capacities with the power to open lines of escape."[58] Knowledge held in the body and gathered through observation, intuition, and physical repetition can offer a powerful challenge to colonial theories and ideologies with little grounding in a material reality. Craft's emphasis on the body—its possibilities, entanglements, and limits—in relation to a world of agential forces that exceed the human suggests an alternative, more egalitarian, configuration of earthly relationships. In Cohen's words, craft "models another posture that might yield safety from the jaws of [self] destruction."[59] This is a posture of strength in humility. It can facilitate a powerful expression of human agency that respects the Earth's own capacities.

On March 16, 1978, *Hōkūleʻa* embarked on a second voyage to Tahiti—this time, with traditional foodstuffs and *lauhala* (hala leaf) sails to facilitate her passage. Without Piailug, Nainoa Thompson served as the vessel's non-instrument navigator, while another crewmember provided back up with modern navigational instruments. At 6:30 p.m. *Hōkūleʻa* departed Ala Wai Harbor in Honolulu, heading southeast amid high winds and waves. The vessel had sailed successfully through similar conditions before, but that night the heavily laden craft proved too difficult for the crew to maneuver effectively. Around midnight, a large wave capsized *Hōkūleʻa* west of Lanaʻi Island, flipping her over completely. All through the night and next morning, the vessel's sixteen-member crew clung to her overturned hulls, drifting farther out to sea while their emergency flares and radio signals went unanswered.[60]

At 10:30 a.m., the crew was growing desperate and reluctantly sent Eddie Aikau, a well-known Hawaiʻian surfer and lifeguard, on his surfboard to get help. In the late hours of March 17, a passenger plane on an altered flight path finally spotted *Hōkūleʻa*'s flares and alerted a rescue team to her crew's condition. Aikau, however, was never seen again. The loss was profound for many Hawaiʻians, including Thompson who counted Aikau among his closest friends and mentors:

> He was a great teacher. He was a lifeguard . . . he guarded life, and he lost his own, trying to guard ours. . . .
>
> After Eddie's death, we could have quit. But then Eddie wouldn't have had his dream fulfilled. He was my spirit. He was saying to me,

"Raise those islands." His tragedy also made us aware of how dangerous our adventure was, how unprepared we were in body and in spirit.[61]

As Thompson describes it, Aikau's death represented an important turning point for the Polynesian Voyaging Society. The loss was as humbling as it was transformative. In reckoning with Aikau's death, *Hōkūleʻa*'s crew had to reckon with their lack of knowledge and experience. To sail in the wake of their ancestors, they needed to gather more of what their ancestors knew. In Piailug's absence, Thompson had tried to teach himself traditional navigation using astronomy books, trips to the planetarium, and courses in ocean sciences at the University of Hawaiʻi. Independent study, however, could not replace the embodied knowledge of a master wayfinder. To become better navigators—and lifeguards—Thompson and his fellow crew members knew they needed Papa Mau.

In the months after Aikau's death, Thompson found Mau in Micronesia and asked him for his help. Though hesitant at first, when Mau heard about *Hōkūleʻa*'s loss, he called Thompson and said, "I will train you to find Tahiti, because I don't want you to die." Over the next two years, Mau helped Thompson "find [his] way to the sea." Every day, they spent time on the water together, as Thompson recalls: "I watched what he watched, listened to what he listened to, and felt what he felt." Thompson had originally wanted to learn, as Mau had, through observation and oral transmission. Yet, he discovered that his upbringing in a text-based education system built on the assumptions of Western science made traditional learning difficult. As a result, a mix of cultural traditions guided Thompson's way to the sea. By following Mau's embodied practice while studying the mathematics of navigation, Thompson strengthened his instincts until he could internalize "all the faces of the ocean."[62] In the spring of 1980, *Hōkūleʻa* and her crew felt ready—"in body and in spirit"—to attempt another voyage to Tahiti.[63]

When Thompson reflects on the value of traditional navigation in a time of radio and satellite technology, he talks a good deal about respect—for the ocean, for the ancestors, and for oneself. In physically retracing ancestral ocean pathways, native islanders rediscover and reinforce the value of ancestral ecological knowledge and lifeways. This process, says Thompson, is important to the overall well-being of Native Hawaiʻians, who he witnesses grappling with the legacies of colonialism, namely, climate disruption, exhausted ecosystems, declining populations, and high rates of poverty and incarceration. Public health scholars Ilima Ho-Lastimosa, Phoebe W.

Hwang, and Bob Lastimosa likewise advocate for the physical and psychological benefits of canoe building. Traditional voyaging vessels, they argue, function as both a "metaphor and method" for empowering Hawai'ian communities. While representational of cultural values and history, canoes also express *ma ka hana ka 'ike*, or "in doing, one learns."[64] Thus, in showing others how to reconnect with their ocean heritage through traditional voyaging, Thompson believes he can help forge a more hopeful, healthy, and ecologically balanced future for Native Hawai'ians. Ultimately, he sees traditional canoe building and sailing as crafts with the capacity to "replace abuse with renewal."[65]

Craft can be understood as a force that gives physical shapes to things—one's course through the world as well as the vessel itself. An ocean-going vessel is itself synonymous with a "craft," the physical manifestation of certain tactile skills and material relationships. Islanders require specific local resources to build canoes: in Hawai'i, a koa tree for the hull, a basalt adz sourced from a volcanic quarry to shape it, coconut fibers for lashings, and pandana leaves for the sails and for making measurements.[66] Traditional canoe building thus depends on the health of local ecosystems and the availability of native plants—an issue of great concern on islands facing sea-level rise and drought, as well as historical resource exploitation and modern real estate development. For example, when the Polynesian Voyaging Society attempted to build a second vessel, *Hawi'iloa*, using only traditional Hawai'ian materials, they could not find a koa tree in their diminished forests large enough for the vessel.[67] Fortunately, Native Alaskans gifted them two giant spruce trees from their forests for the project. In turn, the Polynesian Voyaging Society established a program in Hawai'i to plant eleven thousand koa trees, so that there would be "forests of voyaging canoes" for the next generation.[68]

Traditional voyaging provides a compelling alternative to unfettered environmental exploitation and emission. Ocean-going mobility by traditional means is dependent on winds and currents, as well as on close observations of the natural world. The practice is carbon neutral, but it also underscores the reality of natural limits and the practical value of noticing changes in one's environment. According to a common Hawai'ian saying, *He wa'a he moku; He moku he wa'a*: "a canoe is an island; an island is a canoe." While the canoe sustains its crew, the crew must collectively work to sustain it. As on any island of habitability, natural threats and limitations require crew members to be attentive and resourceful stewards of a shared, finite space.[69]

With this island aphorism in mind, *Hōkūleʻa* embarked on a three-year voyage to circumnavigate Island Earth in 2013, bringing her message of *mālama honua* (caring for our Earth) to 18 countries, 150 ports, and eight Marine World Heritage sites. Along the way, the Polynesian Voyaging Society sought to share their vision for the Mālama Honua Movement in Hawaiʻi: a "perpetual voyage" in environmental stewardship that commenced with a widespread "Promise to PaeʻĀina O Hawaiʻi."[70] This movement emphasizes the human responsibility to act, to put a more compassionate environmental ethic into practice. Throughout their voyage, *Hōkūleʻa*'s crew thus planned to collect and connect examples of *mālama honua* in the hopes of uplifting some of "the world's greatest navigators"—people responding intelligently and compassionately to environmental crises. Of the voyage's goals, Nainoa Thompson said, "We aren't going to change the earth, but we will form a network of people around the world who will change it." When Thompson and his fellow crew members embark on these traditional voyages, they look to their heritage for guidance and to their island neighbors for a shared vision, an ecologically balanced future for which the inhabitants of Island Earth are all responsible. *Hōkūleʻa*'s "perpetual voyage" in the Mālama Honua Movement is both a "metaphor and method" for crafting that future.[71]

A few months before *Hōkūleʻa* set sail on her second voyage to Tahiti in 1980, Papa Mau brought Nainoa Thompson to Lanaʻi Lookout, not far from where the canoe had capsized less than two years before. Mau pointed to the horizon and asked his student, "Can you see the island [Tahiti]?" Of course, Thompson could not physically see Tahiti, which is over two thousand miles from the Hawaiʻian archipelago, but he told his teacher he could picture it in his mind. "Good," said Mau. "Don't ever lose that image or you will be lost." Nainoa understood what Mau meant by the exercise: "[He] was telling me that I had to trust myself and that if I had a vision of where I wanted to go and held onto it, I would get there."[72] No matter how rigorously *Hōkūleʻa*'s crew prepared for the journey ahead, once they left the land, they would be exposed to a turbulent ocean world beyond their control. They would need to be courageous, to trust the strength of their craft—their canoe, their ancestors' knowledge, and their own bodies—to arrive safely at their destination.

"Every Ahupuaʻa Needs a Voyaging Canoe"

In assessing the social and environmental conditions of Oceania nearly twenty-five years ago now, Epeli Hauʻofa lamented: "What kind of teaching is it to stand in front of young people from your own region, people you claim as your own . . . to tell them that their countries are hopeless? Is this not what neocolonialism is all about? To make people believe they have no choice but to depend?"[73] Hau'ofa expounds on a pervasive sense of helplessness in Oceania: inescapable cycles of exploitation, cultural dissolution, and poverty. He asks that island nations across the Pacific realize their collective power and again become active agents of their own histories.

The voyaging revival answers Hau'ofa's call for Pacific mobilization. Not only does the canoe revive a cultural heritage in long-distance ocean voyaging, it confronts and challenges centuries of destructive colonial engagements with the Pacific. Thus, Lastimosa and others conclude that "every ahupua'a [land] needs a voyaging canoe."[74] Canoes strengthen a community. They connect people with the Earth and their heritage. They trace physical and emotional relationships among individuals and nations. They foster a sense of belonging and solidarity. Most importantly, the canoe empowers—bringing courage to Oceania's warriors. Along its passages near and far, the canoe navigates uncertainty, pushes the boundaries of the known world, and offers visions of the future. Both figuratively and physically, the voyaging canoe is a vehicle through which Pacific peoples are centering themselves and their connections to the land and sea, while navigating their own way forward amid climate crisis.

Notes

1. For a discussion of haole and racial politics in Hawai'i, see Judy Rohrer, "Disrupting the 'Melting Pot': Racial Discourse in Hawai'i and the Naturalization of Haole," *Ethnic and Racial Studies* 31, no. 6 (2008): 1110–25. See also Ben Finney's account of the racial tensions on board *Hōkūle'a* during her maiden voyage in *Hokule'a: The Way to Tahiti* (New York City: Dodd, Mead, 1979).
2. Ben Finney, *Voyage of Rediscovery: A Cultural Odyssey through Polynesia* (Berkeley: University of California Press, 1994). See also Fenton Lutunatabua, "Inspiring Islanders: Fenton Lutunatabua," The Coconet, https://www.thecoconet.tv/the-coconettv-series/inspiring-islanders/inspiring-islanders-fenton-lutunatabua-1/.
3. "The Story of Hōkūle'a," Polynesian Voyaging Society, http://www.hokulea.com/voyages/our-story/.
4. *The Navigators: Pathfinders of the Pacific*, directed by Sam Low and Boyd Estus (Documentary Educational Resources, 1983), https://www.youtube.com/watch?v=uxgUjyqN7FU/.
5. Ben Finney, "Myth, Experiment, and the Reinvention of Polynesian Voyaging," *American Anthropologist* 93, no. 2 (1991): 383–404.

6. Nainoa Thompson, "Voyaging and the Revival of Culture and History," http://archive.hokulea.com/ike/intro_ike.html/.
7. In a 1997 speech at the Polynesian Union Conference, Thompson stated, "Hokule'a was built out of scientific inquiry, not culture." See: "Nainoa Thompson," Hawai'ian Voyaging Traditions, http://archive.hokulea.com/index/founder_and_teachers/nainoa_thompson.html/.
8. "The Story of Hōkūle'a," Polynesian Voyaging Society, 2019, http://www.hokulea.com/voyages/our-story/.
9. Christine Graef, "Canoes vs. Coal Ships: Climate Warriors Blockade World's Largest Coal Port," *MPN News*, October 23, 2014, https://www.mintpressnews.com/canoes-vs-coal-ships-climate-warriors-blockade-worlds-largest-coal-port/198033/; Andrea Egan, "Heart of the Spirit: Traditional Fijian Voyaging Bridges Ancestral Wisdom and Renewable Energy, Inspires Youth and Proves Carbon-Neutral Travel Is Possible," *Global Environment Facility*, June 5, 2018, https://www.thegef.org/news/heart-spirit-traditional-fijian-voyaging-bridges-ancestral-wisdom-and-renewable-energy-inspires/.
10. Ilima Ho-Lastimosa, Phoebe W. Hwang MS, and Bob Lastimosa, "Community Strengthening through Canoe Culture: Ho'omana'o Mau as Method and Metaphor," *Journal of Medicine & Public Health* 73, no. 12 (2014): 39–79.
11. Seth Doane, "Climate Refugees: Nations under Threat," *CBSN On Assignment*, August 21, 2017, https://www.youtube.com/watch?v=4MXoUbsswHY/.
12. Carol Farbotko and Heather Lazarus, "The First Climate Refugees? Contesting Global Narratives of Climate Change in Tuvalu," *Global Environmental* Change 22 (2012): 387.
13. Ibid., 389.
14. Epeli Hau'ofa, "A Sea of Islands," *The Contemporary Pacific* 6, no. 1 (1994): 148–61.
15. Ibid.
16. Ibid, 154.
17. Epeli Hau'ofa, "Our Place Within: Foundations for a Creative Oceania," in *We Are the Ocean: Selected Works* (Honolulu: University of Hawai'i Press, 2008), 81.
18. Tammy Haili'Ōpua Baker, Sharon Mazer, and Diana Looser, "'The Vessel Will Embrace Us': Contemporary Pacific Voyaging in Oceanic Theater," *Performance Research* 21, no. 2 (2016): 40.
19. *Waka* is the Maori term for a canoe; ibid., 44.
20. *Waka* is a term also used to refer to elders.
21. Baker, Mazer, and Looser, "Contemporary Pacific Voyaging," 41.
22. Wolfgang Kempf and Elfriede Herman, "Uncertain Futures of Belonging: Consequences of Sea-level Rise in Oceania," in *Belonging in Oceania: Movement, Place-Making and Multiple Identifications*, ed. Elfriede Herman, Wolfgang Kempf, and Toon van Meijil (Oxford: Berghahn Books, 2014).
23. Farbotko and Lazarus, "The First Climate Refugees?"

24. Kempf and Herman, "Uncertain Futures," 192.
25. Carol Farbotko, "Wishful Sinking: Disappearing Islands, Climate Refugees, and Cosmopolitan Experimentation," *Asia Pacific Viewpoint* 51, no. 1 (April 2010).
26. Climate activist Ursula Rakova of the Carteret Islands firmly rejects initiatives to "save the climate refugees." Invoking the canoe, Rakova and her community in the Carteret Islands founded a relocation organization in 2006 called Tulele Peisa, or "Sailing the Waves on Our Own." This nonprofit is organizing to relocate Carteret Islanders to the Bougainville Province in Papua New Guinea over the coming decades. See "Q & A with Ursula Rakova, Climate Activist for the Carteret Islands," *Women Deliver*, https://womendeliver.org/2016/qa-ursula-rakova/.
27. Teresia Teaiwa and Juliette Launiuvao, "Te Wa, the Canoe or the Hokule'a of Kiribati," *Amota Ataneka Merang*, April 2, 2015, https://kiribatidaily.wordpress.com/2015/04/02/te-wa-the-canoe-or-the-hokulea-of-kiribati/.
28. Ibid.
29. According to Marita Davies, building a canoe is an "intimate process" in large part because makers use their hands as a means of measuring the vessel. See "Te Wa, Kiribati's Way to the Water," *Lindsay*, March 12, 2017, http://lindsaymagazine.co/kiribati-te-wa/.
30. Ibid.
31. Ibid.
32. At the time, Australia was the second-largest coal exporter in the world. Today, it is first in the world and accounts for nearly one-third of global coal exports, "Ranking the Top Five Coal Exporting Countries Globally," *NS Energy*, January 24, 2020, https://www.nsenergybusiness.com/features/five-coal-exporting-countries/.
33. Graef, "Canoes vs. Coal."
34. "Coral Bleaching Events," Australian Institute of Marine Science, https://www.aims.gov.au/docs/research/climate-change/coral-bleaching/bleaching-events.html/.
35. Graef, "Canoes vs. Coal."
36. Farbotko and Lazarus, "The First Climate Refugees?," 382.
37. For a more comprehensive discussion of the strategies Pacific climate activists employ to challenge state-centric global climate governance agendas, see Samid Suliman et al., "Indigenous (Im)mobilities in the Anthropocene," *Mobilities* 14, no. 3 (2019).
38. Bronte Hogarth, "Maria Tiimon Is Building Awareness of How Climate Change Affects Low Lying Islands in the Pacific," *1 Million Women*, March 7, 2016, https://www.1millionwomen.com.au/blog/women-youre-voice-maria-timon-building-awareness-how-climate-change-affects-low-lying-islands-pacific/.
39. Lutunatabua, "Inspiring Islanders."

40. "The Pacific Climate Warrior Journey," 350.org, https://world.350.org/pacificwarriors/the-pacific-warrior-journey/.
41. "President's Daughter in Climate Change Protest," *Adventist Record*, October 30, 2014, https://vimeo.com/110522108/.
42. "Pacific Warriors Canoe for Climate Change," Cultural Survival.org, https://www.culturalsurvival.org/news/pacific-warriors-canoe-climate-change/.
43. "The Pacific Climate Warriors," 350.org, https://world.350.org/pacificwarriors/.
44. Melinda Benson and Robin Craig, "The End of Sustainability," *Society and Natural Resources* 27 (2014): 780.
45. "Stand with the Pacific Climate Warriors," 350.org, https://act.350.org/act/pacific_solidarity_petition/.
46. Fenton Lutunatabua, "Climate Impacts and the Climate Warriors of the Island Nation of Kiribati," 350.org, March 28, 2014, https://350.org/climate-impacts-and-the-climate-warriors-of-the-island-nation-of-kiribati/.
47. Lutunatabua, "Inspiring Islanders."
48. Since the 2014 demonstration in Australia, the Pacific Climate Warriors have been active in global climate politics. In 2015, they joined the People's Pilgrimage to Paris ahead of COP21. During COP23, which was hosted by Fiji but held in Germany, the Warriors delivered "The Pacific Warrior Declaration on Climate Change" to leaders. See "The Pacific Warrior Declaration on Climate Change," *Have Your Sei*, https://haveyoursei.org/.
49. *Pathfinders of the Pacific.*
50. "Nainoa Thompson."
51. Margaret Cohen, *The Novel and the Sea* (Princeton: Princeton University Press, 2010), 2.
52. Ibid., 15, 2.
53. Ibid, 56.
54. Ibid, 2.
55. "Nainoa Thompson."
56. Cohen, *The Novel and the Sea*, 58.
57. Ibid, 28.
58. Ibid, 58.
59. Ibid.
60. "1978 Voyage to Tahiti Canceled After Hokule'a Capsizes," Hawai'ian Voyaging Traditions, http://archive.hokulea.com/holokai/1978/voyage_cancelled.html/.
61. "Nainoa Thompson."
62. As one crew member aboard *Hōkūle'a* once reflected, "I think when you're a master like Mau, it's much less of a science than an art." See *Pathfinders of the Pacific.*
63. Ibid.
64. Ilima Ho-Lastimosa, Phoebe W. Hwang, and Bob Lastimosa, "Community Strengthening through Canoe Culture: Ho'omana'o Mau as Method and Metaphor," *Journal of Medicine & Public Health* 73, no. 12 (2014): 398.

65. "Nainoa Thompson."
66. "Plants and Tools Used for Building Canoes," Hawai'ian Voyage Traditions, http://archive.hokulea.com/ike/kalai_waa/plants_and_tools.html/.
67. According to Nainoa Thompson in a 1997 speech to the Polynesian Union Conference, 90 percent of Hawai'i's koa forests were cut down in less than a century. See "Nainoa Thompson."
68. Ibid.
69. Karen Valentine, "He Wa'a He Moku; He Moku He Wa'a"—A Canoe Is an Island; an Island Is a Canoe," *Keola Magazine*, Summer 2016, https://keolamagazine.com/ocean/makalii-voyaging-canoe/.
70. "Promise to Pae'Āina O Hawai'i," Mālama Honua Declarations, April 23, 2014, http://www.hokulea.com/wp-content/uploads/2016/08/Promise-to-Paeaina.pdf/.
71. "The Mālama Honua Worldwide Voyage," Polynesian Voyaging Society, http://www.hokulea.com/worldwide-voyage/.
72. "Nainoa Thompson."
73. Hau'ofa, "A Sea of Islands," 148.
74. Ho-Lastimosa, Hwang, and Lastimosa, "Community Strengthening through Canoe Culture," 398.

PART II

Re-envisioning Ocean Protection

Editors' Introduction to Part II

THE ESSAYS IN PART I explored new oceanic pathways to connectedness and community within human society. Part II will unpack humanity's ongoing relationship to the ocean itself. Western societies have historically envisioned the ocean as a fragmented collection of valuable goods—food in the form of fish, shellfish, and seaweed; illumination in the form of whale oil; warmth in the form of marine mammal fur—and as a global garbage disposal. Even as the Scientific Revolution took hold in these societies, early scientific conceptions of the ocean viewed it as much too large for puny humans to affect, setting the stage for increasing global exploitation facilitated by both the Industrial Revolution's contribution of ever-bigger ships and the engines that made ocean voyaging relatively safe, and parallel seventeenth-century international law developments that grounded the rules governing the ocean in "freedom of the seas."

Science and law have continued to be joint partners in re-envisioning humanity's relationship to the ocean—although, importantly, the breadth of both has expanded considerably since the early years of European colonialism. On the science side, there is no escaping the fact that humans' interactions with the ocean—and with their environment more generally—are taxing the ocean's health. We now know that human beings are more than capable of changing the ocean's physics, its chemistry, its ecosystems, its functions, and, ultimately, its ability to support the planet's biosphere. Moreover, the concept of "scientific knowledge" is expanding to incorporate more than just Western ways of knowing, reinvigorating indigenous visions of humanity's relationship to the sea. At the same time, law has provided more incentives and avenues for protecting the ocean, from increased national control over and a duty to sustainably protect coastal waters under the United Nations Convention on the Law of the Seas to an international goal to protect 10 to 30 percent of the ocean in marine protected areas pursuant to the United Nations Convention on Biological Diversity.

Re-envisioning the Anthropocene ocean requires that we reconceptualize our own relationship to the ocean and the means through which we interact with it. The essays in Part II begin this process by presenting new visions of a protected and productive ocean from Western science, indigenous knowledge, and law. However, in so doing, they also spotlight contradictory valuations of the ocean and its bounty—ocean as the source of valuable goods versus ocean as culture and life support system—that have shaped humanity's historical and contemporary relationships with the ocean. These essays demonstrate the need to shift how we value the ocean so that we can re-envision both the reasons for ocean protection and the means of that protection.

Jeremy B. C. Jackson has spent decades illuminating the long-term impacts of human activities on ocean ecology—especially overfishing and pollution—through the methodologies of Western science—observe, measure, test, explain. He summarizes these studies to suggest new paradigms for reducing human use and increasing human benefit from the ocean.

As much of the world is increasingly coming to realize, however, Western science does not provide the only way of knowing—or protecting—the world's complex ecosystems, including those in the ocean. Thomas Swensen demonstrates historical clashing valuations of the ocean in his recounting of how the Karluk Village on Kodiak Island, Alaska, attempted to protect the Karluk River Basin, including the coastal waters it feeds, from Russian and then American exploitation. Although local in focus, this story exemplifies the (often violent) displacement of indigenous coastal cultures, their visions of the ocean, and their place-based wisdom in managing their local marine territories. As Christopher Finlayson will note in Part III, throughout the world both the surviving indigenous communities and the westernized societies that subsumed them are making new efforts to revitalize these native visions of ocean management and protection.

Jackson and Swensen show that the law has an important role to play in shaping and effectuating any re-envisioning of the ocean. The last three essays in Part II scrutinize law's potential to structure a re-envisioned and more mutually protective relationship between humanity and the ocean. Robin Craig explores both President Barack Obama's use of the Antiquities

Act to create the Northeast Canyons and Seamounts Marine National Monument and the litigation that followed. Her essay illuminates the ways old laws can re-envision ocean protection and proves that valuation conflicts are alive and well in the United States. Nathaniel Broadhurst, in turn, focuses on the recurring presidential attempts to formulate a national ocean policy for the United States, emphasizing President Trump's executive order on the subject and its revision of President Obama's desire to increase protection for the ocean. Finally, Abigail Benesh describes a relatively new legal phenomenon, the Rights of Nature movement, and explores how the law might effectively recognize personhood rights in the ocean and its ecosystems. Re-creating the ocean as a legal person with its own legal rights to exist and flourish would offer a completely new re-envisioning of the ocean in law, putting it on an equal footing with human beings for the first time in human history.

CHAPTER 6

Humanity's Changing Relationship with the Ocean

JEREMY B. C. JACKSON

VAST, BOUNTIFUL, AND FEARSOME, the oceans throughout history provided seemingly inexhaustible fish for people brave and skillful enough to exploit them. When fish catches declined, fishers sailed farther and farther from home to meet their needs.[1] This geographic expansion began more than two thousand years ago to meet the gastronomic needs of the coastal Greek and Roman Empires by the construction of new fermented fish sauce factories that extended beyond the Mediterranean to the Atlantic coasts of northern Spain and France.[2] Nowadays, the entire global ocean is accessible as large factory ships and the magic of refrigeration have allowed fishers to venture out for months or years at a time.[3] New ways of fishing using more efficient and diverse kinds of gear have increased catches ever more, with little understanding of the incremental reduction of fish stocks.[4] No one but a few scientists worried about how long the bounty could last until suddenly, in the middle of the twentieth century, everything changed. Mini-wars were fought over cod, exclusive economic zones were established to keep foreigners away, and fisheries kept collapsing with tragic consequences for entire communities, as epitomized by the precipitous failure of Newfoundland cod when thirty thousand fishers and fish-plant workers forever lost their jobs.[5]

Humanity has also used the oceans as the ultimate sewer for our garbage, excrement, and chemicals, especially since the advent of intensive agriculture and urban growth. Coastal pollution most obviously began in the stench of estuaries like the Thames and New York Harbor, which by the nineteenth century had become serious hazards to human health.[6] Soon afterwards, entire semi-enclosed seas like the Baltic and Adriatic, Chesapeake Bay, and embayments of the Mississippi Delta were so polluted by excess nutrients and organic matter that oxygen levels declined, and fish kills became commonplace.[7] More recently, industrial pollution of toxic chemicals and greenhouse gases from burning fossil fuels have extended to the farthest reaches of the oceans and the atmosphere. Scientists long warned about these pollutants, but now their effects can no longer be ignored as toxins like mercury accumulate dangerously in tuna and swordfish,[8] plastics litter the oceans,[9] and carbon emissions from fossil fuels acidify and heat up ocean water, intensify extreme storms, and cause sea-level rise that threatens the homes and livelihoods of more than a billion people.[10]

Long out of sight and mind, we are only beginning to understand that entire ocean ecosystems are gravely threatened, with dire implications for human wellbeing.[11] I will first review the extraordinary scope of the ongoing changes in the oceans, with emphasis on the discoveries of the last twenty years. The basic points are simple. Humans are now the ultimate apex predators in the oceans, dominating food webs in all marine ecosystems. Humans are also the primary agents of habitat change in the ocean, causing profound disruptions in ecosystem structure and function. Lastly, climate change exacerbates all of these problems to make everything worse.

I will then consider how we can do better for the oceans and for ourselves. First and foremost is to decarbonize the global economy within the next twenty years because everything else depends on that.[12] I will return to this point at the end, but my emphasis here is on three kinds of more immediate actions and policies that are already working or showing signs of doing so. First, we need to recognize the extraordinary diversity of marine species as living beings rather than mere commodities; this is an essential step towards understanding the different roles that species play in the maintenance of healthy ecosystems for the future. Second, we need to leave much of the ocean alone to stabilize resources and ultimately reap greater resources and profit. Third, we need to intervene to try to repair ocean ecosystems by integrating policies for both the land and oceans, treating them together, as related systems, rather than apart.

Human Impacts on the Ocean

Overfishing

Humans dominate ocean ecosystems every bit as much as they do land ecosystems.[13] The most extensive data are for coastal ecosystems, including estuaries, marsh and mangrove wetlands, seagrass meadows, kelp forests, and coral reefs, where, in spite of great differences in their inhabitants, the dominant predators were historically large animals, including some combination of killer whales, sharks, seals, crocodiles, predatory fishes like tunas and sharks, and seabirds.[14] Nowadays, most of these animals are so severely depleted as to be ecologically extinct. Humans have taken their place as the dominant predators at almost all trophic levels above the zooplankton—and humans even dominate the zooplankton in Antarctica, where there is a major fishery for krill without anything close to the necessary ecological data and an adequate stock assessment to know what is sustainable.[15]

The biomass of preferred fishery species has diminished by an order of magnitude or more,[16] as best demonstrated for the groundfish of the northwest Atlantic[17] but equally affecting coral reefs,[18] kelp forests,[19] estuaries and coastal seas,[20] and even the high seas.[21] Fisheries scientists attacked these conclusions, but a detailed USA National Research Council assessment strongly confirmed the extreme depletion of large animal ocean biomass.[22] Arguments have continued,[23] but it is now generally accepted that two-thirds of global fisheries are overfished and getting worse, while many of the remaining better-managed fisheries are not yet sufficiently recovered to be economically viable.[24]

Unsurprisingly, given all of the above, global fish catches are declining in spite of greatly increased effort and subsidies.[25] The losses are greatest for large scale industrial fisheries, whereas artisanal catches are still slowly increasing. Risks of biological extinction are also increasing for large animals.[26] Caribbean monk seals are already lost,[27] and their Hawaiian and Mediterranean counterparts are gravely threatened.[28] Killer whales are rapidly diminishing, especially the newly recognized species that depend on highly specific overfished prey like salmon.[29] Caribbean sea turtles have declined in abundance a hundredfold.[30] Caribbean crocodiles are threatened to endangered throughout most of their range and are hybridizing with critically endangered island species.[31] To imagine how abundant American crocodiles must have been, one needs to go to northwestern Australia, where they are still abundant and dangerous.[32] Sharks are globally threatened, with losses of numerous species exceeding 90 percent or more.[33] Sharks were already so rare on Caribbean reefs in the 1960s that I personally saw only a few dozen over nearly fifty years of fieldwork there. That's in comparison to the countless hundreds I saw diving on Micronesian reefs in just two weeks in 1969, although Indo-Pacific sharks are now also gravely depleted.[34] SCUBA-diving marine ecology would have been a perilous business in pristine ecosystems!

Habitat Destruction and Its Consequences

There are more than five hundred coastal hypoxic "dead zones" worldwide, an astonishing exponential increase since the 1960s,[35] resulting from massive increases in nutrient runoff from intensive agriculture made possible by cheap nitrogen fertilizer manufactured from petroleum.[36] Runoff of excess nitrogen fuels population explosions of phytoplankton (microscopic algae and microbes) far beyond the capacity of zooplankton and other suspension feeders to consume them. The excess phytoplankton die and sink to the seafloor, where they are metabolized by microbes, a process that consumes most or all the oxygen in bottom waters (hypoxia or anoxia). Animals that cannot swim away die from asphyxia, except for a very few species that can survive extremely low oxygen conditions, with potentially great losses in fisheries.

The structural integrity of coastal marine habitats from the tropics to the temperate zone is dependent on the abundance of a small number of structurally dominant species of mangroves, saltmarshes, seagrasses, kelps, and reef corals that stabilize sediments and provide critical shoreline protection from storms.[37] These are also important sites of carbon deposition and sequestration[38] and are important nursery habitats for fisheries.[39] Wholesale losses of these species because of coastal development and climate change effectively kills the environment, reducing biological structural stability and complexity from three-dimensional marine forests to featureless level bottoms of sediment, thereby eliminating the ecosystem services enumerated above. Global losses have been alarming, reaching 50 percent for mangroves[40] and 30 percent for seagrasses.[41] Patterns for kelp forests are highly varied, with most global ecoregions showing significant declines but a few regions increasing in extent.[42] Global declines in living coral cover on reefs is also highly variable but commonly exceeds 50 percent throughout the Caribbean and Indo-Pacific.[43]

Other increasingly widespread forms of anthropogenic habitat change are more immediately destructive in reducing habitat complexity and biodiversity.[44] The most damaging include dynamite fishing on coral reefs, which blows corals to bits and kills fishes indiscriminately;[45] seabed trawling for shrimp, scallops, and groundfish, which transforms biodiverse underwater forests into depauperate level bottoms of mud;[46] and deep seafloor mining, which, if it is allowed to proceed, will inevitably destroy seafloor ecosystems for decades and possibly centuries.[47] Container ship traffic is also increasing almost exponentially and carries the double risk of fatal collisions with endangered whales and sound pollution that is dangerous for all cetaceans.[48]

Seismic oil and gas exploration cause even more severe sound pollution that sometimes causes mass mortalities of cetaceans.[49]

Introductions of exotic species are also increasing because of expanding ship traffic discharging at sea ever-increasing volumes of ballast water that contains larval stages of invertebrates and fishes, plankton, and pathogens.[50] The data are mostly circumstantial, but the first outbreaks of mass mortality of the sea urchin *Diadema antillarum* occurred next to the Caribbean entrance of the Panama Canal,[51] and the first widespread outbreaks of coral diseases in the Caribbean were recorded from nearby Colombia and adjacent Netherlands Antilles.[52] Coral diseases are exacerbated by global warming (see below), but the first Caribbean disease outbreaks occurred two decades before the first reports of coral bleaching due to extreme warming events.[53] Introductions of exotic species also occur as a result of deliberate or accidental release from aquaria, which, for example, is how Indo-Pacific lionfish colonized the Caribbean, with devastating effects on native fish populations.[54]

Anthropogenic Climate Change

Impacts to the ocean from climate change caused by the burning of fossil fuels are both direct and indirect and include rising average temperatures, extreme heating events, declining oxygen, ocean acidification, outbreaks of disease, and intensification of extreme storms.

Sea surface temperatures are rising globally, but disproportionately, with the greatest increases in polar seas and semi-enclosed basins in the temperate zones, such as the Gulf of Maine.[55] The latitudinal limits of myriad species are rapidly increasing in response,[56] as in the case of the Humboldt squid, whose northern limit shifted from southern California to the Gulf of Alaska in just a few decades as a result of a combination of climate change and overfishing that reduced the abundance of predators.[57] Most species' range shifts are more gradual but pervasive, with great implications for fisheries.[58] For example, optimal conditions for Atlantic and Barents Sea cod are moving northward, out of traditional fishing grounds and into different international jurisdictions,[59] further exacerbating the consequences of historical overfishing.[60] Tropical reef corals are also migrating towards higher latitudes, most strikingly along the southwest coast of Australia, where kelp forests are dying off and being replaced by subtropical species, including reef corals.[61]

As oceans continue to warm, species characteristic of colder polar conditions have nowhere else to migrate and are at risk of extinction. Arctic species and entire ecosystems are increasingly threatened by the loss of summer sea ice.[62] The iconic example is the increasing starvation of polar bears, which historically fed on seals captured at breathing holes.[63] Populations are plummeting,[64] and starving bears are showing up around human settlements, where they forage on garbage and potentially whatever else they can catch.[65] Other effects on polar food webs are still poorly understood, but the collapse of Antarctic krill, for example, would have grave impacts on the baleen whales that feed upon them.[66]

Global warming is also causing increases in the magnitude and frequency of extreme heating events, wherein sea surface temperatures may rise 2–3° C above normal maxima in just a few months.[67] Consequences for reef corals can be catastrophic.[68] Healthy reef corals exist in symbiosis with dinoflagellates within their tissues which are critical to coral nutrition and calcification.[69] Extreme heat breaks down this symbiosis, and the corals evict the symbiont, which leaves them ghostly white, hence bleached. Such coral bleaching is commonly fatal to the corals unless the symbiosis is reestablished within a matter of weeks. Mass bleaching events are increasingly frequent and severe, raising questions about the very survival of coral reefs. The most recent extreme example was in 2015–2016, when most corals along the northern Great Barrier Reef bleached and died,[70] and similar mass bleaching and mortality occurred across the Pacific.[71] Another example of sudden sea temperature increases is the enormous blob of hot water that appeared in the northeast Pacific in 2014[72] which was associated with collapses in species abundance and outbreaks of diseases.[73]

Oxygen concentrations are declining in the open ocean because the warming of surface waters makes them lighter, which in turn slows down the vertical mixing of the oceans, a runaway process that decreases the rate of oxygen transport to the deep sea and upwelling of nutrients to the sea surface as lighter surface waters continue to accumulate.[74] The process is especially striking in the equatorial Pacific and in the Arctic Ocean where the cover of summer sea ice is rapidly decreasing.[75] Sea ice is highly reflective, dispersing heat back into the atmosphere, whereas seawater absorbs heat, setting up a positive feedback that is effectively irreversible. Reduced nutrient upwelling and declining oxygen are strongly associated with decreases in the open ocean productivity that is the basis for high seas fisheries.[76]

The ocean is also becoming more acidic as increasing amounts of carbon dioxide in the atmosphere due to the burning of fossil fuels are dissolved in seawater. As a result, average ocean pH has declined by 0.1 units over the past century.[77] The biological consequences of this ocean acidification on physiology, growth, and development are still incompletely understood although what we do know is of great concern. The most obvious impacts are on organisms that form their skeletons of calcium carbonate, which is more easily dissolved under more acidic conditions. Acidification is already affecting shellfish aquaculture industries in Washington State, where pH is steadily declining.[78] Aquaculturists have been forced to raise vulnerable juvenile clams and oysters under less acidic conditions in aquaria on land before placing them in the ocean.[79] Reef corals are also vulnerable to ocean acidification. Corals grown under present-day more acidic conditions grew 15 percent more slowly than corals where pH was maintained experimentally at historically less acidic conditions.[80] The combined threat of acidification and rising temperatures pose an existential threat to the future survival of corals and coral reefs unless efforts are rapidly intensified to abandon fossil fuels for wind, solar, and other forms of green energy.

Climate change also sets off cascading series of indirect effects that magnify its impact. Well-known terrestrial examples include the advance of parasites and diseases that are moving poleward with warming temperatures, as in the case of pine borers and ticks in northern forests and tropical diseases expanding northward into the southern United States.[81] Similarly, in the oceans, the impact of coral diseases has greatly increased, especially in connection with mass bleaching events.[82] Outbreaks of coral diseases are especially impactful on polluted reefs[83] and those where overfishing has resulted in population explosions of fleshy algae, which have been shown experimentally to increase the vulnerability of corals to disease.[84] In contrast, disease outbreaks are comparatively rare on unpolluted reefs in marine protected areas with abundant grazing fishes.[85] Lobsters along the northeast coast of North America are also more vulnerable to shell wasting disease as waters warm, effectively wiping out the fishery in Long Island Sound.[86]

Climate change is also increasing human risk through the increased destructiveness of extreme storms, coupled with sea-level rise and storm surge.[87] The year 2017 was the first year that three Category 4 or 5 hurricanes struck the United States in a single year (the previous record was one), with catastrophic damage and loss of life.[88] The Intergovernmental Panel on Climate Change estimates that sea levels will rise by about one meter by 2100,[89] but the models are based on what we know, and every year we keep learning more and more about increasing melting rates in Greenland and Antarctica. Consequently, estimates by numerous prominent experts hover closer to two or more meters of sea-level rise by the end of the century. The majority of essential global infrastructure, including ports, energy plants, cement production, transportation hubs, and megacities, are within the danger zone of sea-level rise.

Reversing the Tide

Now that the enormity of harmful human impacts on the oceans is so clearly documented, attention is increasingly shifting from ever more detailed documentation of the problems to increasing focus on possible solutions.[90] Most efforts to date have been local measures to address specific problems within areas of national jurisdiction, although there is increasing momentum for global governance of the high seas through the United Nations. Important developments include changes in public perceptions of marine life as people learn more about species and their endangerment through ecotourism and environmental media, establishment of marine protected areas to stabilize and rebuild populations and ecosystems, and reductions in pollution and habitat destruction.

Learning about Species and What They Do

A century ago, bird watchers fighting to stop the slaughter of herons and egrets for women's hats were pioneers in the early rise of the conservation movement, marked by organizations such as the newly formed Audubon Society.[91] Similarly, it's not just important for tourism that increasing numbers of people pay good money to see whales up close in the wild and to SCUBA dive with sharks. Besides the thrill of witnessing these creatures' power and grace, whale and shark watchers learn about the lives and behavior of these magnificent animals and how they fit into ocean ecosystems, which, in turn, leads to increased support for their protection.

Horror at the slaughter of whales was a major factor in the establishment of the International Whaling Convention and its implementing Commission in 1946. Despite persistent opposition from a few countries, this treaty and the whaling moratorium that the Commission imposed have resulted in dramatic recoveries of most whale species.[92] In addition to the ethical issues inherent in the mass slaughter of such magnificent animals, we now know that the great whales were once (and increasingly are now again) vitally important "ecosystem engineers," as predators of massive amounts of fish and invertebrates, prey for other large predators, highly mobile reservoirs of carbon and nutrients, and, as carcasses, sources of energy and habitat in the deep sea.[93]

Similar public concerns about losses of other marine mammals were a driving factor in the enactment of the United States Marine Mammal Protection Act in 1972, which prohibits the killing, harm, harassment, or collection of any marine mammal in US territorial waters or by US citizens anywhere else. It also forbids importation of any marine mammal products or parts, subject to exceptions for old products like nineteenth-century scrimshaw artifacts or fossil marine mammal ivory. Populations of most marine mammals have varyingly recovered, although their comparative success is strongly associated with their life histories, habitat requirements, and geographic range.[94] Depletion of essential forage as a result of overfishing also inhibits recovery.[95] One obvious manifestation of success is the greatly increased abundance of seals along the east and west coasts of the US, where their activities and real or perceived impacts on fisheries are not always welcome.[96] The resurgence of seals has also led to increases in great white sharks near shore,[97] restoring a degree of balance to marine food webs while generating new questions about perceived risks to humans and potential impacts on endangered species.[98]

Increased tourist revenues have also led to the banning of shark fishing on coral reefs by entire nations because the sharks are vastly more lucrative alive than dead. Economic analysis for the government of Palau demonstrated that diver tourism provides 39 percent of total GDP and that 21 percent of divers come principally to dive with sharks. The approximately one hundred sharks in prime shark dive sites are each worth about US$180,000 per year in tourist revenue, or US$1.9 million during their lifetimes, versus about $110 each for their fins and meat.[99] Shark diving is a burgeoning global industry that is not without its environmental concerns, although, done responsibly, the net conservation value appears to be generally positive.[100]

New studies of the remarkable behavior and migrations of ocean species are also generating public support for increased protections. Electronic tagging of thousands of individuals of different species of Pacific whales, seabirds, sea turtles, tunas and other large fish, and sharks has revealed striking transoceanic migrations of some species versus others that travel much smaller distances.[101] Bluefin tuna, for example, move back and forth across the Atlantic and Pacific, hanging out for up to a year or more in the same general location before moving on. In contrast, eastern Pacific great white sharks move back and forth between the California coast, where they feed on burgeoning seal populations, and an area of deep ocean halfway between Baja California and Hawaii dubbed the "White Shark Café," where they feed on vertically migrating fishes and invertebrates.[102] Over two hundred of the sharks have been tagged and followed for up to twenty years; their movements and histories are available at http://whitesharkcafe.org/.

Marine aquaria, an ever-expanding library of guidebooks to coastal marine life, and approximately fifteen million active SCUBA divers and snorkelers in the United States alone[103] are also increasing public awareness of maintaining healthy oceans. That number is still dwarfed by the roughly sixty million US birdwatchers,[104] but the increasing numbers of people being drawn to the ocean is a critical factor in their growing awareness of threats to the ocean and support for increased actions and protections.[105]

Marine Protected Areas

Marine protected areas (MPAs) are an increasingly popular and effective conservation strategy for biodiversity and habitat protection when effectively financed, administered, and enforced,[106] but unprotected "paper parks" can do more harm than good by lulling people into thinking everything is fine when it is not.[107] MPAs are also controversial from the perspective of fisheries management,[108] with some arguing that MPAs are the most effective tool available,[109] versus those who believe that management tools such as catch shares and gear restrictions are more effective than simple area closures.[110]

Cabo Pulmo in the southern Sea of Cortez is one of the most spectacular success stories of an effectively enforced MPA.[111] Cabo Pulmo was designated as a Mexican marine national park in 1995 on the basis of its coral populations, although it was severely overfished. Protections did not become effective, however, until local villagers self-organized to enforce protection of the entire park as a no-take area in the late 1990s. Fish biomass was less than one metric ton per hectare in 1999, comparable to other unprotected areas or paper parks throughout the Gulf of California. Biomass increased over the next ten years of absolute protection to about 4.5 metric tons while all other areas failed to increase, an extraordinary example of the power of local leadership for conservation.[112] Biomass and diversity have fluctuated up and down since 2009 in large part through reorganization of the relative abundance of different fishes as the community evolves towards a more natural composition that includes greater populations of schooling fishes as well as more abundant corals.[113] Now the greatest potential threat to Cabo Pulmo is its notorious success, which is attracting burgeoning numbers of tourists and development.[114]

A network of nine well-enforced no-take MPAs and two partial-take MPAs was established around four of the northern Channel Islands off

California in 2003 and revisited ten years later.[115] Biomass of preferred fisheries species approximately doubled within MPAs at three of the four islands, but nontargeted species showed little response. Biomass of targeted species also increased by about one quarter outside the reserves, possibly because of a spillover effect. Similar results were obtained in the Cowcod Conservation Areas established in the southern Channel Islands in 2001, where abundances of six of eight targeted species and four of seven non-targeted rockfish species increased regionally from 1998 to 2013.[116] Rising temperatures during the study are a complicating factor. Nevertheless, 75 percent of targeted species but none of the nontargeted species increased inside compared to outside of the MPAs, even controlling for environmental factors. These results suggest that well-chosen and enforced MPAs can compensate for fishing effort, restoring a more natural species mixture within their boundaries.

In spite of strong evidence for their effectiveness, establishment of very large marine protected areas within nations' exclusive economic zones has increased the area of ocean protected by MPAs to a mere 3.5 percent, with only 1.6 percent under strong protection.[117] Meanwhile, most ocean ecosystems are hemorrhaging as most of the major fishing fleets continue to expand their operations on a truly global scale.[118] This state of affairs may be changing, however, as the international community finally begins to seriously consider international governance of the high seas, defined as areas beyond national jurisdiction. The first major achievement was the agreement to establish the world's largest marine protected area by the twenty-five member nations constituting the Commission for the Conservation of Antarctic Marine Living Resources.[119] The agreement protects all wildlife and bans fishing for overfished krill and Patagonian and Antarctic Toothfish in six hundred thousand square miles of the Ross Sea for thirty-five years. Much more will have to be done, however, to preserve other marine areas around Antarctica where other populations of these species are threatened by overfishing and rapid climate change, with ripple effects on the marine mammals and penguins that depend upon them.[120]

One of the most exciting new developments is the realization that closing the high seas to fisheries is a win for both conservation and the long-term sustainability of fisheries that are dwindling globally. This is because nearly 98 percent of global seafood production comes from exclusive economic zones (EEZs) of individual nations and aquaculture, with a small proportion from fresh water. In contrast, the high seas catch is only 2.4 percent, mostly of luxury species of tuna and billfishes, and their commercial value is even less.[121] Moreover, most high seas fisheries are heavily dependent on government subsidies by a small number of wealthy countries that can afford the enormous costs.[122] Closure of the high seas to fishing would therefore have great economic and social benefits in addition to environmental protection of fish stocks and the remarkable long-distance migration routes of marine megafauna documented by Block and colleagues.[123] Most compellingly, the overwhelming majority of high seas fishery species are also major components of fisheries within national EEZs, which means that closure of the high seas to fishing would produce a vast MPA where commercially important species could prosper, reproduce, and spill over into EEZs, where potential catches would increase.[124] Further advantages would include simplification of policing the rampant problem of pirate fishing and transfers at sea.[125]

Another driving factor in the United Nations negotiations is the threat of deep-sea mining of rare minerals that is now technically feasible but at demonstrably catastrophic environmental cost.[126] Large-scale mining has not yet begun but is being debated at the United Nations under intense lobbying pressure from the would-be mining conglomerates.

Ecological Restoration

While too commonly overshadowed by bad news, concerted actions to reduce pollution and protect keystone species have resulted in many recoveries of marine populations and ecosystems.[127] Installation of modern sewage systems and reductions in nutrient runoff have varyingly improved water quality, reduced excess planktonic productivity and toxic algal blooms, and restored seagrass meadows, salt marshes, and fisheries in many estuaries around the world.[128] The Billion Oyster Project is a remarkable ongoing citizen science program under the auspices of the New York Harbor School to establish one billion live oysters in New York Harbor by 2035, with the ecological goals of reestablishing oyster reefs, improving water quality through the filtration of the oysters, and rebuilding nearshore ecosystems.[129]

These efforts demonstrate that even greater progress could be achieved in stabilizing coastal ecosystems if adequate measures are taken to eliminate or greatly reduce runoff of pollutants, including most importantly agricultural nutrient pollution that feeds the dead zones and pollutes coral reefs.[130] However, serious efforts to do so have not yet materialized because the farmers don't have to pay to reduce their pollution. There is also a problem of scale in semi-enclosed seas like the Baltic, because nutrient buildups in sediments are already so great that simply reducing runoff of nutrients may not suffice.

A very different sort of success story stems from the banning of fish pots around Bermuda in 1990, where fish populations had collapsed because of overfishing and coral reefs were degraded as a result of diverse impacts. Fishes greatly rebounded after the ban and were rapidly dominated by schools of large parrotfish.[131] Since then, abundances have remained high except for the large predatory fish that remain overfished. Coral populations have steadily increased, apparently as a result of the control of algal populations by the abundant parrotfish. Caribbean coral reefs are generally extremely overfished, but the few places where both fishing and pollution are effectively controlled—such as Bermuda, Bonaire, and the Flower Gardens reefs in the Gulf of Mexico—continue to uniquely support high coral and reef fish abundance comparable to that throughout the Caribbean several decades before.[132] How long such local protection can help protect these reefs from rising temperatures is uncertain but they will inevitably succumb, just as so much of the Great Barrier Reef,[133] unless drastic actions are taken to shift away from fossil fuels.

Navigating the Future in the Brave New Ocean

Despite all of these important accomplishments, no major nation has managed to enact the kinds of comprehensive policies necessary to address the unsustainability of the modern global economy that is driving ecosystem collapses both on the land and in the oceans and that threatens human well-being. Nature is a complex system and much of that system as we know it is irreversibly breaking down.[134] This is especially true of the ocean, which is so vulnerable to impacts from the land in addition to the consequences of overexploitation and climate change.

We need to accept that environmental perturbations in one place almost inevitably have repercussions down the line because everything is so strongly interconnected, be it the aforementioned agricultural pollution in the US cornbelt causing the dead zone in the Gulf of Mexico or the effects of runoff and overfishing on outbreaks of disease affecting reef corals. We also need to accept that there are no quick fixes. For example, huge energy and investment in projects to restore populations of corals in Florida and on the Great Barrier Reef are making much progress in terms of the technical details of raising, breeding, and growing corals. But they are also absurdly expensive and small scale, not to mention that putting the corals back into the same nutrient-polluted environments and expecting them to somehow survive is folly.[135] More fundamentally, they are bandaids to address the symptoms of ocean decline rather than addressing the fundamental root causes of the ocean crisis: global warming, overfishing, and land-based pollution.[136]

Nevertheless, we are finally making progress. The most encouraging developments involve large-scale efforts to decarbonize the global economy that are beginning to gain traction despite political intransigence, not least because green energy is a financially better option than heavily subsidized fossil fuels in addition to its obvious advantages for human health and the environment.[137] California, the fifth-largest global economy, is committed to be carbon neutral by 2045 and is well on track. Electric cars are becoming a more practical alternative to gasoline and diesel. Increasingly destructive extreme weather events combined with indisputable evidence of sea-level rise are raising public alarm, especially as insurance companies are abandoning coverage of low-lying areas. The momentum of all these changes seems irreversible and the outstanding question is how rapidly opposition can be overcome to speed things up.

Perhaps most importantly, we need fundamentally new paradigms to guide our relationship to the oceans, both for the sake of the oceans and for our own sustainability and well-being.[138] Increasing focus on closing the high seas to fishing is a promising example. Long thought to be a preposterous impossibility, the idea is gaining scientific and political traction precisely because it makes sense not only for the survival of open ocean ecosystems but also for the continued productivity of oceanic fisheries within individual nations' exclusive economic zones. Similar arguments apply to reforming industrial agriculture to drastically reduce downstream pollution while reducing runaway problems of massive soil erosion, loss of pollinators, superweeds, and pollution of drinking water that are a major threat to agricultural sustainability and human health.[139] Changing the conversation in these newly positive ways helps to redefine our goals towards tractable solutions.[140]

Notes

Conversations with Terry Hughes, Jennifer Jacquet, Nancy Knowlton, Loren McClenachan, and Bob Steneck continue to influence my thinking about the ocean crisis and paths forward, some more optimistically than others. The Smithsonian Tropical Research Institute nurtured me for thirteen magical years to follow my instincts and the Scripps Institution of Oceanography allowed me to stick my neck out even when it made them slightly nervous. To all I am very grateful.

1. W.J. Bolster, *The Mortal Sea* (Harvard University Press, Cambridge, MA: 2014); Callum Roberts, *The Unnatural History of the Sea*, 2nd ed. (Island Press, Washington, DC: 2009); D.H. Cushing, *The Provident Sea* (Cambridge: Cambridge University Press, 1988).
2. Declan Henesy, "Fish sauce in the Ancient World," *Ancient History Encyclopedia* (2018), https://www.ancient.eu/article/1276/.
3. D. Tickler, J.J. Meeuwig, M.-L. Palomares, Daniel Pauly, & D. Zeller. 2018. Far from home: Distance patterns of global fishing fleets. *Science Advances* 4: eaar3279, https://www.science.org/doi/10.1126/sciadv.aar3279/.
4. Bolster, *supra* note 1; Roberts, *supra* note 1; Cushing *supra* note 1.
5. John Kennedy. 1997. "At the Crossroads: Newfoundland and Labrador Communities in a Changing International Context." *The Canadian Review of Sociology and Anthropology* 34 (3): 297–318; and Michael Harris, *Lament for an Ocean: The Collapse of the Atlantic Cod Fishery, a True Crime Story* (Toronto: McClelland & Stewart, 1998).
6. J. Waldman, *Heartbeats in the Muck: The History, Sea Life, and Environment of New York Harbor* (revised ed.; Empire State Editions, 2012); M.J. Atrill, ed., *A Rehabilitated Estuarine Ecosystem: The Environment and Ecology of the Thames Estuary* (New York: Springer, 1998).
7. T.B.H. Reusch, J. Dierking, H.C. Andersson, E. Bonsdorff, J. Carstensen, M. Casini, et al. 2018. The Baltic Sea as a time machine for the future coastal ocean. *Science Advances* 4: DOI: 10.1126/sciadv.aar8195; Nancy N. Rabalais, R.E. Turner, B.K. Sen Gupta, E. Platon, & M.L. Parsons. 2007. Sediments tell the history of eutrophication and hypoxia in the northern Gulf of Mexico. *Ecological Applications* 17 (supplement): S129–S143; H.K. Lotze, H.S. Lenihan, B.J. Bourque, R.H. Bradbury, R.G. Cooke, M.C. Kay, S.M. Kidwell, M.X. Kirby, C.H. Peterson, & Jeremy B.C. Jackson. 2006. Depletion, degradation, and recovery potential of estuaries and coastal seas. *Science* 312: 1806–1809.
8. A.T. Schartup, C.P. Thackray, Qureshi, C. Dassuncao, Gillespie, A. Hanke, & E.M. Sunderland. 2019. Climate change and neurotoxicant in marine predators. *Nature* 572: 648–650, https://doi.org/10.1038/s41586-019-1468-9/.
9. J.R. Jambeck, R. Geyer, C. Wilcox, Y.R. Siegler, M. Perryman, A. Andrady, R. Narayan, & K.L. Law. 2015. Plastic waste inputs from land into the ocean. *Science* 347: 768–771.
10. Intergovernmental Panel on Climate Change. 2019: Summary for Policymakers. In: *IPCC Special Report on the Ocean and Cryosphere in a Changing Climate* (H.-O. Pörtner, D.C. Roberts, V. Masson-Delmotte, P. Zhai, M. Tignor, E. Poloczanska, K. Mintenbeck, M. Nicolai, A. Okem, J. Petzold, B. Rama, N. Weyer, eds.) [hereinafter 2019 IPCC Ocean Report).
11. Jeremy B.C. Jackson. 2008. Ecological extinction and evolution in the brave new ocean. *Proceedings of the National Academy of Sciences USA* 105, supplement 1: 11458–11465.

12. J.-P. Gattuso, A. Magnan, R. Bille, W.W.L. Cheung, E.L. Howes, F. Joos, et al. 2015. Contrasting futures for ocean and society from different anthropogenic CO_2 emissions scenarios. *Science* 349: aac4722-1.
13. 2019 IPCC Ocean Report, *supra* note 10; Springmann, M., Clark, M., Mason-D'Croz, D. *et al.* Options for keeping the food system within environmental limits. *Nature* 562 (2018): 519–525; P.M. Vitousek, H.A. Mooney, Jane Lubchenco, & J.M. Melillo. 1997. Human domination of earth's ecosystems. *Science* 277: 494–499.
14. H.K. Lotze & Boris Worm. 2009. Historical baselines for large animals. *Trends in Ecology and Evolution* 24: 254–262; Jackson, *Ecological Extinction, supra* note 11; Roberts, Callum. *The Unnatural History of the Sea.* Washington, DC: Island Press, 2007; Jeremy B.C. Jackson, M.X. Kirby, W.H. Berger, K.A. Bjorndal, L.W. Botsford, B.J. Bourque, R.H. Bradbury, et al. 2001. Historical overfishing and the recent collapse of coastal ecosystems. *Science* 293: 629–638; Jeremy B.C. Jackson. 2001. What was natural in the coastal oceans. *Proceedings of the National Academy of Sciences USA* 98: 5411–5418.
15. Jennifer Jacquet, Daniel Pauly, D. Ainley, S. Holt, P. Dayton, & Jeremy B.C. Jackson. 2010. Seafood stewardship in crisis. *Nature* 467: 28–29.
16. Jackson, *Ecological Extinction, supra* note 11; Jackson et al., *supra* note 14.
17. Bolster, *supra* note 1; V. Christensen, S. Guenette, G.H. Heymans, C.J. Walters, R. Watson, D. Zeller, & Daniel Pauly. 2003. Hundred-year decline of North Atlantic predatory fishes. *Fish and Fisheries* 4: 1–24.
18. L. McClenachan. 2008. Documenting loss of large trophy fish from the Florida Keys with historical photographs. *Conservation Biology* 23: 636–643; Jackson, *supra* note 14; Jeremy B.C. Jackson. 1997. Reefs since Columbus. *Coral Reefs* 16 (supplement): S23–S32.
19. K.A. Krumhansi, D.K. Okamoto, A. Rasseiler, M. Novak, J.J. Bolton, K.C. Cavanaugh, et al. 2016. Global patterns of kelp forest change over the past half-century. *Proceedings of the National Academy of Sciences USA* 113: 13785–13790.
20. Lotze et al., *supra* note 7.
21. J.K. Baum, R.A. Myers, D.G. Kehler, Boris Worm, S.J. Harley, & P.A. Doherty. 2003. Collapse and conservation of shark populations of the Northwest Atlantic. *Science* 299: 389–392; Daniel Pauly, V. Christensen, J. Dalsgaard, R. Froese, & F. Torres Jr. 1998. Fishing down marine food webs. *Science* 279: 860–863.
22. National Research Council, National Academy of Sciences, *Dynamic changes in marine ecosystems: Fishing, foodwebs, and future options* (Washington, DC: National Academy Press, 2006).
23. R. Hilborn & D. Ovando. 2014. Reflections on the success of traditional fisheries management. *ICES Journal of Marine Science* 71: 1040–1046; Boris Worm, R. Hilborn, J.K. Baum, T.A. Branch, J.S. Collie, C. Costelleo,

et al. 2009. Rebuilding global fisheries. *Science* 325: 578–585; T.E. Essington, A.H. Beaudreau, & J. Wiedenmann. 2006. Fishing through marine food webs. *Proceedings of the National Academy of Sciences USA* 103: 3171–3175.

24. C. Costello, D. Ovando, T. Clavelle, C.K. Strauss, R. Hilborn, M.C. Melnychuk, et al. 2016. Global fishery prospects under contrasting management regimes. *Proceedings of the National Academy of Sciences USA* 113: 5125–5129; Boris Worm. 2016. Averting a global fisheries disaster. *Proceedings of the National Academy of Sciences USA* 113: 4895–4897.
25. Daniel Pauly & D. Zeller. 2016. Catch reconstructions reveal that global marine fisheries catches are higher than reported and declining. *Nature Communications* 7: 10244, doi: 10.1038/ncomms10244.
26. D.J. McCauley, M.L. Pinsky, Steven R. Palumbi, J.A. Estes, F.H. Joyce, & R.H. Warner. 2015. Marine defaunation: Animal loss in the global ocean. *Science* 347: 1255641:1–7.
27. L. McClenachan & A.B. Cooper. 2008. Extinction rate, historical population structure and ecological role of the Caribbean monk seal. *Proceedings of the Royal Society B* 275: 1351–1358.
28. J.D. Baker, A.L. Harting, T.A. Wurth, & T.C. Johanos. 2010. Dramatic shifts in Hawaiian monk seal distribution predicted from divergent regional trends. *Marine Mammal Science* 27: 78–93; A.A. Karamanlidid, P. Dendrinos, P.F. de Larrinoa, A.C. Gucu, W.M. Johnson, C.O. Kirac, & R. Pires. 2015. The Mediterranean monk seal *Monachus monachus*: Status, biology, threats, and conservation priorities. *Mammal Review* 46: 92–105.
29. K.L. Ayres, R.K. Booth, J.A. Hempelmann, K.L. Koski, C.K. Emmons, R.W. Baird, et al. 2012. Distinguishing the impacts of inadequate prey and vessel traffic on an endangered killer whale (Orcinus orca) population. *PloS One* 7: e36842. https://doi.org/10.1371/journal.pone.0036842/.
30. Loren McClenachan, Jeremy B.C. Jackson, & M.J.H. Newman. 2006. Conservation implications of historic sea turtle nesting beach loss. *Frontiers in Ecology and Environment* 4: 290–296.
31. J.B. Thorbjarnarson. 2010. American crocodile *Crocodylus acutus*. In S.C. Manolis & C. Stevenson, eds., *Crocodiles. Status Survey and Conservation Action Plan*, Third Edition: 46–53 (Crocodile Specialist Group: Darwin); J.B. Thorbjarnarson, F. Mazzotti, E. Sanderson, F. Buitrago, M. Lazcano, K. Minkowski, et al. 2006. Regional habitat conservation priorities for the American crocodile. *Biological Conservation* 128: 25–36.
32. Y. Fukuda, C. Manolis, & K. Appel. 2014. Management of human-crocodile conflict in the Northern Territory, Australia: Review of crocodile attacks and removal of problem crocodiles. *The Journal of Wildlife Management* 78: 1239–1249; Y. Fukuda, G. Webb, C. Manolis, R. Delaney, M. Letnic, G. Linder, & P. Whitehead. 2011. Recovery of saltwater crocodiles following unregulated hunting in tidal rivers in Northern Territory, Australia. *The Journal of Wildlife Management* 75: 1253–1266.

33. Baum et al., *supra* note 21; N.K. Dulvy, J.K. Baum, S. Clarke, L.V.J. Compagno, E. Cortes, A. Domingo, et al. 2008. You can swim but you can't hide: The global status and conservation of oceanic pelagic sharks and rays. *Aquatic Conservation: Marine and Freshwater Ecosystems*, doi: 10.1002/aqc.
34. T.B. Letessier, D. Moullot, P.J. Bouchet, L. Vigliola, M.J. Fernandes, C. Thompson, et al. 2019. Remote reefs and seamounts are the last refuges for marine predators across the Indo-Pacific. *PloS Biology* 17(8): e3000366. https://doi.org/ 10.1371/journal.pbio.3000366/.
35. R.J. Diaz & R. Rosenberg. 2008. Spreading dead zones and consequences for marine ecosystems. *Science* 321: 926–929; D. Breitburg, L.A. Levin, A. Oschlies, M. Gregoire, F.P. Chavez, D.J. Conley, et al. 2018. Declining oxygen in coastal waters. *Science* 359: eaam7240 (2018) 1–11.
36. Reusch et al., *supra* note 7; J.E. Cloern, P.C. Abreu, J. Carstensen, L. Chauvaud, R. Elmgren, J. Grall, et al. 2016. Human activities and climate variability drive fast-paced change across the world's estuarine-coastal ecosystems. *Global Change Biology* 22: 513–529; Rabalais et al., *supra* note 7; J.E. Cloern. 2001. Our evolving conceptual model of the coastal eutrophication problem. *Marine Ecology Progress Series* 210: 223–253.
37. G. Guanell, K. Arkema, P. Ruggiero, & G. Verutes. 2016. The power of three: Coral reefs, seagrasses and mangroves protect coastal regions and increase their resilience. *PloS One*: https://doi.org/10.1371/journal.pone.0158094/; E.B. Barbier, S.D. Hacker, C. Kennedy, E.W. Koch, A.C. Stier, & B.R. Silliman. 2011. The value of estuarine and coastal ecosystem services. *Ecological Monographs* 81: 169–193.
38. E. McLeod, G.L. Chmura, S. Bouillon, R. Salm, M. Bjork, C.M. Duarte, C.E. Lovelock, W.H. Schlesinger, & B.R. Silliman. 2011. A blueprint for blue carbon: Toward an improved understanding of the role of vegetated coastal habitats in sequestering CO_2. *Frontiers in Ecology and Environment* 9: 552–560.
39. O. Aburto-Oropeza, E. Ezcurra, G. Danemann, V. Valdez, J. Murray, & E. Sala. 2008. Mangroves in the Gulf of California increase fishery yields. *Proceedings of the National Academy of Sciences USA* 105: 10456–10459; I. Nagelkerken & G. van der Velde. 2004. Relative importance of interlinked mangroves and seagrass beds as feeding habitats for juvenile reef fish on a Caribbean island. *Marine Ecology Progress Series* 274: 153–159.
40. S.S. Romanach, D.L. DeAngelis, H.L. Koh, S.Y. The, R. Sulaiman, R. Barizan, & L. Zhai. 2018. Conservation and restoration of mangroves: Global status, perspectives, and prognosis. *Ocean and Coastal Management* 154: 72–82.
41. M. Waycott, C. Duarte, T.J.B. Carruthers, R.J. Orth, W.C. Dennison, S. Olyarnik, et al. 2009. Accelerating loss of seagrass across the globe threatens coastal ecosystems. *Proceedings of the National Academy of Sciences USA* 106: 12377–12381.
42. Krumhansi et al., *supra* note 19.

43. Jeremy B.C. Jackson, M.K. Donovan, K.L. Cramer, & W. Lam, eds. *Status and trends of Caribbean coral reefs: 1970–2012* (Global Coral Reef Monitoring Network, IUCN, Gland, Switzerland, 2014); G. De'ath, K.E. Fabricius, H. Sweatman, & M. Puotinen. 2012. The 27-year decline of coral cover on the Great Barrier Reef and its causes. *Proceedings of the National Academy of Sciences USA* 109: 17995–17999; Terry P. Hughes, J.T. Kerry, M. Alvarez-Noriega, J.G. Alvarez-Romero, K.D. Anderson, A.H. Baird, et al. 2017. Global warming and recurrent mass bleaching of corals. *Nature* 543: 373–378.
44. McCauley et al., *supra* note 26.
45. E.E. DeMartini & J.E. Smith. Effects of fishing on the fishes and habitat of coral reefs. In C. Mora, ed., *Ecology of fishes on coral reefs* (Cambridge: Cambridge University Press, 2015), 135–144.
46. W. Watling & E.A. Norse. 1998. Disturbance of the seabed by mobile fishing gear: A comparison to forest clear cutting. *Conservation Biology* 12: 1180–1197; R.O. Amoroso, C.R. Pitcher, A.D. Rijnsdorp, R.A. McConnaughey, A.M. Parma, P. Suuronen, et al. 2018. Bottom trawl fishing footprints on the world's continental shelves. *Proceedings of the National Academy of Sciences USA* 115: E10275–E10282.
47. D.C. Dunn, C.L. Van Dover, R.J. Etter, C.R. Smith, L.A. Levin, T. Morato, et al. 2018. A strategy for the conservation of biodiversity on mid-ocean ridges from deep-sea mining. *Science Advances* 4: eaar4313.
48. McCauley et al., *supra* note 26.
49. D.P. Nowacek, C.W. Clark, D. Mann, P.J.O. Miller, H.C. Rosenbaum, J.S. Golden, et al. 2015. Marine seismic surveys and ocean noise: Time for coordinated and prudent planning. *Frontiers in Ecology and Environment* 13: 378–386.
50. J.B. Geller, J.A. Darling, & J.T. Carlton. 2010. Genetic perspectives on marine biological invasions. *Annual Review of Marine Science* 2: 367–393; H. Seebens, N. Schwartz, N. Schupp, & B. Blasius. 2016. Predicting the spread of marine species introduced by global shipping. *Proceedings of the National Academy of Sciences USA* 113: 5646–5651.
51. H.A. Lessios, D.R. Robertson, & J.D. Cubit. 1984. Spread of *Diadema* mass mortality through the Caribbean. *Science* 226: 335–337.
52. E. Weil & C.S. Rogers. Coral diseases in the Atlantic-Caribbean. In Z. Dubinsky & N. Stambler, eds., *Coral reefs: An ecosystem in transition* (Netherlands: Springer, 2011), 465–491.
53. Jackson et al., *supra* note 43.
54. M.A. Albins & M.A. Hixon. 2013. Worst case scenario: Potential long-term effects of invasive predatory lionfish (*Pterois volitans*) on Atlantic and Caribbean coral-reef communities. *Environmental Biology of Fishes* 96: 1151–1157.
55. K.E. Mills, A.J. Pershing, C.J. Brown, Y. Chen, F.-S. Chiang, D.S. Holland, S. Lehuta, J.A. Nye, J.C. Sun, A.C. Thomas, & R.A. Wahle. 2013: Fisheries management in a changing climate: Lessons from the 2012 ocean heat wave in the northwest Atlantic. *Oceanography* 26: 191–195.

56. M.T. Burrows, D.S. Schoeman, L.B. Buckley, P. Moore, E.S. Poloczanska K.M. Brander, et al. 2011. The pace of shifting climate in marine ecosystems. *Science* 334: 652–655; M.L. Pinsky, Boris Worm, M.J. Fogarty, J.L. Sarmento, & S.A. Levin. 2013. Marine taxa track local climate velocities. *Science* 341: 1239–1242.
57. L.D. Zeidburg & B.H. Robison. 2007. Invasive range expansion by the Humboldt squid, *Dosidicus gigas*, in the eastern North Pacific. *Proceedings of the National Academy of Sciences USA* 104: 12948–12950; J.S. Stewart, E.L. Hazen, S.J. Bograd, J.E.K. Byrnes, D.G. Foley, W.F. Gilly, B.H. Robison, & J.C. Field. 2014. Combined climate- and prey-mediated range expansion of Humboldt squid (*Dosidicus gigas*), a large marine predator in the California Current system. *Global Change Biology* doi: 10.1111/gcb.12502.
58. Cheung, William W.L., Vicky W.Y. Lam, Jorge L. Sarmiento, Kelly Kearney, Reg Watson, and Daniel Pauly. "Projecting global marine biodiversity impacts under climate change scenarios." *Fish and fisheries* 10, no. 3 (2009): 235–251.
59. Pershing, Andrew J., et al. "Slow adaptation in the face of rapid warming leads to collapse of the Gulf of Maine cod fishery." *Science* 350 (October 2015): 80–81; Fossheim, M., Primicerio, R., Johannesen, E., et al. Recent warming leads to a rapid borealization of fish communities in the Arctic. *Nature Climate Change* 5 (2015): 673–677.
60. Jackson et al., *Historical overfishing*, *supra* note 14; K.M. Brander. 2018. Climate change not to blame for cod population decline. *Nature Sustainability* 1: 262–264.
61. T. Wernberg, S. Bennett, R.C.B. Babcock, T. de Bettignies, K. Cure, M. Depczynski, et al. 2016. Climate-driven regime shift of a temperate marine ecosystem. *Science* 353: 169–172.
62. E. Post, U.S. Bhatt, C.M. Bitz, J.F. Brodie, T.L. Fulton, M. Hebblewhite, et al. 2013. Ecological consequences of sea ice decline. *Science* 341: 519–524.
63. C.D. Hamilton, K.M. Kovacs, R.A. Ims, J. Aars, & C. Lydersen, 2017. An Arctic predator–prey system in flux: Climate change impacts on coastal space use by polar bears and ringed seals. *Journal of Animal Ecology* 86: 1054–1064.
64. J.F. Bromaghin, T.L. McDonald, I. Stirling, A.E. Derocher, E.S. Richardson, & E.V. Regehr. 2015. Polar bear population dynamics in the southern Beaufort Sea during a period of sea ice decline. *Ecological Applications* 25: 634–651.
65. I. Stanley-Becker. 2019. A mass invasion of polar bears is terrorizing an island town. Climate change is to blame. *The Washington Post*: https://www.washingtonpost.com/nation/2019/02/11/mass-invasion-polar-bears-is-terrorizing-an-island-town-climate-change-is-blame/?noredirect=on/.
66. E. Seyboth, K.R. Groch, L. Dall Rosa, K. Reid, P.A.C. Flores, & E.R. Secchi. 2016. Southern right whale (*Eubalaena australis*) reproductive success is influenced by krill (*Euphausia superba*) density and climate. *Scientific Reports* 6: 28205 | DOI: 10.1038/srep28205.

67. Mills et al., *supra* note 55; E. Di Lorenzo & N. Mantua. 2016. Multi-year persistence of the 2014/15 North Pacific marine heatwave. *Nature Climate Change* 6: 1042–1047.
68. O. Hoegh-Guldberg, P.J. Mumby, A.J. Hooten, R.S. Steneck, P. Greenfield, E. Gomez, et al. 2008. Coral reefs under rapid climate change and ocean acidification. *Science* 318: 1737–1742.
69. Nancy Knowlton. 2001. The future of coral reefs. *Proceedings of the National Academy of Science* 98 (10): 5419–5425.
70. Hughes et al., *supra* note 43.
71. Terry P. Hughes, K.D. Anderson, S.R. Connolly, S.F. Heron, J.T. Kerry, J.M. Lough, et al. 2018. Spatial and temporal patterns of mass bleaching of corals in the Anthropocene. *Science* 359: 80–83.
72. Di Lorenzo & Mantua, *supra* note 67.
73. C.D. Harvell, D. Montecino-Latorre, J.M. Caldwell, J.M. Burt, K. Bosley, A. Keller, et al. 2019. Disease epidemic and a marine heat wave are associated with the continental-scale collapse of a pivotal predator (*Pycnopodia helianthoides*). *Science Advances* 5: eaau7042.
74. D. Roemmich & J. McGowan. 1995. Climatic warming and the decline of zooplankton in the California Current. *Science* 267: 1324–1326; S. Schmidtko, L. Stramma, & M. Visbeck. 2017. Decline in global oceanic oxygen content during the past five decades. *Nature* 542: 335–339.
75. Post et al., *supra* note 62.
76. Jeffrey J. Polovina et al., "Ocean's Least Productive Waters are Expanding," *Geophysical Research Letters* 35, no. 3 (February 2008), doi:10.1029/2007GL031745.
77. J.C. Orr, V.J. Fabry, O. Aumont, L. Bopp, S.C. Doney, R.A. Feely, et al. 2005. Anthropogenic ocean acidification over the twenty-first century and its impact on calcifying organisms. *Nature* 437: 681–686.
78. J.T. Wootton, C.A. Pfister, & J.D. Forester. 2008. Dynamic patterns and ecological impacts of declining ocean pH in a high-resolution multi-year dataset. *Proceedings of the National Academy of Sciences USA* 105: 18848–18853; J.A. Ekstrom, L. Suatoni, S.R. Cooley, L.H. Pendleton, G.G. Waldbusser, A. Cinner, et al. 2015. Vulnerability and adaptation of US shellfisheries to ocean acidification. *Nature Climate Change* 5: 207–214.
79. Alan Barton, G.G. Waldbusser, R.A. Feely, S.B. Weisberg, J.A. Newton, B. Hales, et al. 2015. Impacts of coastal acidification on the Pacific Northwest shellfish industry and adaptation strategies implemented in response. *Oceanography* 28 (2): 146–159, https://doi.org/10.5670/oceanog.2015.38/.
80. R. Albright, L. Caldeira, J. Hosfelt, L. Kwiatkowski, J.K. Maclaren, B.M. Mason, et al. 2016. Reversal of ocean acidification enhances net coral reef calcification. *Nature* 531: 362–365.
81. D.W. Williams & A.M. Liebhold. 2002. Climate change and the outbreak ranges of two North American bark beetles. *Agricultural and Forest*

Entomology 4: 87–99; S. Altizer, R.S. Ostfeld, P.T.J. Johnson, S. Kutz, & C.D. Harvell. 2013. Climate change and infectious diseases: From evidence to a predictive framework. *Science* 341: 514–519.

82. A. Croquer & E. Weil. 2009. Changes in Caribbean coral disease prevalence after the 2005 bleaching event. *Diseases of Aquatic Organisms* 87: 33–43; J. Miller, E. Muller, C. Rogers, R. Waara, A. Atkinson, K.R.T. Whelan, M. Patterson, & B. Witcher. 2009. Coral disease following massive bleaching in 2005 causes 60 percent decline in coral cover on reefs in the US Virgin Islands. *Coral Reefs* 28: 925–937.
83. B.E. Lapointe, R.A. Brewton, L.W. Herren, J.W. Porter, & C. Hu. 2019. Nitrogen enrichment, altered stoichiometry, and coral reef decline at Looe Key, Florida Keys, USA: A 3-decade study. *Marine Biology* 166: 108–111.
84. Terence Hughes. 2007. Phase shifts, herbivory, and the resilience of coral reefs. *Current Biology* 17: 360–365; K.L. Barott & F.L. Rohwer. 2012. Unseen players shape benthic competition on coral reefs. *Trends in Microbiology* 20: 621–628; D.S. Beatty, J.M. Valayil, C.S. Clements, K.B. Ritchie, F.J. Stewart, & M.E. Hay. 2019. Variable effects of local management on coral defenses against a thermally regulated bleaching pathogen. *Science Advances* 5: eaay1048.
85. Jackson et al., *supra* note 43; J.R. Zanefeld, D.E. Burkepile, A.A. Shantz, C.E. Pritchard, R. McMinds, J.P. Payet, et al. 2016. Overfishing and nutrient pollution interact with temperature to disrupt coral reefs down to microbial scales. *Nature Communications* 7: 11833.
86. M.L. Groner, J.D. Shields, D.F. Landers Jr., J. Swenarton, & J.M. Hoenig. 2018. Rising temperatures, molting phenology, and epizootic shell disease in the American lobster. *The American Naturalist* 192: E163–E177.
87. K. Emanuel. 2005. Increasing destructiveness of tropical cyclones over the past 30 years. *Nature* 436: 686–688; R. Marsooli, N. Lin, K. Emanuel, & K. Feng. 2019. Climate change exacerbates hurricane flood hazards along US Atlantic and Gulf coasts in spatially varying patterns. *Nature Communications* 10: 3785.
88. Jeremy B.C. Jackson & Steven Chapple. *Breakpoint: Reckoning with America's Environmental Crises* (New Haven: Yale University Press, 2018).
89. 2019 IPCC Ocean Report, *supra* note 10.
90. C.M. Duarte, S. Agusti, E. Barbier, G.L. Britten, J.C. Castilla, J,-P. Gattuso, R.W. Fulweiler, Terry P. Hughes, Nancy Knowlton, C.E. Lovelock, H.K. Lotze, M. Predragovic, E. Poloczanska, Callum Roberts, & Boris Worm. 2020. Rebuilding marine life. *Nature* 580: 39–51; Nancy Knowlton. 2020. Ocean optimism: Moving beyond the obituaries in marine conservation. *Annual Reviews of Marine Science* 13: 479–499; Y. Malhi, J. Franklin, N. Seddon, M. Solan, M.G. Turner, C.B. Field, & Nancy Knowlton. 2020. Climate change and ecosystems: Threats, opportunities and solutions. *Philosophical Transactions of the Royal Society* B 375: 1–8.

91. J. Price. 2000. *Flight Maps*. Basic Books.
92. H.K. Lotze, M. Coll, A.M. Magera, C. Ward-Paige, & L. Airoldi. 2011. Recovery of marine animal populations and ecosystems. *Trends in Ecology and Evolution* 26: 595–605.
93. J. Roman, J.A. Estes, L. Morissette, C. Smith, D. Costa, J. McCarthy, J.B. Nation, et al. 2014. Whales as ecosystem engineers. *Frontiers in Ecology and the Environment* 12: 377–385.
94. H.K. Lotze, J.M. Flemming, & A.M. Magera. 2017. Critical factors for the recovery of marine mammals. *Conservation Biology* 31: 1301–1311.
95. M.G. Burgess, G.R. McDermott, B. Owashi, L.E. Peavey Reeves, T. Clavelle, D. Ovando, et al. 2018. Protecting marine mammals, turtles, and birds by rebuilding global fisheries. *Science* 359: 1255–1258.
96. K.M. Cammen, D.B. Rasher, & R.S. Steneck. 2019. Predator recovery, shifting baselines, and the adaptive management challenges they create. *Ecosphere* 10: e02579.
97. G.B. Skomal, J. Chisholm, & S.J. Correia. 2012. Implications of increasing pinniped populations on the diet and abundance of white sharks off the coast of Massachusetts. In M.L. Domeier, ed., *Global perspectives on the biology and life history of the white shark* (CRC Press, Boca Raton, FL): 405–418.
98. J. Roman, M.M. Dunphy-Daly, D.W. Johnston, & A.J. Read. 2015. Lifting baselines to address the consequences of conservation success. *Trends in Ecology & Evolution* 30: 299–302; J.K. Carlson, M.R. Heupel, C.N. Young, J.E. Cramp, & C.A. Simpfendorfer. 2019. Are we ready for elasmobranch conservation success? *Environmental Conservation* 46: 264–266.
99. G.M.S. Vianna, M.G. Meekan, D. Pannell, S. Marsh, & S. Meeuwig. 2012. Socio-economic value and community benefits from shark-diving tourism in Palau: A sustainable use of shark populations. *Biological Conservation* 145: 267–277.
100. A.J. Gallagher, G.M.S. Vianna, Y.P. Papastamatiou, C. Macdonald, T.L. Guttridge, & S. Hammerschlag. 2015. Biological effects, conservation potential, and research priorities of shark diving tourism. *Biological Conservation* 184: 365–379.
101. B.A. Block, S.L.H. Teo, A. Walli, A. Boustany, M.J.W. Stokesbury, C.J. Farwell, et al. 2005. Electronic tagging and population structure of bluefin tuna. *Nature* 434: 1121–1127; B.A. Block, I.D. Jonsen, S.J. Jorgensen, A.J. Winship, S.A. Shaffer, S.J. Bograd, et al. 2011. Tracking apex marine predator movements in a dynamic ocean. *Nature* 475: 86–90.
102. S.J. Jorgensen, N.S. Arnoldi, E.E. Estess, T.K. Chapple, M. Ruckert, S.D. Anderson, & B.A. Block. 2012. Eating or meeting? Cluster analysis reveals intricacies of white shark (*Carcharodon carcharias*) migration and offshore behavior. *PloS One* 7: e47819.
103. Diving Equipment & Marketing Association, *DEMA Home Page*, www.dema.org/.

104. E. Carver. 2013. Birding in the United States: A demographic and economic analysis. Addendum to the 2011 survey of fishing, hunting, and wildlife-associated recreation. *US Fish and Wildlife Service Report 2011-1*: 1–16.
105. H.K. Lotze, H. Guest, J. O'Leary, A. Tuda, & D. Wallace. 2018. Public perceptions of marine threats and protection from around the world. *Ocean and Coastal Management* 152: 14–22.
106. S.E. Lester, B.S. Halpern, K. Grorud-Colvert, J. Lubchenco, B.I. Ruttenberg, S.D. Gaines, S. Airame, R.R. Warner. 2009. Biological effects within no-take marine reserves: a global synthesis. *Marine Ecology Progress Series* 384: 22–46; Jane Lubchenco & K. Grorud-Colvert. 2015. Making waves: The science and politics of ocean protection. *Science* 350: 382–383.
107. D.A. Gill, M.B. Mascia, G.N. Ahmadia, L. Glew, S.E. Lester, M. Barnes, et al. 2017. Capacity shortfalls hinder the performance of marine protected areas globally. *Nature* 543: 665–669.
108. L.H. Pendleton, G.N. Ahmadia, H.I. Browman, R.H. Thurstan, D.M. Kaplan, & V. Bartolino. 2018. Food for thought: Debating the effectiveness of marine protected areas. *ICES Journal of Marine Science* 75: 1156–1159.
109. E.g., E. Sala & S. Giakoumi. 2018. Food for thought: No-take marine reserves are the most effective protected areas in the ocean. *ICES Journal of Marine Science* 75: 1166–1168.
110. E.g., R. Hilborn, 2018. Food for thought: Are MPAs effective? *ICES Journal of Marine Science* 75: 1160–1162.
111. O. Aburto-Oropeza, B. Erisman, G.R. Galland, A. Mascarenas-Osorio, E. Sala, & E. Ezcurra. 2011. Large recovery of fish biomass in a no-take marine reserve. *PloS One* 6: e23601.
112. A.N. Rife, B. Erisman, A. Sanchez, & O. Aburto-Oropeza. 2013. When good intentions are not enough: Insights on networks of "paper park" marine protected areas. *Conservation Letters* 6: 200–212; O. Aburto-Oropeza, E. Ezcurra, J. Moxley, A. Sanchez-Rodriguez, I. Mascarenas-Osorio, C. Sanchez-Ortiz, B. Erisman, & T. Ricketts. 2015. A framework to assess the health of rocky reefs linking geomorphology, community assemblage, and fish biomass. *Ecological Indicators* 52: 353–361.
113. O. Aburto-Oropeza, pers. comm.
114. A. Langle-Flores, P. Ocelik, & P. Perez-Maqueo. 2017. The role of social networks in the sustainability transformation of Cabo Pulmo: A multiplex perspective. *Journal of Coastal Research* 77: 131–142.
115. J.E. Caselle, A. Rassweiler, S.L. Hamilton, & R.R. Warner. 2015. Recovery trajectories of kelp forest animals are rapid yet spatially variable across a network of temperate marine protected areas. *Scientific Reports* 5: 14102.
116. A.R. Thompson, D.C. Chen, L.W. Guo, J.R. Hyde, & W. Watson. 2017. Larval abundances of rockfishes that were historically targeted by fishing increased over 16 years in association with a large marine protected area. *Royal Society Open Science* 4: 170639.

117. Lubchenco & Grorud-Colvert, *supra* note 106.
118. Tickler et al., *supra* note 3.
119. B.C. Howard. 2016. World's largest marine reserve created off Antarctica. https://www.nationalgeographic.com/news/2016/10/ross-sea-marine-protected-area-antarctica/.
120. C.M. Brooks, D.G. Ainley, P.A. Abrams, P.K. Dayton, R.J. Hofman, J. Jacquet, & D.B. Siniff. 2018. Watch over Antarctic waters. *Nature* 558: 177–180.
121. U.R. Sumaila, V.W.Y. Lam, D.D. Miller, L. Teh, R. Watson, D. Zeller, et al. 2015. Winners and losers in a world where the high seas is closed to fishing. *Scientific Reports* 5: 8481; L. Schiller, M. Bailey, J. Jacquet, & E. Sala. 2018. High seas fisheries play a negligible role in addressing global food security. *Science Advances* 4: eaat8351.
122. E. Sala, J. Mayorga, C. Costello, D. Kroodsma, M.L.D. Palomares, Daniel Pauly, U.R. Sumaila, & D. Zeller. 2018. The economics of fishing the high seas. *Science Advances* 4: eaat2504.
123. N. Queiroz, N.E. Humphries, A. Couto, M. Vedor, I. da Costa, A.M.M. Sequeira, et al. 2019. Global spatial risk assessment of sharks under the footprint of fisheries. *Nature* 572: 461–466.
124. Sumaila et al., *supra* note 121.
125. K. Boerder, N.A. Miller, & Boris Worm. 2018. Global hotspots of transshipment of fish catch at sea. *Science Advances* 4: eaat7159.
126. Dunn et al., *supra* note 47.
127. Lotze et al., *supra* note 92.
128. J.M.P. Vaudry, J.N. Kremer, B.F. Branco, & F.T. Short. 2010. Eelgrass recovery after nutrient enrichment reversal. *Aquatic Botany* 93: 237–243; Waldman, *supra* note 6; Cloern et al., *supra* note 36; H. Greening, A. Janicki, & E.T. Sherwood. 2016. Seagrass recovery in Tampa Bay, Florida (USA). In Christopher M. Finlayson, G.R. Milton, & N.C. Davidson, eds., *The Wetland Book II: Distribution, Description, and Conservation* (Springer Science and Business): 495–506.
129. Billion Oyster Project, *Our Vision*, https://billionoysterproject.org/.
130. Rabalais et al., *supra* note 7; F.J. Kroon, P. Thornburn, B. Schaffelke, & S. Whitten. 2016. Towards protecting the Great Barrier Reef from land-based pollution. *Global Change Biology* 22: 1985–2002; Reusch et al., *supra* note 7; Lapointe et al., *supra* note 83.
131. B.E. Luckhurst. 1994. A fishery-independent assessment of Bermuda's coral reef fish stocks by diver census following the fish pot ban—a progress report. *Proceedings of the 46th Gulf and Caribbean Fisheries Institute:* 309–323.
132. Jackson et al., *supra* note 43.
133. Hughes et al., *supra* note 43.
134. Bill McKibben. *The End of Nature* (New York: Random House, 1989); Bill McKibben. *Falter: Has the human game begun to play itself out?* (New York: Henry Holt & Company, 2019); Vitousek et al., *supra* note 13; Jackson, *supra* note 11.

135. T.H. Morrison, Terry P. Hughes, W.N. Adger, K. Brown, J. Barnett, & M.C. Lemos. 2019. Save reefs to rescue all ecosystems. *Nature* 573: 333–336.
136. R.H. Bradbury & R.M. Seymour. 2009. Coral reef science and the new commons. *Coral Reefs* 28: 831–837.
137. Jackson & Chapple, *supra* note 88.
138. Duarte et al., *supra* note 90; Malhi et al., *supra* note 90.
139. Jackson & Chapple, *supra* note 88.
140. Knowlton, *supra* note 90.

CHAPTER 7

A Reservation of Water

THOMAS MICHAEL SWENSEN

THE ROLLING HILLS of Kodiak Island's southern side around the Karluk River Basin are an ancestral home for both the Alutiiq people and the Kodiak bear.[1] Together the bears and the islanders have long lived near the many rivers of the Karluk area because they are age-old fish-spawning waters for five species of Pacific Ocean salmon, as well as for freshwater steelhead and dolly varden. The Karluk River, the largest among these channels of water, propagates from a range of small mountains that encircles Karluk Lake like a clasped fist to produce springtime runoff from melting winter snow. There, hiding in the mix of the brush and tall yellow grass, the Kodiak bears pick berries in the spring season, then travel down the mountains and hills to catch salmon at the river. The fish in the river have drawn bears and humans alike since before recorded history. Archaeologists have uncovered at the sides of Karluk Lake numerous remnants of Native homes and campsites alongside the skeletons of bears that date back seven thousand years.[2]

Downstream from the lake, at the mouth of the Karluk River, federal administrators in the Department of the Interior incorporated the Karluk Village as a tribal government in 1939. This formal certification was the first time that villagers had any political independence since Russians entered the region in the seventeenth century. The village was on an island over a thousand miles away from the union, as part of the Alaska Territory, which was the largest United States' noncontiguous territorial possession. Before statehood the Alaskan village of Karluk took advantage of new twentieth-century laws that were at first directed toward the Native nations located in the United States union in lower North America. The new tribal government, implemented by Interior Secretary Harold L. Ickes and Interior Department Solicitor Nathan R. Margold, employed the federal policies of the Indian New Deal of 1934 to create a reservation from land surrounding the village. These federal administrators helped the village extend the reservation boundaries to include a tract of the ocean up to three thousand feet from the shoreline.[3] The move to section off water for tribal ownership sought to limit access to the salmon at the mouth of the river from longtime industrial competitors. The river has served as one of the most profitable commercial salmon fishing sites in North America since the Russians claimed the Kodiak area in the eighteenth century. The villagers wanted to regain dominion over these resources after lacking the means to do so for so long.

This essay investigates the problems encountered by the Karluk Village's tribal government when they asserted water rights in the Alaska Territory. Less than a century after annexation by Russia, the village's legal actions as a tribal government revealed how the United States refused to apply the federal Indian policies that supported governance in the villages on Kodiak. Federal law as applied to Native American governments in Lower North America drew from Western European practices of claiming in the Western Hemisphere. Russia, unlike these Western European countries, conquered regions in Eurasia and in the Americas without recognizing what are called the occupational rights of villagers. When the United States purchased the Russian American territory from the tsar in 1867, the nation adhered to the Russian custom of not distinguishing village occupational land rights during the initial administrative period before statehood. In the 1930s, however, twenty years before Alaska joined the union, the United States arranged for a number of villages to organize as tribal governments. Out of the hundreds of Alaskan villages, the Department of the Interior raised a small number to the status of a tribal government and from these administrators helped measure out only four reservations. The Karluk Village's claim to water made visible the exceptional nature of United States legal policies towards people of annexed lands outside the contiguous part of the nation.

The chapter contributes to this volume by focusing on how the village lost their control of a portion of the sea through laws implemented for the public good. The reservation of water history illustrates how the people of Kodiak have experienced the Anthropocene in a manner distinct from their Russian and U.S. governors. I argue that the archive shows that re-envisioning ocean protection means keeping in mind how policies made by administrators affect local control in annexed territories. The overarching themes that emerge from this chronology suggest that there are questions about how universal practices, laws, and regulations governing the ocean relate to subject peoples.

For on Kodiak Island before Alaskan statehood, the Karluk Village attempted to incorporate an adjacent stretch of the ocean as their own property.[4] As a valuable resource for commercial fishing, the Karluk leaders made the move to close public access, which spurred widespread resistance to the reservation of water. In response, many branches of the government, and private industry, turned to a law that forbids such a land allocation to be made for Karluk. The history of this reservation of water fits into the chronology ascribed to federal Indian water claims but also to the exceptional aspects of United States federal law in territorial annexations outside the then United States union in Lower North America.

Unique to the Alaska Territory as an American political entity was having the status of a non–Western European colony before purchase by the United States. All other regions in the Americas faced initial annexation by European countries, such as Spain, England, or Portugal. When these countries began usurping the Americas, they employed an international law called the Doctrine of Discovery in order to invalidate another's possession of land. The doctrine laid out that if selections were not claimed by Western European Christian monarchs, then the controlling country held the right to own the land. The doctrine originated from the Treaty of Tordesillas in 1494, which commanded that Western Europeans could only annex non-Western non-Christian regions of the world. Much later, the United States Supreme Court, in a series of three decisions from 1823 through 1832 called the Marshall Trilogy, affirmed the nation's relations to Native nations through the Doctrine of Discovery. Chief Justice John Marshall, writing for the majority in all three cases, traced how European countries took over lands belonging to foreign nations in the Americas, during Europe's period of extensive overseas exploration that ran from the fifteenth through the seventeenth centuries. Within that time the nation established a practice of distinguishing occupational title for the original nations guided by policies of colonial administration.

The Alaska Territory's Russian past modified how the United States initially dealt with the villagers in places like Karluk after the purchase by passing them over for individual or collective property rights. During the Russian era the people of the Kodiak region held no rights beyond the will of Russian residents who forced them into labor. For over a century, managers of sea otter hunting parties whipped and kidnapped islanders to make them follow their directions. From 1733 to 1867 the tsar bound southern regional villagers in places like Kodiak to make continued tribute payments of fur for their survival. This imbursement method called *iasak* was an alternative to Western Europe's practices of economic unfreedom like chattel slavery or the Mita system in South America.[5] The tsar used iasak throughout the Russian territory while Western Europeans relied on the pope's proclamation to legitimize the annexation of the Americas. This Doctrine of Discovery was a Western notion about the ethics of colonial authority that claimed to create a legal relationship between Europeans, and then with the postcolonial American states, with pre-existing political organizations in the Western Hemisphere. The tsar sold the North American portion of Russia to the United States in 1867, and that region came to be known as Alaska, the longest held North American colony separate from the metropole. Then the region's political transition between Russia and the United States marked an opportunity for the federal government to cease extending occupational title to territorial possessions as the nation had done in previous expansions.

The assertion of a reservation of water eighty years after the Alaska Purchase was emblematic of Russia's history in North America meeting with United States Indian policies. Since Russia forged no treaties or other agreements with villages they eventually passed on to the United States, the nation treated these villagers without any occupational rights at the beginning of their territorial annexation. However, through court cases and congressional acts, federal law came to apply to these territorial villagers in Alaska as residents of the United States, eventually as citizens, then later as members of federally recognized tribal governments. In the decades following the Alaska Purchase, the United States naturalized the villagers as part of the new federal territory while, with sluggish speed, they constructed laws to govern the vast region. At that time, administrators flouted the idea that villagers held any independence from the federal government but closed off the possibility of national citizenship until 1917. After wholesale obligatory citizenship in 1924, the federal government also began applying Western-styled laws and policies, drawing from the Doctrine of Discovery, onto villages and their residents. Laws derived from the doctrine had remained an enduring and guiding principal for the federal government as the United States expanded across the lower portion of North America in the nineteenth century, but in Alaska the government applied them with an uneven geographic scope. For instance, down in the United States tribal governments with reservations retained a reserved right to water unlike the multitude of villages in Alaska like Karluk.

The extension of the Indian Reorganization Act in Alaskan villages after 1936 to include Alaska villagers proclaimed:

> Groups of Indians in Alaska not heretofore recognized as bands or tribes, but having a common bond of occupation, or association, or residence within a well-defined neighborhood, community, or rural district may organize to adopt constitutions and bylaws and to receive charters of incorporation and Federal loans under section 16, 17, and 10 of the Act of June 18, 1934 (48 Stat. 984).[6]

Under the law, Karluk met this requirement as a rural community, which then redirected how relations between the village and the federal government proceeded for decades to come and also marked a line between Russia and United States approaches to territorial acquisition. The Russian empire was less concerned about social conditions than about enduring the profitable sea otter industry off the labor of individuals like the Karluk villagers. Both governments dictated how life in Karluk unfolded and disregarded the village's desire for self-rule.

Before the twentieth century the United States used the Doctrine of Discovery to allow people who faced the coming of the nation to retain aspects of governance over lands they held preceding the nation's annexation. The federal policy was that an aboriginal title allowed people to possess an occupational right over their long-held lands in the face of development. This part of legal practice lent them a set of reserved rights for distinct areas.[7] The Winters doctrine, established in 1903, grew as the fundamental manner for tribal governments to possess a degree of reserved water rights, but the government did not apply these to the villages of the Alaska Territory.[8] The villages' histories with Russia and with the United States produced limitations on their abilities to request the rights offered through the Winters doctrine.[9] The people on Kodiak were a non-western people who were first colonized by a non-Western, Eastern Orthodox country before the United States erected a rule of law there. The Russian government contracted their dominion over the region to what became the Russian American Company, which deprived villagers of their ability to govern themselves. These United States laws, stemming from Russian annexation, produced relationships in Alaska different than those in the nation's earlier expansions. For instance, the Treaty of Cession between Russia and the United States spelled out that Alaska Native villages might fall under the laws the United States employs on the tribes [sic].[10] "The uncivilized tribes will be subject to such laws and regulations as the United States may, from time to time," the treaty reads, "adopt in regard to aboriginal tribes of that country."[11] This language of the law promised no rights outside of United States authority and stood inactive until passage of the Act of May 17, 1884. The law established federal services in the region and promised aboriginal people they could continue to live in their occupied territories without mentioning the possession, or termination, of Aboriginal title rights that may at a future point be dealt with by congress. "Indians or other persons in said district," the law said, "shall not be disturbed in the possession of any lands actually in their use or occupation or now claimed by them." With this passage the federal government pushed the question of Native title to an unforeseeable postponed date, proclaiming that, "the terms under which such persons may acquire title to such lands is reserved for future legislation by Congress."[12] The government made no commitment to work within the principals of the Doctrine of Discovery as the country had done when appropriating other parts of North America. They entered into no treaties with villagers nor any other type of formal agreements that registered their political separateness. Instead, the passage licensed an ambiguity about the ownership of lands by conditional aims that the federal government might consider at another time. The cession agreement reserved the government's entitlement to assert a ward-guardian relationship with the people of the Alaskan villages whom federal administrators deemed unable to transition into perceived "civilized" lives.

Imperative to how private land ownership evolved in the Alaska Territory was the court case *United States v. Seveloff* that denied demarking the region as "Indian Country" in 1873.[13] Indian Country was a legal term denoting that organizations indigenous to the region other than the United States held territorial authority in North America. Federal law defines Indian Country as "land within the limits of an Indian reservation under the jurisdiction of the United States government" as well as "dependent Indian communities" or "Indian allotments still in trust, whether they are located within reservations or not."[14] In the contiguous United States, tribal governments in Indian Country possessed a power to rule as they wished with limited federal intervention in their affairs. In *United States v. Seveloff*, however, the federal territorial judge ruled that the Alaska colony failed to exist as Indian Country, and therefore he denied Aboriginal title for Alaska Native villages because a previous law described Indian Country as not extending beyond the Pacific coast. The law, called the Trade and Intercourse Act of 1834, ascribed that Indian Country failed to exist beyond the boundaries of the Oregon Territory, and therefore the judge ruled that

Alaska was not Indian Territory.[15] This marked the Alaska villagers as being outside the frontier of national expansion, a non-Western people of a foreign state. Contrary to twentieth-century popular notions about Alaska as the "Last Frontier," the administrative law executed there from the nineteenth century forward handled the people and geography of the Alaska Territory as a former part of another state, not as an "Indian Country." When the 1884 Alaska Act set up the Department of Alaska, the federal government began administering a judicial review showing that region was less a frontier than a stable region under the nation's rule.[16] The court in the Seveloff case ruled that the framers of the Alaska Act would have said the region was such an Indian Country in 1867 if they believed it to be.[17] As an exception, less than a decade later the *Waters v. Campbell* case asserted that the Alaska colony was Indian Country only in the sense that it was forbidden to sell alcohol there.[18] The federal government wished to control the sale of liquor but refused to extend the provisions of Indian Country to the villages.

The administrators applied the Oregon legal code to the possession until the 1884 Organic Act which provided for federal territorial governance that regulated the region under a new territorial court. In the case *Kie v. United States* of 1886, the court held that Alaska failed to meet the criteria for Indian Country. The judge surmised that "it [is not] at all probable that the aborigines of Alaska can or will be considered as dependent or domestic nations, or people having any title to the Boil of the country, to be extinguished by the United States, as were the Indian tribes north and west of the Ohio river."[19] If Alaska was not an "Indian Country" then why could not villagers hold fee-simple title to their lands? The case *Tee-Hit-Ton Indians v. United States* would confront this issue in 1955.[20] The judicial reviews up to this point refused to apply any legal traditions onto Alaska, let alone those set up by the Marshall Trilogy. In an unusual case from the beginning of the twentieth century, the Allotment Act of May 17, 1906, allowed villagers to make individual land claims along the Karluk basin and throughout Alaska in lot sizes under 160 acres, but not as one applied the occupational title that the courts had ruled was inherent to Native Americans. These selections had to be vacant of residents, and "unappropriated" by others, in order to qualify.[21] The law conveyed no large-scale rights to villages as tribal governments in Alaska but instead worked with the intent of the Dawes Allotment era to dismantle villagers from their communities into individual parcels.

In contrast to the Allotment era policies in Alaska, an unexpected case arose that showed that the United States had no intention to apply policies grown from the Marshall Trilogy on Alaskan Territorial villagers. The case

Alaska Pacific Fisheries v. United States in 1918, based around the Annette Island reservation, diverged from the historical legal path of villages in Alaska in that the people of the reservation came from outside the boundaries of the Alaska Territory.[22] In 1891, Congress made an 85,000-acre reservation for a Tsimshian group that migrated across from Canada for religious reasons to the Alaska Territory four years prior led by an Anglican missionary named William Duncan. On permission of the United States, they settled on Annette Island with an assigned reservation. The court found in *Alaska Pacific Fisheries v. United States* that the tribal government did possess rights to the water implicated to the conveyance of a reservation that was formed by congressional action with the Karluk leaders.

Unlike the Metlakatla Indian community nine hundred miles away in southeastern Alaska, the Karluk tribal government's reservation in southern Kodiak formed through the Indian Reorganization Act of the Indian New Deal.[23] This law was the first time Karluk was able to assert authority over both Kodiak Island villagers and foreigners since the coming of the Russians in the eighteenth century. In the 1740s, the people of southern Alaska fell under the rule of the tsar who allowed Russian managers to make islanders work after hundreds lost their lives at their hands during the Awa'uq Massacre or Refuge Rock Massacre of 1784.[24] The village of Karluk was established a few years after in 1787 as a Russian trading outpost by Russian Evstrat Delarov after the slaughter at Refuge Rock on the other side of the island. Delarov was a head manager for an enterprise that later consolidated with other sea otter hunting operations in southern Alaska into the Russian American Company.[25] During the Russians' rule over Kodiak, at the turn of the nineteenth century, the Karluk river grew as one the best for red salmon fishing in North America with Russians commanding villager fishing activities. This was part of an organized global market, not an unexplored frontier. Up until that time, islanders lived on both sides of the river while the Russians created a trading post at the river's mouth. With Russians running the enterprise, the river cultivated a reputation for an abundance of fish that continued into the twentieth century as the marine economy swelled in southern Alaska. When the United States bought the island, fishers continued to use the Karluk River as a source of industry.

In the late nineteenth century, the J. P. Morgan Company directed the production of a safe and quick canning technique that allowed for the profitable commercial fishing industry to increase in productivity along the Pacific coast. Non-Native fishers assumed the sole command over the mouths of rivers, like that of the Karluk River, in southern Alaska with fish caches and wheels.[26] At the end of the nineteenth century, the river was the site of the first fish hatchery in Alaska.[27] By the 1930s, under federal authority, overfishing brought the robust salmon industry in Karluk to a nadir driving many processors northwest around the big island to spots in Larsen Bay. Companies further advanced the salmon fishing in other locations in the Kodiak Archipelago, yet the decline in harvesting at Karluk failed to end the site's participation in marine commerce. The Karluk region still commanded a profitable section of the red salmon market that lured seiners to throw their purse nets in the waters surrounding the village.

In response, the fishers hustled boats into the area to control all the waters of the Kodiak region. By June 1946, Fish and Wildlife agents followed the reservation laws by patrolling the waters of Karluk reservation, checking to see if fishers had the tribe's permission to work there. The dispute was a test for the political strength of Alaska villages as tribal governments in Alaska. The case explored whether the tribal government operated outside the authority of the territorial government like tribes did in the union. The case did this by asking if the creation of the water reserve was a violation of the public's right by lending a certain class of people, certified as a tribal government, rights over the rest of the public? The wording of the law read that "no exclusive or several right of fishery shall be granted therein, nor shall any citizen of the United States be denied the right to take, prepare, cure, or preserve, fish or shellfish in any area of the waters of Alaska where fishing is permitted by the Secretary of Commerce." The White Act was clear that citizens held the ability to use waterways for their benefit, but could the federal government impose regulations on the access to water? The law outlined the opposite, proclaiming, "The right herein given to establish fishing areas and to permit limited fishing therein shall not apply to any creek, stream, river, or other bodies of water in which fishing is prohibited by specific provisions of this Act, but the Secretary of Commerce through the creation of such areas and the establishment of closed seasons may further extend the restrictions and limitations imposed upon fishing by specific provisions of this or any other Act of Congress."[28] Was the Department of the Interior's Indian New Deal, an act by Congress, that allowed Karluk to make a legal boundary around part of the ocean for their own governance? Canneries filed suit against the head of the Department of the Interior arguing that the Karluk's reservation of water violated the White Act. Private business interests argued that the department's authority under the Indian New Deal did not allow it to take these waters out of public ownership and subject them to the tribe's independence. The significance here was that Alaska was

Figure 7.1. From the Amelia Elkinton Collection. Photograph of Karluk River and canneries taken sometime between 1905 and 1931. Alaska and polar regions collections, Elmer E. Rasmuson Library, University of Alaska, Fairbanks (UAF-1974-175-117). Used by permission.

not a state in the union and the Karluk Village was an international site of industry.

The United States Department of the Interior, representing the interests of Karluk, went on to pursue an unsuccessful case for the reservation of water at both the district and circuit levels; then it reached the U.S. Supreme Court. The court delivered an opinion that contradicted the *Alaska Pacific Fisheries v. United States* finding of the village's state of water ownership in 1916. The fisheries argued that tribal regulation of the water was against the territorial White Act of 1924 that forbade the private use of public lands. Federal laws like the White Act sought to entice people to move and invest in the Alaska Territory, and the Department of the Interior's attempt to lend villages tribal authority over natural resources went against this support for enterprise. For at the end of the nineteenth century, the only tribal government with a reservation was the Metlakatla Indian Community, created by the Congress in Alaska, and whose members came from outside the Alaska Territory. In contrast to this reservation, by the twentieth century Alaska made up 20 percent of the national geography but the hundreds of villages, which existed long before the United States' presence, lacked clear and broad legal standing with the federal government. After

Figure 7.2. "An 80,000-pound haul, Karluk, 1901." From Alaska State Library Collections and the Wickersham State Historic Site Photographs, 1882–1930s (ASL-P277-008-065). Used by permission.

the introduction of the Indian Reorganization Act into Alaska, four out of the five major cases about Alaska's people that went to the United States Supreme Court involved tribal governments set up by the IRA.

In *Hynes v. Grimes Packing Company*, Karluk brought to the court how they had worked toward regulating the nontribal business activity near the mouth of the river.[29] Under their appeal the water was to be included in the property holdings of the tribal government as part of their reservation. The basis for the claim was that the U.S. Department of the Interior controlled both the Bureau of Indian Affairs and the Fish and Wildlife Service, which managed Alaska's resources through the White Act of 1924. The territorial law spelled out how operations involving territorial water were to function. The act allowed for demarcating fish and other game reserves but stipulated that the directive barred limiting public reserved land and water use to distinct classes of people. Kent McNeil points out the origination of how the case came before the court. He writes: "By a 1943 Order, the Secretary of the Interior designated a certain area, including tidelands and coastal waters, as an Indian reservation for the native inhabitants of Karluk. Then in 1946, the Secretary amended the Alaska Fisheries General Regulations to prohibit commercial fishing in the coastal waters included in the reservation." Yet, at the same time federal administrators bolstered Karluk's right to regulate, regional canneries made the case that such laws negatively affected their ability to maintain profitability. McNeil shows that the industries "that depended on fish from these waters claimed that this prohibition would substantially impact their businesses," and continues to say that they "asked the Alaska District Court to enjoin enforcement of the prohibition on the grounds that the order and amendment to the regulations were invalid. The district court granted the injunction, and the Ninth Circuit Court of Appeals affirmed, leading to a certiorari application to the Supreme Court."[30]

Economist Steve Colt explains the importance of the White Act to the region's political future. He writes, "The White Act of 1924 was the result of the ensuing political battle and remained the foundation of all further federal regulation through 1959. The act prescribed an escapement goal of 50% of all Salmon and allowed the Secretary of Commerce to regulate all aspects of Alaska salmon fishing except for access to the fishery."[31] With that access for the public, Colt explained, the law spelled out that "[n]o exclusive right of fishery shall be granted." The House version of the bill abolished fish traps, but the senators from Washington blocked that provision.[32] This case intervened in the legal development of Alaska by claiming that tribe was a separate governmental entity from the Alaska territory and therefore holding powers and abilities outside of federal public law. In fact, the language spelled out that in "Section 2 of the extending act, set out at the beginning of this opinion, 337 U.S. 91, 69 S. Ct. 972, supra, gives no power to the Secretary to dispose finally of federal lands." The law specified that the Department of the Interior was not able to allocate parts of Alaska that limited the public's admittance. Section three of the White Act then directed:

> [I]t shall be unlawful to erect or maintain any dam, barricade, fence, trap, fish wheel, or other fixed or stationary obstruction, except for purposes of fish culture, in any of the waters of Alaska at any point where the distance from shore to shore is less than one thousand feet, or within five hundred yards of the mouth of any creek, stream, or river into which salmon run, excepting the Karluk and Ugashik Rivers, with the purpose or result of capturing salmon or preventing or impeding their ascent to the spawning grounds, and the Secretary of Commerce is hereby authorized and directed to have any and all such unlawful obstructions removed or destroyed.[33]

In contrast to public rights, the court noted that the Indian Reorganization Act was brought into existence in order to convey property to tribal governments, not to "create reservations of any kind." Led by Justice Stanley Forman Reed, the court's majority opinion concluded that the water reservation was invalid because the Reorganization Act supplied no avenue to the creation of reservations of this sort. The court further argued that the Reorganization Act failed to grant absolute ownership of lands to tribal governments, opining that the law "gives no power to the Secretary to dispose finally of federal lands."[34] *Hynes* proved contrary to the *Alaska Pacific Fisheries v. United States* case that upheld an implied congressional intent to develop tribal fishing.[35] With the Reorganization Act, the Reed Court followed in the tradition of positioning village as incongruent with public affairs. Reed did so by contending that the reservation of water granted exceptional privileges to the village, a class of people, to the detriment of the public, the non-Native fishers, in Alaska. The ruling figured Karluk villagers as a special class of people within the territorial public without the extraconstitutional rights to land afforded to the Tsimshian Metlakatla Indian community in the *Alaska Pacific Fisheries v. United States* case. The court ignored the precedents laid out for Native national rights in the Marshall Trilogy in refusing to concede that Karluk Village owned the land. In allowing members of the industry to bring such a charge in opposition to the tribe, the court treated residents of the Native villages as though they lacked long-term proprietorship of land. The court was unwilling to apply Western law as it pertains to Native Americans in the union onto a region that was once part of Russia. The Alaska Territory was not a state, nor a lawless frontier, and the villagers at Karluk held no claims to land through the Reorganization Act.

In comparison, the law under the Winters doctrine when Congress reserved land for use as a tribal reservation in the union in lower North America, also reserved the volumes of water necessary to allow the tribal government to maintain and grow their society.[36] The argument Justice Reed presented against Karluk was that Congress failed to provide the ground of the reservation, based on the Indian New Deal, beyond the village's use and occupation of their homelands. In the absence of a treaty, Reed saw no evidence of tribal ownership in the congressional dispositions for the Karluk Village. Justice Reed relied on the Indian New Deal to render the Karluk Tribe a "special class of people," instead of an independent people, for the purpose of regulating privatized industrial fishing. For the *Hynes v. Grimes Packing Company* court, the denial of tribal governance can be read against how the Winters doctrine recognized reservation tribes as in possession of water rights. The court when ruling in Winters confirmed that the tribal government held rights to water as part of their resources. The ruling noted that the reservation included an "implied reservation of water" that came with the land upon the reservation's designation. Justice Reed, in divergence from Winters, employed how he believed the Indian Reorganization Act policies failed to promise such power. Reed thus presented the denial of an occupational title, overriding the tribe's inherent sovereignty, as a decision made on behalf of the public good. In this case, the Reorganization Act's creation of villages as tribal governments served to subordinate the village's claims to that of a "public interest" even though Native villagers constituted a significant part of the territorial public. McNeil explains that "the Court held that the order including the coastal waters was valid, but found the amendment to the regulations limiting commercial fishing in those waters to the native inhabitants of the reservation and their licensees to be invalid (it was on the latter issue that four judges dissented, as they viewed the regulations as valid except to the extent that they permitted the Karluk Indians to grant authority to others to fish commercially in the reservation waters)."[37]

Justice Reed attested that "[t]here is no language in the various acts, in their legislative history, or in the Land Order 128, from which an inference can be drawn that the Secretary has or has claimed power to convey any permanent title or right to the Indians in the lands or waters of Karluk Reservation."[38] In the dissenting opinion, Justice Wiley Blount Rutledge Jr. wrote that in creating the reservation, "the necessary effect was to forbid others to enter the area for purposes inconsistent with the reservation's objects, thus making persons so entering trespassers and subject to such remedies as the law may afford to prevent or redress their wrongful entry."[39] The Karluk Village would endure as a tribe without a reservation of water finding no redress in U.S. law after two hundred years of foreign annexation.

The *Hynes v. Grimes Packing Company* case fit into the federal government's historical treatment of Alaska villages in the twentieth century by limiting the scope of their ownership to be consistent with other members of the then-federal territory.[40] The groups of people whom the United States called Alaska Natives, which included the Karluk villagers, held a path to land rights that continued to diverge from the nation's established relations with Native nations in the lower section of North America. Federal law classified the villagers of the Alaska Territory as part of the greater Alaskan public, not separate members of countries that existed before Russian arrival with a set of autonomous governance practices. The 1952 case *Arizona v. California* provides an example of the distinctions between Alaska villages and the Native nation in the then United States. Adjudicated just a few years after the *Hynes v. Grimes Packing Company* ruling, *Arizona* involved the state of Arizona suing California in a fight over a division of water from the Colorado River. During the case, the federal government asserted federal rights to the river, which included those of five tribes—through Winters rights—who also claimed to possess a right to the Colorado River. The Supreme Court's majority again opined that whenever the government creates a reservation for a tribal government, there was an "implied reservation of water rights" allowing the tribal governments the water required to operate the reservation in the present and for future growth.

In the court's view, tribal governments held this water as part of their title whether they employed the right or not, therefore affirming that Congress formed reservations, or the ability to allocate reservations, with the intention to make them habitable and productive. Regardless of whether water rights are addressed explicitly in treaties, the court presumed that tribes have a right to enough water to satisfy the purpose of the reservation in both present and future endeavors. This was a more singular argument than the Winters doctrine in that a tribal government's water rights were reserved, not observed by first use as they are in the Doctrine of Prior Appropriation in the Western United States. Though federal water law with Native nations in lower North America met distinct challenges throughout the twentieth century, the court continued to recognize forms of water authority for tribes in the contiguous part of the nation, as in *United States v. New Mexico*, a ruling that affirmed a tribal government's reserved right to water, though for a limited purpose.[41] This reserved right granted to tribal governments, however, plays little part in the legal history for villagers in Alaska. Historian Stephen Haycox sees that even though the Department of the Interior tried to create a reservation culture in Alaska, as seen with the Karluk Village, the movement failed over time until after the 1989 Exxon oil spill.[42]

Yet, Alaska villagers continued to assert further claims to land and water. Six years following *Hynes v. Grimes Packing Company*, the 1955 Supreme Court case involving an Alaska Native village, called *Tee-Hit-Ton v. United States*, reflected the earlier opinion. The judgment denied the village complainants the right to bring their charge of land right violation against the federal government to the Supreme Court.[43] The court refused village members' class action suit against the government's creation of the Tongass National Forest, denying their right to bring before the bench a Fifth

Amendment Taking case against the federal government. The majority opinion concluded that the village petitioners were without any claim because Congress never recognized their ownership as to enable the Supreme Court's province.

In the *Tee-Hit-Ton* majority opinion, Justice Stanley Reed again derided village ownership of land in the Alaska Territory. He emphasized that "every American school boy" knew that Europeans conquered Native Americans long before the Tee-Hit-Ton claim passed to his bench. He said this despite the facts that few non-Native citizens lived in the Alaska Territory at the time compared to the Native population and that the nation had fought no wars with Alaska Natives. In fact, the Russians entered into few wars with the people in the expanse of Northern Alaska and were not strong enough to quell the Tlingit clans, even after their victory in the Battle of Sitka in 1804. The United States bought the region, and there was no national conquest of Karluk outside of these court proceedings. What "every American school boy knew" was based upon a financial transaction with Russia that failed to include military actions or wholesale encroachment. By using the mythology of United States expansion, his words pushed against the general idea that the Alaska Territory was distinct from the union. The United States' 1867 purchase of Alaska created the nation's first long-term colonial possession while marking a policy change regarding Native peoples on the national mainland compared with other national possessions. The *Tee-Hit-Ton* and *Hynes* rulings signified the end of broad dealings with villagers in Alaska as separate governing bodies from that time forward.[44]

The conclusion to be drawn from the reservation of water in the history of the Alaska Territory's development is that the federal government wrestled with how to place this post-Russian colony into the nation. Unlike all other regions of the Americas, this was a place only miles from the large continent of Eurasia. The judicial opinions that developed Alaska Native relationships with the United States did so in a manner distinct from federal relations with American Indian tribes in the contiguous part of the nation.

In comparison to the canon of American Indian water law in Lower North America, *Hynes v. Grimes Packing Company* foreshadowed how the United States continued to deal with villagers in Alaska less as part of Native nations than as once members of non-western foreign nations whose country became part of the United States. In doing this, the government worked around the implementation of policies derived from the Doctrine of Discovery. After *Tee-Hit-Ton*, the Alaska State Constitution of 1959 proclaimed water as a public benefit for all residents, like the White

Act. Within a decade following statehood, a company located a massive petroleum deposit beneath Alaska's North Slope in 1968, the same year that the federal government paid the settlement to the Tlingit Haida tribal council. Because of the eagerness of companies to extract oil, the new state and the federal government were compelled to enter into a settlement with the multiple indigenous tribes in Alaska to end possible legal action concerning Aboriginal title.

The result was Congress passing the Alaska Native Claims Settlement Act of 1971.[45] This public law extinguished further Native claims to Aboriginal title within the state of Alaska in exchange for parcels of land that are taxable if developed by the Native corporations that the law also created. The common saying about the settlement was that the deal conveyed "mountain tops and lake bottoms," not water rights, to Native-operated corporations, like my own Koniag, Inc., or Leisnoi, Inc. The law released indigenous claims to land for a cash settlement that Alaska Natives would incorporate into thirteen regional and over two hundred village-based for-profit corporations, such as my own Leisnoi, Inc. The land holdings of the corporations were held tax-free unless the corporation developed the acreage in any way. Gone was the prospect of creating tribal reservations of water, with the exception of the Tsimshian Metlakatla Indian community on Annette Island in southeast Alaska. The Alaska Native Claims Settlement disbanded the Karluk Tribe until 1994, when the Department of the Interior chose to recognize all Alaska Native Villages as tribal governments. The ruling in *Hynes v. Grimes Packing Company* failed to build upon the Doctrine of Discovery or the history of American Indian reserved water rights that were first recognized by the U.S. Supreme Court in *Winters v. United States* in 1908. A Russian legacy of territorial acquisition lingered in the outcome. The result helped lay the foundation for how ANCSA would regard Native property.

The Alaska Native Claims Settlement avoided language that specified indigenous rights to subsistence lifestyles, even though thousands of Native residents rely upon hunting, gathering, and fishing to sustain their households. In 1980, Congress passed the Alaska National Interest Lands Conservation Act (ANILCA), which protected rural resident subsistence practices.[46] The state government argued that the federal law interfered with the state constitution that grants equal access to fish and game to all residents. Over decades, a series of cases informally known as the Katie John cases sought to specify Native rights to water, and other public lands, providing access for the sake of subsistence living. Katie John was an Athabascan activist who argued for her rights to fish on the Copper River, as her family had for generations. By 2014, after her passing, the subsistence right for rural residents was confined to federal lands, not state or private parcels. The laws came to reflect the history of denying Native control over resources by giving all rural residents rights to subsistence practices.

The turbulent legal history surrounding the Karluk Village waters and lands continued past the Alaska Native Claims Settlement Act. In 1980, numerous southern Kodiak villages, including Karluk, merged lands with the Native regional corporation Koniag, Inc., after the claims settlement. Many villagers considered the integration to be a mistake, and some leaders then separated their villages from Koniag, Inc. Nonetheless, the regional corporation retained the land at the sides of the Karluk River that the tribe had held as its reservation before the settlement.[47] *Hynes v. Grimes* possesses an intractable hold on the longtime inhabitants of these North Pacific shores. The reservation of water pursued by the Karluk Village was one of the first Supreme Court rulings that set Alaska outside the principles of the Doctrine of Discovery. However, federal law failed to extinguish the environmental history of people, fish, or bears. Today amid the constraints imposed by the case, villagers continue to use the waters of the Karluk region. This essay has reflected on Karluk, the salmon, and the Ocean to advocate that the re-envisioning of Anthropocene waters means changing patterns of administration. In imaging how to protect the Ocean, one could consider how the realization of governance would affect others.

Notes

1. Known as the *Taquka-aq* in the island's indigenous Alutiiq language and in Latin as *Ursus arctos middendorffi*.
2. Knecht, Richard A. "The late prehistory of the Alutiiq people: Culture change on the Kodiak Archipelago from 1200–1750 AD." PhD Diss., University of Aberdeen, 1995.
3. 50 C.F.R. § 208.23 (Supp. 1946).
4. Philp, Kenneth R. "The New Deal and Alaskan Natives, 1936–1945." *The Pacific Historical Review* 50, no. 3 (1981): 309–327, 318.
5. Black, Lydia. *Russians in Alaska, 1732–1867*. University of Alaska Press, 2004, 6.
6. Pub. L. 74-538, 49 Stat. 1250 (1936).
7. In 1971 Congress extinguished Native Alaska's claims to aboriginal title in the state of Alaska with the Alaska Native Claims Settlement Act.
8. *Winters v. United States*, 207 U.S. 564 (1908).
9. The exception to this history is the Metlakatla Indian Community on Annette Island.
10. Treaty of Cession 15 Stat. 539 (1867).

11. Treaty of Cession, Article 3.
12. Act of May 17, 1884, Sec. 8, 23 Stat. 26.
13. United States v. Seveloff, 1 Alaska Fed. Rpts. 64 (D. Or. 1872).
14. 18 U.S.C. §1151, US Code.
15. Twenty-Third Congress. Sess. I. Ch. 161 (1834).
16. Forty-Eighth Congress. Sess. I. Chs. 52, 53 (1884).
17. Harring, Sidney L. *Crow Dog's case: American Indian sovereignty, tribal law, and United States law in the nineteenth century*. Cambridge University Press, 1994. 214–216.
18. Waters v. Campbell. 29 F. Cas. 411 (C.C.D. Or. 1876).
19. Kie v. United States. 27 F. 351 (C.C.D. Or. 1886).
20. 348 U.S. 272 (1955).
21. (34 Stat. 197).
22. Alaska Pacific Fisheries v. United States, 248 U.S. 78 (1918).
23. Pub. L. No. 73-383, 48 Stat. 984 (1934).
24. Haakanson, Sven. "Written Voices Become History." *Being and Becoming Indigenous Archaeologists.* Ed. George Nicholas. (2010): 220.
25. Black, *Russians in Alaska*. 108
26. Clark, John H., Andrew McGregor, Robert D. Mecum, Paul Krasnowski, and Amy M. Carroll. "The commercial salmon fishery in Alaska." *Alaska Fishery Research Bulletin* 12, no. 1 (2006): 1–146 (63).
27. Ibid., 68.
28. Transcript. Sixty-Eighth Congress. sess. I. ch. 272 (1924). 465. "[in margin] Restriction in limited fishing areas."
29. Hynes v. Grimes Packing Co., 337 U.S. 86 (1949).
30. McNeil, K. 2019. How the New Deal Became a Raw Deal for Indian Nations: Justice Stanley Reed and the Tee-Hit-Ton Decision on Indian Title. *Am. Indian L. Rev.*, 44, p. 1.
31. Steve Colt. 1999. "Salmon fish traps in Alaska." Report published by the Institute of Social and Economic Research at the University of Alaska, https://scholarworks.alaska.edu/handle/11122/12509/, p. 12.
32. Ibid.
33. White Act sec. 3.
34. *Hynes*, 337 U.S. 86 (1949).
35. 248 U.S. 78 (39 S.Ct. 40, 63 L.Ed. 138).
36. *Winters*, 207 U.S. 564 (1908).
37. McNeil, 27.
38. Reed Opinion, *Hynes v. Grimes Packing Co.*, 337 U.S. 86 (1949).
39. Wiley Blount Rutledge Jr. Opinion. *Hynes v. Grimes Packing Co.*, 337 U.S. 86 (1949).
40. My own village's land was usurped when a religious organization turned in the plat to the developing Kodiak regional government in the early twentieth century.
41. 438 U.S. 696 (1978).

42. Stephen W. Haycox. *Alaska: an American colony*. University of Washington Press, 2020, 253.
43. Tee-Hit-Ton Indians v. United States, 348 U.S. 272 (1955).
44. The federal government ended treaty making with the Indian Appropriation Act of March 3, 1871. By the end of the twentieth century, the federal government did begin nation-to-nation relationships with Alaska Native tribal governments.
45. 43 U.S.C. 1601 et seq.
46. December 2, 1980, Public Law 96-487, 94 Stat. 2371.
47. Naomi Klouda. "Karluk Tribe keeps up fight over ownership of river." Alaska Journal of Commerce. Oct. 18, 2017. Available at: www.alaskajournal.com/2017-10-18/karluk-tribe-keeps-fight-over-ownership-river.

CHAPTER 8

Re-envisioning the Value of Marine Spaces in Law

Massachusetts Lobstermen's Association v. Ross

ROBIN KUNDIS CRAIG

IN THE UNITED STATES, as is true in many coastal nations, the law governing human use of the ocean tends to view—and value—the marine realm as a series of discrete goods and services that benefit humans, ranging from fish to offshore oil and gas to tourism and recreation to shipping lanes.[1] Legal protections for the ocean, similarly if more perversely, also work to fragment any coherent vision of the marine environment. Thus, for example, the earliest and still most common legal protections that countries established for the ocean, both domestically and internationally, apply to marine species, either individually, as is the case for most endangered species protection regimes, or in limited groups, such as whales, seals, or tunas. In the United States, about forty-five federal statutes protect marine species, generally on a species-by-species basis, and include the Anadromous Fish Conservation Act of 1965,[2] Atlantic Salmon Convention Act of 1982,[3] Atlantic Striped Bass Conservation Act,[4] Atlantic Tuna Convention Act of 1975,[5] Eastern Pacific Tuna Licensing Act of 1984,[6] Fur Seal Act Amendments of 1983,[7] International Dolphin Conservation Program Act,[8] Northern Pacific Halibut Act of 1982,[9] South Pacific Tuna Act of 1988,[10] Sponge Act,[11] and the Whaling Convention Act of 1949.[12] The two more general federal statutes that protect marine species, the Marine Mammal Protection Act of 1972,[13] and the Endangered Species Act (ESA) of 1973,[14] focus their protections on species already in crisis and thus also underscore the species-specific approach to protecting living marine resources. Indeed, marine species only receive the Endangered Species Act's protections on a species, subspecies, or distinct population segment basis.[15]

This traditional legal approach—whether for exploitation or for protection—thus envisions the ocean to be a collection of discrete component parts rather than a complex system of systems or networked series of functional places—in contemporary parlance, social-ecological systems or SESs. This piecemeal approach also leads to significant fragmentation in how governments manage the ocean. As the U.S. Commission on Ocean Policy summarized in 2004 for the United States, at the federal level, "more than 55 congressional committees and subcommittees oversee some 20 federal agencies and permanent commissions in implementing at least 140 federal ocean-related statutes."[16] In addition, 38 coastal states, multiple state agencies, and multiple coastal tribes play important roles in marine governance, particularly along the nation's coasts. Despite the fact that the United States controls more ocean than any other country in the world, it has no overarching and comprehensive vision for its marine territory, and the various federal "agencies interact with one another and with state, territorial, tribal, and local authorities in sometimes haphazard ways."[17]

This fragmented vision of the ocean has resulted in a legal system that promotes extractive and commercial uses of the ocean. This promotion is particularly true for marine fishing, with federal laws often creating significant blinders to the effects of fishing and overfishing on the ocean's overall health. Thus, for example, many federal fishing statutes in the United States evidence an underlying consumption-promoting policy, such as the Central, Western, and South Pacific Fisheries Development Act,[18] Fish and Seafood Promotion Act of 1986,[19] Fisheries Financing Act of 1996,[20] and the Fisherman's Protective Act of 1967.[21] This pro-fisheries policy is also evident in the most general and most important of the federal fisheries statutes, the 1976 Magnuson-Stevens Fishery Conservation and Management Act.[22] Although Congress recognized in this act that a national program for the conservation and management of the United States' fishery resources is necessary to prevent overfishing, to rebuild overfished stocks, to ensure conservation, and to realize the full potential of the nation's fishery resources,[23] it did not enact a comprehensive ecosystem-based regulatory regime to achieve those goals. Instead, geographically, the Magnuson-Stevens Act leaves fisheries regulation in the first three miles of ocean largely to the states, then divides management in the federal Exclusive Economic Zone—that two-hundred-nautical-mile-wide band of ocean jurisdiction accorded to coastal nations under international law—among eight regional Fisheries Management Councils, which enact fishery management plans for each troubled fishery within their respective jurisdictions. Moreover, even after Congress enacted the Sustainable Fisheries Act of 1996[24] to address continued problems of overfishing and bycatch (the incidental catching of nontarget species in commercial fishing), federal fishery management plans tend to remain focused on each fishery and each individual stock of fish,[25] and the statutory goal for fisheries management remains "optimum yield," that is, "the maximum sustainable yield from such fishery which will provide the greatest overall benefit to the Nation."[26]

In contrast, there is no federal mandate to protect marine ecosystems in the United States. The most obvious candidate, the National Marine Sanctuaries Act,[27] contains no such mandate, does not require that sanctuaries focus on ecosystems or biodiversity, and does not even require that sanctuaries be fully protective. Instead, the act notes that areas of the marine environment can be nationally or internationally significant because of their "conservation, recreational, ecological, historical, scientific, educational, cultural, archeological, or esthetic qualities,"[28] and it allows the Secretary of Commerce, acting through the National Oceanic and Atmospheric Administration (NOAA), to designate National Marine Sanctuaries for any of these reasons, or because "the area is of special national significance due to . . . its resource or human use values."[29] The Flower Garden Banks National Marine Sanctuary in the Gulf of Mexico contains offshore oil and gas wells and platforms, while three National Marine Sanctuaries—Monitor, Mallows Bay–Potomac River, and Thunder Bay—exist primarily to protect historic shipwrecks.

Against this background, political and legal investment in establishing marine protected areas (MPAs) and systems of MPAs represents a potentially transformative re-envisioning of the ocean. MPAs legally set aside a specific area of the ocean and restrict at least some uses of that area. The most protective MPAs, generally referred to as marine reserves, significantly restrict or prohibit all resource extraction from the area—especially fishing. The legal act of establishing a marine reserve expressly to protect and enhance marine biodiversity thus operationally re-envisions the ocean and its importance in two ways. First, it shifts the focus of marine protection from charismatic (whales) or economically important (tuna) species to the full web of marine biodiversity. For example, Robert Wilder has noted that

> [p]rotecting marine biodiversity requires a different sort of thinking than has occurred so far. Common misperceptions about what is needed abound, such as a popular view that biodiversity policy ought to focus on the largest and best-known animals. But just as on land,

> biodiversity at sea is greatest among smaller organisms such as diatoms and crustacea, which are crucial to preserving ecosystem function. Numerous types of plants such as mangrove trees and kelps have equally essential roles but are often overlooked entirely. We look away from the small, slimy, and ugly, as well as from the plants, in making marine policy. The new goal must be to consider the ecological significance of all animals and plants when providing policy protections and to address all levels of the genome, species, and habitat.[30]

Second, it re-values the ocean's importance from being a source of goods (fish, oil) to being a complex global system that provides humanity with multiple benefits and primary life support. Biodiversity-based marine recreation like snorkeling and SCUBA diving and marine biophilia's contribution to human psychological well-being are among the most obvious of these other values. However, marine scientists continually discover that marine biodiversity is also critical to a plethora of other ecosystem services, including atmospheric oxygen production, carbon sequestration, marine ecosystem adaptation to climate change, and some aspects of climate regulation.

As a result, the proposed or actual designation of a new MPA often provides the occasion for very public discussions about the value of marine spaces or ecosystems. When the new MPA is a marine reserve, those public discussions often turn into fights and lawsuits that pit the value of extractive use (generally fishing, but also occasionally oil and gas) against the values that a protected system can provide, such as recreation and tourism, biodiversity protection, or species restoration. In other words, MPA and marine reserve designations often serve to spectacularly illuminate previously obscure conflicting views of the Anthropocene ocean generally and, more specifically, differing societal valuations of particular marine spaces.

These conflicts are common and global and raise a variety of values, precluding any universally "correct" view of MPAs and marine reserves, even when governments' biodiversity goals are clear. Where fishers have depended on an area for many years, the creation of a marine reserve certainly feels like—and may actually be—an at least temporary economic deprivation, even when fishers eventually benefit from increased numbers of larger fish outside the reserve. Thus, when the United Kingdom in January 2016 announced "eight new Marine Conservation Zones around the South West, from Land's End to Foreland Point in North Devon," fishers claimed that "[t]he growth of Marine Conservation Zones is killing Britain's fishing industry" and petitioned the government to stop.[31] Marine reserve establishment can also threaten subsistence fishing and privilege (or at least appear to privilege) the nonindigenous tourist sector. For example, a few months after the UK protests, Native Hawaiian fishers on the Big Island of Hawaii protested on these grounds "the establishment of the Kaupulehu Marine Reserve, the island's first initiative to put a reef off-limits to fishing," which sought to impose a ten-year moratorium on *all* taking of fish while a subsistence plan was being drafted for Kaupulehu Bay coastline.[32]

Nor does the successful establishment of a marine reserve necessarily accomplish a re-visioning of the ocean's value away from exploitative use. In the late 1990s, for example, opposition from fishing interests derailed NOAA's attempt to protect 20 percent of the Florida Keys National Marine Sanctuary in various kinds of marine reserves.[33] NOAA then used an extended public process to designate the Dry Tortugas Ecological Reserve, completed in 2001. However, the process worked in part because NOAA actively reframed the reserve's value from marine ecosystem and biodiversity protection to enhancement of fisheries. In less than five years, therefore, marine reserves in the sanctuary "evolved from being highly suspicious ecosystem- and biodiversity-focused management tools, generally opposed by commercial and recreational fishers, to being important tools in restoring admittedly over-exploited fisheries, with fishers actively participating in the creation of a large new reserve."[34]

Occasionally, however, a marine reserve designation and the resulting conflicts with fishers spotlight even more basic legal issues with respect to re-envisioning the ocean and its value to humanity—such as who has legal authority to impose a new vision and how legally ephemeral that new vision might be. This chapter tells the story of the Northeast Canyons and Seamounts National Marine Sanctuary, whose designation in 2016 by President Barack Obama created not just the fairly normal political conflicts with fishers but also multiple legal conflicts about presidential authority to re-envision the value of marine spaces to the United States.

The Northeast Canyons and Seamounts Marine National Monument

On September 15, 2016, using the Antiquities Act of 1906,[35] President Barack Obama established the Northeast Canyons and Seamounts Marine National Monument by presidential proclamation.[36] Under the Antiquities Act, "[t]he President may, in the President's discretion, declare by public proclamation historic landmarks, historic and prehistoric structures, and other objects of historic or scientific interest that are situated on land owned or controlled by the Federal Government to be national monuments."[37] This statute thus gives the president considerable authority to establish unilaterally protected areas—national monuments—throughout the United States.

This newest marine national monument encompasses 4,913 square miles (about the size of Connecticut)[38] of sensitive ocean ecosystems and is located approximately 130 miles southeast of Cape Cod in federal ocean waters. Under the United Nations Convention on the Law of the Sea, coastal nations can claim authority over a 200-nautical-mile-wide EEZ and 200 nautical miles of continental shelf.[39] While the United States is not a party to that treaty, it views the convention's provisions on ocean jurisdiction as customary international law and has claimed jurisdiction over both its continental shelf and its EEZ. Under U.S. domestic law—namely, the Submerged Lands Act—the coastal states have primary jurisdiction over the first three miles of ocean,[40] but the federal government controls everything in the rest of the EEZ.[41] The Northeast Canyons and Seamounts Marine National Monument encompasses both the water column and the seafloor, but, as its name suggests, it focuses on bottom features—specifically, three underwater canyons in an area that covers approximately 941 square miles and four seamounts in a separate area that encompasses 3,972 square miles (see Figure 8.1).[42]

According to Presidential Proclamation 9496, the new National Marine Monument contained multiple "objects of . . . scientific interest" worthy of presidential protection. For example, in these waters, the Atlantic Ocean meets the continental shelf in a region of great abundance and diversity as well as stark geological relief. The waters are home to many species of deep-sea corals, fish, whales, and other marine mammals. Three submarine canyons and, beyond them, four undersea mountains lie in the waters approximately 130 miles southeast of Cape Cod. This area (the canyon and seamount area) includes unique ecological resources that have long been the subject of scientific interest.[43]

Specifically, the objects of historical and scientific interest "are the canyons and seamounts themselves, and the natural resources and ecosystems in and around them."[44] As NOAA describes this system in more detail:

> The submarine canyons and seamounts create dynamic currents and eddies that enhance biological productivity and provide feeding grounds for seabirds; pelagic species, including whales, dolphins,

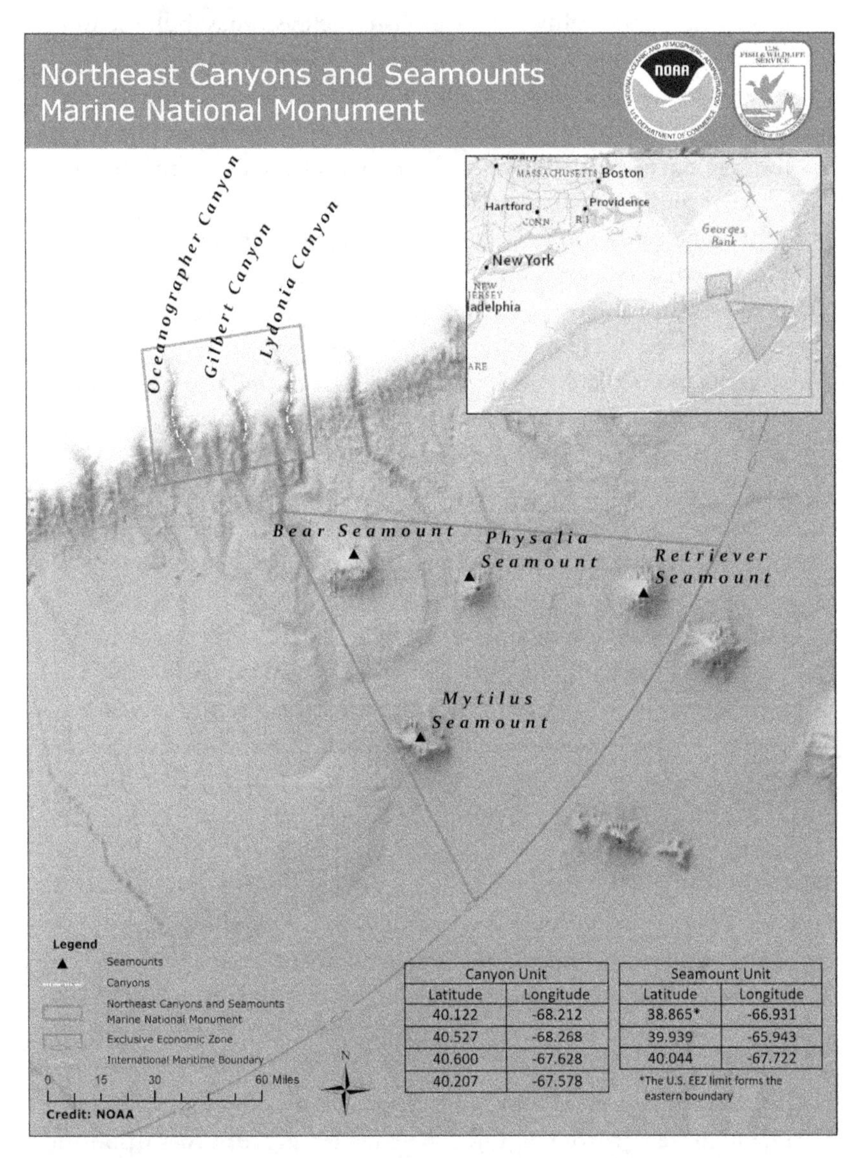

Canyon Unit	
Latitude	Longitude
40.122	-68.212
40.527	-68.268
40.600	-67.628
40.207	-67.578

Seamount Unit	
Latitude	Longitude
38.865*	-66.931
39.939	-65.943
40.044	-67.722

Figure 8.1. Map of the Northeast Canyons and Seamounts Marine National Monument, showing its location off the coast of Massachusetts. Created by President Barack Obama through the Antiquities Act, it consists of approximately 4,913 square miles (12,724 square kilometers) and is located about 130 miles east-southeast of Cape Cod. SOURCE: https://www.fisheries.noaa.gov/resource/map/northeast-canyons-and-seamounts-marine-national-monument-map-gis-data.

> and turtles; and highly migratory fish, such as tunas, billfish, and sharks. More than ten species of shark, including great white sharks, are known to utilize the feeding grounds of the canyon and seamount area. Additionally, surveys of leatherback and loggerhead turtles in the area have revealed increased numbers above and immediately adjacent to the canyons and Bear Seamount.
>
> Marine birds concentrate in upwelling areas near the canyons and seamounts. Several species of gulls, shearwaters, storm petrels, gannets, skuas, and terns, among others, are regularly observed in the region, sometimes in large aggregations. Recent analysis of geolocation data found that Maine's vulnerable Atlantic puffin frequents the canyon and seamount area between September and March, indicating a previously unknown wintering habitat for those birds.[45]

These wonders deserve legal protection, moreover, because they are at risk from climate change and fishing. Thus:

> These canyons and seamounts are home to at least 54 species of deep-sea corals, which live at depths of at least 3,900 meters below the sea surface. The corals, together with other structure-forming fauna such as sponges and anemones, create a foundation for vibrant deep-sea ecosystems, providing food, spawning habitat, and shelter for an array of fish and invertebrate species. These habitats are extremely sensitive to disturbance from extractive activities.[46]

As a result, the Proclamation asserts, "it is in the public interest to preserve the marine environment, including the waters and submerged lands, in the area to be known as the Northeast Canyons and Seamounts Marine National Monument, for the care and management of the objects of historic and scientific interest therein."[47]

While the monument was "subject to valid existing rights," President Obama withdrew it from all forms of "sale, leasing, or other disposition under the public land laws," including patents under the federal mining laws and "disposition under all laws relating to development of oil and gas, minerals, geothermal, or renewable energy."[48] The Secretary of Commerce, acting through NOAA, and the Secretary of the Interior, acting through the U.S. Fish & Wildlife Service, received joint authority to manage the new monument, which they were to do through a joint management plan prepared by September 2019. However, the proclamation also directed the secretaries to prohibit six activities and to regulate seven more, while leaving the Armed Forces exempt from most restrictions.[49]

Fishing was one of the activities that the proclamation prohibited and regulated. In general, the proclamation prohibited "[r]emoving, moving, taking, harvesting, possessing, injuring, disturbing, or damaging, or attempting to remove, move, take, harvest, possess, injure, disturb, or damage, any living or nonliving monument resource, except as provided under regulated activities below."[50] To emphasize the point, the proclamation explicitly prohibited commercial fishing, "except for the red crab fishery and the American lobster fishery," which were subject instead to a seven-year phaseout before becoming completely forbidden.[51]

Thus, in President Obama's vision, the Northeast Canyons and Seamounts Marine National Monument would be a marine reserve in federal waters, established to protect unusual oceanic geological features and the ecosystems that they support. As is common with marine reserves, however, the proclamation's phaseout of two commercial fisheries and eventual prohibition of all fishing sparked a protest among the affected fishers. This protest transcended the political level to become two legal battles—specifically, the federal court litigation of *Massachusetts Lobstermen's Association v. Ross* and *Conservation Law Foundation v. Trump*.

The Antiquities Act as a Legal Mechanism for Protecting Marine Space

Law is one of the fundamental mechanisms for structuring societies. As such, law can significantly shape how citizens and governments envision multiple aspects of their lives, whether by articulating cultural norms, promoting some perspectives and activities, or prohibiting some perspectives and activities. Thus, as discussed above, federal law applicable to the United States' ocean promotes a species-by-species, resource-by-resource, and activity-by-activity perspective on the marine realm, fragmenting that realm geographically, topically, and among regulatory authorities.

Nevertheless, it is a rare law—even when that law takes the form of a federal statute—that imposes completely unavoidable blinders or tunnel vision on either government actors or the general public. Most law, whether coming from legislatures, regulatory agencies, or courts, leaves "wiggle room" that can allow new visions of reality to emerge.[52] In legalese, this wiggle room often derives from governmental discretion or textual ambiguity, or both.

The Antiquities Act that President Obama used to create the Northeast Canyons and Seamounts Marine National Monument is an example of a federal statute that might provide sufficient wiggle room—deriving from both discretion and textual ambiguity—to allow presidents (and hence the nation as a whole) to re-envision the ocean. Like President Obama, presidents have used the Antiquities Act in the ocean primarily to protect ecosystems and biodiversity, overriding the fragmented vision of the ocean created through federal marine law. In addition, use of the Antiquities Act in the ocean brings the United States' marine realm into its system of protected public lands. While this development may not be desirable in all respects—a domesticated ocean will not be to everyone's taste—the transformation has at least the salutary effect of re-envisioning the ocean as a set of important and public *places* rather than as a series of goods and services to be exploited for private profit.

Under the Antiquities Act, "[t]he President may, *in the President's discretion*, declare by public proclamation historic landmarks, historic and prehistoric structures, and other objects of historic or scientific interest that are situated on *land* owned or controlled by the Federal Government to be national monuments."[53] The reference to "land" in the Antiquities Act indicates that its use to create *marine* national monuments involved a spark of presidential imagination and creativity from the beginning. It also signals a presidential urgency to protect marine spaces. As noted, other legal vehicles for establishing marine protected areas in federal ocean waters do exist, especially the National Marine Sanctuaries Act.[54] Enacted in 1972, the National Marine Sanctuaries Act has allowed NOAA—with the occasional extra boost from Congress—to establish fourteen National Marine Sanctuaries total, ranging from the Hawaiian Islands Humpback Whale National Marine Sanctuary to the Stellwagen Bank National Marine Sanctuary off the coast of New England. Nevertheless, the National Marine Sanctuaries Act designation process is long and cumbersome. For example, in July 2019 the nation added its newest sanctuary to the system, the Mallows Bay–Potomac River National Marine Sanctuary, located in Maryland about forty miles south of Washington, DC. As NOAA notes, "Mallows Bay–Potomac River National Marine Sanctuary protects and interprets the remnants of more than 100 World War I-era wooden steamships—known as the 'Ghost Fleet'—and other maritime resources and cultural heritage dating back nearly 12,000 years."[55] However, the designation process took five years after Maryland governor Martin O'Malley nominated the site in 2014, and Mallows Bay–Potomac River is the first new National Marine

Sanctuary since NOAA designated Thunder Bay National Marine Sanctuary in Lake Huron in 2000. Designation delays have plagued the National Marine Sanctuaries Act for at least three decades now. The Florida Keys National Marine Sanctuary required congressional intervention to complete the designation process in 1990, while President George W. Bush short-circuited the sanctuary designation process for the Northwestern Hawaiian Islands in favor of a Marine National Monument, which he could establish through presidential proclamation in 2006. Notably, moreover, the last two National Marine Sanctuaries do not protect truly *marine* environments, and they are two of three sanctuaries that focus on protecting shipwrecks rather than ecosystems.

While reaching for the Antiquities Act to protect the ocean might have required an initial spark of presidential creativity, its use to protect marine spaces is not new. Both Republican and Democratic presidents have used their Antiquities Act authority to protect marine spaces since 1925. In that year, President Calvin Coolidge established the Glacier Bay National Monument in Alaska,[56] which is now Glacier Bay National Park. The designated monument included fjord waters, intertidal glaciers, and the marine life of Alaska's Inside Passage, in addition to the fjord cliffs and terrestrial resident animals. President Coolidge's use of the Antiquities Act much later reached the U.S. Supreme Court in a case where the State of Alaska tried to claim Glacier Bay for itself.[57] As the court recounted, Glacier Bay had been the subject of *multiple* presidents' Antiquities Act interest.

After President Coolidge invoked the Antiquities Act to create Glacier Bay National Monument, President Franklin D. Roosevelt, in 1939, issued a proclamation expanding the monument to include all of Glacier Bay's waters, extending the monument's western boundary three nautical miles out to sea. In 1955, President Dwight D. Eisenhower issued a proclamation slightly altering the monument's boundaries but leaving the bay's waters within them. In 1980, Congress designated the monument as part of Glacier Bay National Park and Preserve and expanded the resulting reservation's boundaries. For present purposes, however, the important point is that by the time Alaska achieved statehood in 1959, the Glacier Bay National Monument had already existed for thirty-four years as a federal reservation.[58] Thus, the Supreme Court concluded, these presidents had effectively used the Antiquities Act to protect not just the terrestrial features and wildlife of Glacier Bay, but also its submerged lands and associated ecosystems.

President Franklin D. Roosevelt was particularly keen to use the Antiquities Act to protect marine environments. Besides expanding the Glacier

Bay National Monument's boundaries to three nautical miles out to sea in 1939, as the Supreme Court recounted, he also established the Fort Jefferson National Monument in Florida in 1935[59] to protect a deep-water fort, which Congress redesignated as the Dry Tortugas National Park in 1992,[60] and established the Channel Islands National Monument in 1938 to protect most of Anacapa and Santa Barbara Islands off the coast of California.[61] President Truman expanded this last monument in 1949 to encompass "the areas within one nautical mile of the shoreline of Anacapa and Santa Barbara Islands."[62]

President Truman's expansion occasioned the U.S. Supreme Court's most extensive discussion of the Antiquities Act's applicability to the marine environment, in part because of its unusual legal timing. In 1947, the court had decided "that the Federal Government rather than the state has paramount rights in and power over [the three-nautical-mile-wide belt of ocean], an incident to which is full dominion over the resources of the soil under that water area, including oil."[63] Therefore, when President Truman expanded the Channel Islands National Monument in 1949 to include one nautical mile of ocean, there was no question that the president could do so under the Antiquities Act.[64] However, in 1953, Congress enacted the Submerged Lands Act and transferred the first three miles of submerged lands to the states,[65] while confirming later that same year in the Outer Continental Shelf Lands Act that the federal government controls the submerged lands from three miles out.[66] As a result, while the U.S. Supreme Court in 1978 did not question President Truman's authority to use the Antiquities Act to protect submerged lands in the first place, it concluded that, "[b]ecause the United States' claim to the submerged lands and waters within one mile of Anacapa and Santa Barbara Islands derives solely from the doctrine of 'paramount rights' announced in this Court's 1947 *California* decision, we hold that, by operation of the Submerged Lands Act, the Government's proprietary and administrative interests in these areas passed to the State of California in 1953."[67]

Presidents Coolidge, Roosevelt, Truman, and Eisenhower applied the Antiquities Act to the ocean somewhat as an afterthought, focusing more on the terrestrial lands they were protecting. However, in 1961, President John F. Kennedy evolved the Antiquities Act's use for marine protection when he designated Buck Island Reef National Monument, becoming the first president to exercise his authority under the Antiquities Act primarily to protect marine resources—that is, to use the Antiquities Act to create what has come to be known as a Marine National Monument. President Kennedy's proclamation explained that he intended this monument's designation to protect both the island and its surrounding coral reef ecosystem because "its adjoining shoals, rocks, and undersea coral reef formations possess one of the finest marine gardens in the Caribbean Sea."[68] Following expansions by President Gerald Ford (1975) and President Bill Clinton (2001), today the Buck Island Reef National Monument protects the 176-acre tropical island, rare marine life, coral reef ecosystems, and historic shipwrecks.[69] The reef, however, remains the feature of central importance:

> Buck Island's Barrier Reef's underwater scene taxes human perceptions with the abundant variety of shape, patter, color, texture, and movement. It's [*sic*] barrier reef ranks among the Caribbean's best. It's [*sic*] thick branching elkhorn corals push their sheer mass to 30-foot heights. Like fortress walls corals rise off the sea floor and dominate the underwater world.[70]

Thus, from early on, presidents have used the Antiquities Act to create Marine National Monuments that focus on protecting marine ecosystems.

After President Kennedy, presidents of both political parties have used the Antiquities Act to create twelve additional national monuments that protect important marine resources, both cultural and ecological. As with Channel Islands and Buck Island Reef, however, for decades these Marine National Monuments remained relatively close to shore. President Jimmy Carter, for example, proclaimed three national monuments in Alaska, all in December 1978: the Bering Land Bridge National Monument;[71] Kenai Fjords National Monument;[72] and Misty Fjords National Monument.[73] The Bering Land Bridge now lies largely below the Chukchi Sea, Bering Sea, and Bering Strait but served as one of the migration routes to North America for plants, animals, and humans during ice ages, when a receding ocean made it passable.[74] This monument was 2,590,000 acres when President Carter created it and explicitly included the submerged lands and waters within its border.[75] However, in 1980 Congress modified the designation through the Alaska National Interest Lands Conservation Act (ANILCA),[76] and it has been the far more terrestrial Bering Land Bridge National Preserve ever since. "Kenai Fjords National Monument borders the Gulf of Alaska and includes the Harding Icefield and extensions of mountain peaks out into the sea."[77] Again, President Carter explicitly reserved the submerged lands and waters within the monument and explicitly mentioned sea otter recovery as a reason for protecting this area.[78] In ANILCA, Congress made

Kenai Fjords a National Park.[79] "Misty Fjords is an unspoiled coastal ecosystem containing significant scientific and historical features unique in North America."[80] The 2,285,000-acre coastal national monument again reserved submerged lands and waters.[81] In ANILCA, Congress reduced its acreage slightly and redesignated the monument a wilderness in the national forest system.[82]

President Clinton, similarly, designated two coastal national monuments. The California Coastal National Monument encompasses the entire 840 miles of California's coast and all terrestrial features existing above the high-tide line out to 12 nautical miles.[83] While the monument clearly exists to protect coastal ecosystems, it does not include ocean waters or submerged lands. (Notably, President Obama followed this model when he established the San Juan Islands National Monument in Puget Sound, Washington, in 2013.[84]) In 2014, President Obama expanded the monument to protect the Garcia River estuary,[85] then, days before leaving office in January 2017, expanded the monument again to add six additional terrestrial areas.[86] In contrast, President Clinton's Virgin Islands Coral Reef National Monument is located "in the submerged lands off the island of St. John in the U.S. Virgin Islands, [and] contains all the elements of a Caribbean tropical marine ecosystem,"[87] and hence is a full Marine National Monument. The designating proclamation also prohibits all extractive uses[88] and hence creates a marine reserve.

Thus, presidents through President Clinton found new and creative ways to protect coastal marine ecosystems through the Antiquities Act. However, and perhaps surprisingly, it was President George W. Bush who first used the Antiquities Act to establish immense marine reserves—marine protected areas that actually match the physical scale of ocean ecosystems. In 2006, he designated the Papahānaumokuākea (Northwestern Hawaiian Islands) Marine National Monument,[89] followed in 2009 by the Marianas Trench Marine National Monument in the Northern Marianas Islands and Guam;[90] the Pacific Remote Islands Marine National Monument, encompassing Wake, Baker, Howland, and Jarvis Islands, Johnston Atoll, Kingman Reef, and Palmyra Atoll;[91] and the Rose Atoll Marine National Monument off American Samoa.[92] (Closer to shore, President Bush also established the World War II Valor in the Pacific National Monument to protect Pearl Harbor, Hawaii, and the ships sunk during the Japanese attacks on December 7, 1941.)[93] From the beginning, the Papahānaumokuākea Marine National Monument protected the 1,200-nautical-mile chain of islands and atolls that make up the Northwestern Hawaiian Islands and encompassed "139,793

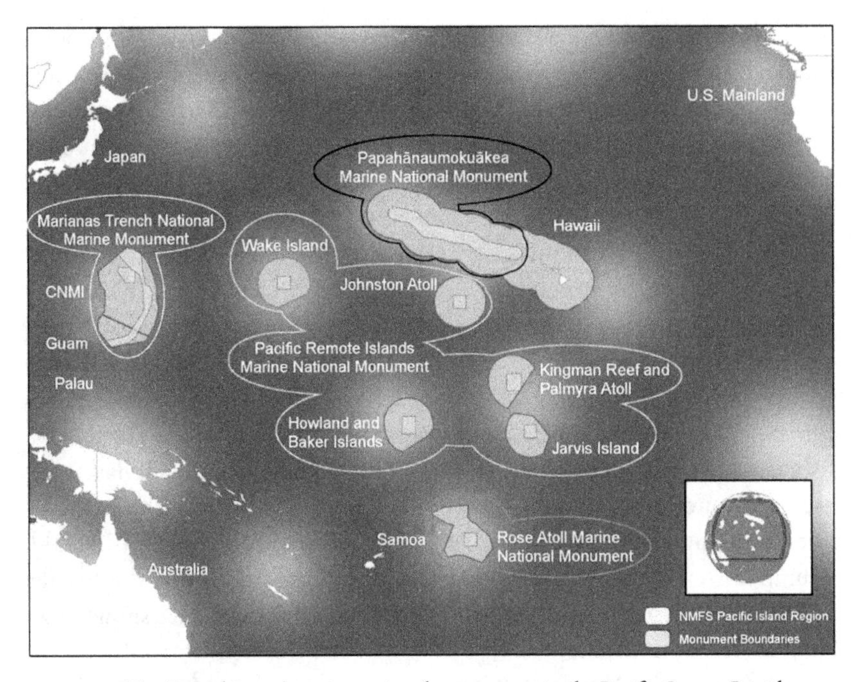

Figure 8.2. The United States' marine national monuments in the Pacific Ocean. President George W. Bush established all four of these marine national monuments through the Antiquities Act—the Papahānaumokuākea Marine National Monument in 2006 and the other three in 2009. After expansions in 2014 and 2016, these four marine national monuments now protect 1,182,717 square miles (3,063,223 square kilometers) of tropical coral reef ecosystems. Image courtesy of NOAA. SOURCE: https://commons.wikimedia.org/wiki/File:Eez_monument_4_6_2011.png.

square miles of emergent and submerged lands and waters."[94] In 2016, however, President Obama expanded the monument to the full extent of the United States' EEZ to protect the extended ecosystem, Native Hawaiian culture and history, and memorials of World War II.[95] The expanded monument "encompasses 582,578 square miles of the Pacific Ocean (1,508,870 square kilometers)—an area larger than all the country's national parks combined."[96] The Marianas Trench Marine National Monument follows the Marianas Trench, the world's deepest marine canyon, while the Rose Atoll and Pacific Remote Islands Monuments extended approximately 50 miles into the ocean, although in 2014 President Obama expanded the latter to the limits of the EEZ around many of its islands (see Figure 8.2).

With President Bush and President Obama, therefore, the Antiquities Act became a tool for legally protecting large marine ecosystems, encompassing a greater variety of habitats and species as well as indigenous human history. Moreover, these two presidents created marine reserves, giving these areas fairly stringent legal protection. President Bush, for example, phased out commercial fishing in the Papahānaumokuākea Marine National Monument, just as President Obama proposed to do in the Northeast Canyons and Seamounts. Finally, these two presidents protected different kinds of marine habitats—not just charismatic tropical coral reefs but also seamounts, marine canyons, and the deep Marianas Trench.

Notably, fisheries interests opposed these Pacific Marine National Monuments. Claiming that the expanded territories around the Remote Pacific Islands and Papahānaumokuākea Marine National Monuments accounted for 20 percent of their annual catch (10 percent from each monument), the fishers and the Western Pacific Fisheries Management Council have lobbied continuously to reopen these marine reserves to fishing. The council's latest attempt came on May 8, 2020, in a direct letter to President Donald Trump. However, President Trump did not, ultimately, open the Pacific marine national monuments to fishing, and scientists have since shown that they do not harm fishing."[97]

In creating the Northeast Canyons and Seamounts Marine National Monument, President Obama carried both the concept of this larger, highly protective Antiquities Act designation and the attendant conflicts with existing fisheries into the open waters of the Atlantic Ocean. The Atlantic Ocean off New England has long been a lot busier with human activity—especially commercial human activity—than the far more remote waters of the Pacific islands. As a result, it did not take long for fishermen's associations to challenge this newest of Marine National Monuments.

Challenging President Obama's Vision, Part 1: *Massachusetts Lobstermen's Association v. Ross*

Despite the number of national monuments that include marine resources and the two U.S. Supreme Court cases essentially upholding presidential use of the Antiquities Act in the marine realm, the federal courts had never squarely addressed whether the open ocean can be "land" "owned or controlled by the federal government" for the purposes of the Antiquities Act until the Massachusetts Lobstermen's Association, the Atlantic Offshore Lobstermen's Association, the Long Island Commercial Fishing Association, the Garden State Seafood Association, and the Rhode Island Fishermen's

Alliance decided to challenge the Northeast Canyons and Seamount Marine National Monument's designation. On March 17, 2017, these plaintiffs filed a complaint in the U.S. District Court for the District of Columbia in Washington, DC, arguing that presidents cannot use the Antiquities Act to establish marine national monuments. The Massachusetts Lobstermen's Association further asserted that the monument "will deplete the value of some of the lobstermen's permits—a key part of these small businesses' value—put more pressure on the fisheries left open to fishermen, and impact coastal businesses that depend on a productive lobster industry, including marinas, bait dealers, mechanics, processors, and restaurants."[98] According to the Atlantic Offshore Lobstermen's Association, "[t]he monument designation will displace over 11,000 lobster traps used by" its members:

> These traps are hauled in weekly, year-round and are thus an important source of employment and income for the industry. The association estimates the impact on the industry will be $3 million. The displacement of these traps will cause severe disruption to the industry and the environment. It will increase conflicts with other gear as lobstermen invade other fisheries. Although the lobster fishery in the Gulf of Maine/Georges Bank is healthier than the Southern New England lobster fishery, this displacement will put further pressure on that fishery.[99]

Similarly, "The Long Island Commercial Fishing Association's trawl and longline fishermen have been injured by the monument declaration, which forbids them from fishing in the area. Previously, this was an important area for New York's fluke, whiting, squid, swordfish, and tuna fishermen. . . . The association estimates that the loss to New York fishermen alone will be $1.6 million per year. But these impacts are further multiplied when you consider impacts to shoreside businesses related to the fishermen, like marinas and restaurants."[100] The Rhode Island Fishermen's Alliance projected that its fishers would lose $3 million per year in income because of the monument's designation.[101]

The plaintiffs in *Massachusetts Lobstermen's Association v. Ross* asserted three legal arguments. First, they contended that submerged lands and the ocean are not "lands" for purposes of the Antiquities Act. Second, they argued, the federal government does not own or control the outer continental shelf or the EEZ. Finally, plaintiffs argued that President Obama did not reserve the smallest possible area compatible with the Northeast Canyons and Seamounts Marine National Monument's purposes, as the Antiquities Act requires.

Both the U.S. District Court for the District of Columbia in 2018 and a unanimous panel at the U.S. Court of Appeals for the D.C. Circuit in 2019 rejected these arguments and granted the federal defendants' motion to dismiss. While presidential designations under the Antiquities Act *are* judicially reviewable,[102] U.S. Supreme Court precedent, presidential practice, and the Antiquities Act's plain meaning all support the president's authority to protect submerged lands and the water associated with them.[103] Nor did the National Marine Sanctuaries Act implicitly repeal that authority in the ocean because, while "both Acts address environmental conservation in the oceans[,] they do so in different ways and to different ends."[104] Federal "control," in turn, did not have to be absolute control, and the federal government exercised sufficient control over the nation's EEZ and outer continental shelf to fit within the Antiquities Act.[105] Finally, given that the Northeast Canyons and Seamounts Marine National Monument protects not only the geological features but also the ecosystems associated with them, the monument's size was justified.[106]

After the D.C. Circuit denied their petition for rehearing on February 28, 2020, the Massachusetts Lobstermen's Association plaintiffs filed a petition for certiorari with the U.S. Supreme Court on July 27, 2020. Multiple amici filed briefs regarding this petition, and the Court extended the time for responses until October 30, 2020. Nevertheless, the Supreme Court denied certiorari on March 22, 2021,[107] ending—for now—the legal challenges to marine national monuments. Nevertheless, in a rare comment appended to the Supreme Court's formal denial of certiorari, Chief Justice John Roberts noted:

> The creation of a national monument is of no small consequence. As part of managing the Northeast Canyons and Seamounts Marine National Monument, for example, President Obama banned almost all commercial fishing in the area with a complete ban to follow within seven years.... According to petitioners—several commercial fishing associations—the fishing restrictions would not only devastate their industry but also put severe pressure on the environment as fishing would greatly expand in nearby areas outside the Monument.[108]

Thus, Chief Justice Roberts made clear that the conflict in valuing the ocean is far from over. Notably, however, recent science indicates that the monument's establishment had no effect on fishing.[109]

Challenging President Obama's Vision, Part 2: President Trump Superimposes a Different Vision

In the meantime, President Trump also directly challenged President Obama's vision, inducing in the process a second round of Antiquities Act litigation involving the Northeast Canyons and Seamounts Marine National Monument. On June 5, 2020, President Trump issued a new proclamation for the monument.[110] "This proclamation lifts the prohibition on commercial fishing, an activity that is subject to the Magnuson-Stevens Fishery Conservation and Management Act (Magnuson-Stevens), 16 U.S.C. 1801 et seq., and other applicable laws, regulations, and requirements. This proclamation does not modify the monument in any other respect."[111] Specifically, President Trump found—ostensibly as part of efforts to provide economic relief to fishers injured by the COVID-19 pandemic—"that a prohibition on commercial fishing is not, at this time, necessary for the proper care and management of the Northeast Canyons and Seamounts Marine National Monument, or the objects of historic or scientific interest therein."[112] While the monument did reopen to fisheries, the fisheries industry's victory was short-lived. On October 8, 2021, President Joe Biden restored full protection to the monument, including the phase-out of commercial fisheries.[113] Fishermen challenged this restoration in 2022,[114] and so the monument is once again in litigation as this book goes to press.

The litigation has not been one-sided, however. Within two weeks of President Trump opening the monument to fishing, the Conservation Law Foundation, Natural Resources Defense Council, and Center for Biological Diversity filed suit in the U.S. District Court for the District of Columbia to challenge the president's Antiquities Act authority to roll back a national monument's protections. On September 11, 2020, the district judge stayed this lawsuit, *Conservation Law Foundation v. Trump* (now *Conservation Law Foundation v. Biden*),[115] until the U.S. Supreme Court granted or denied certiorari in *Massachusetts Lobstermen's Association v. Ross*. Since the Supreme Court's decision *not* to hear *Massachusetts Lobstermen's Association*, the district judge has delayed further action in light of the Biden administration's review of the controversial national monuments, including Northeast Canyons and Seamounts.

Conservation Law Foundation could have had broad implications for presidential authority respecting *all* national monuments. Moreover, *Conservation Law Foundation* completes the process of re-envisioning Antiquities Act Marine National Monuments as federal public "lands" by directly tying the ocean Antiquities Act litigation to the terrestrial Antiquities Act litigation challenging President Trump's reduction of the Bears Ears National Monument in Utah.[116] The legal issue in the two cases is almost precisely the same: can President Trump reduce the size (Bears Ears) or protections (Northeast Canyons and Seamounts) of an Antiquities Act National Monument that another president created? However, when President Biden restored the monument's protections in October 2021, the Conservation Law Foundation voluntarily dismissed its case.

Conclusion

The Northeast Canyons and Seamounts Marine National Monument litigation is important in its own right and may define how presidents can and cannot wield the Antiquities Act for the rest of the twenty-first century and beyond. However, viewed through a wider lens, it also provides a snapshot of an ocean re-envisioning in process. From this wider view, President Obama's designation of the Northeast Canyons and Seamounts Marine National Monument, in conjunction with his expansion of two of President Bush's Pacific Marine National Monuments to the limits of the United States' jurisdiction, represents the culmination of presidential creativity dating back to 1925 that proclaimed the marine realm to consist of *places*, not just fragmented goods and services to be used and extracted for private profit.

In this legal context, marine national monuments are literally *land*scapes that happen to have water on top of them—but those submerged landscapes support important coastal and open ocean ecosystems and biodiversity that are in fact critical to the survival of terrestrial life, including human beings. The earliest national monuments involving the ocean made terrestrial and marine connectivity explicit, extending island and coastal features into the waters with which they interacted. However, in using the Antiquities Act to create extremely large but also extremely remote—for most Americans, at least—marine national monuments, Presidents Bush and Obama have effectively challenged all Americans—not just fishers—to articulate their dependence upon the ocean as well as to re-envision the ocean as places and ecosystems rather than fragmented parts.

The discretion and ambiguities within the Antiquities Act allowed this re-envisioned ocean to emerge out of federal law. Regardless of how long that vision lasts as a matter of law, the snapshots of marine national monuments will remain in cultural memory, perhaps eventually provoking Congress to finally adopt a more comprehensive and permanent legal regime for the ocean.

Notes

This research was made possible, in part, through generous support from the Albert and Elaine Borchard Fund for Faculty Excellence and the University of Utah Global Change and Sustainability Center. The author participated as a *pro bono* law professor amicus in the district court and court of appeals phases of the litigation discussed.

1. Robin Kundis Craig, "Re-Valuing the Ocean in Law: Exploiting the Panarchy Paradox of a Complex System Approach," 41 *Stanford Environmental Law Journal* 3 (2022): 19–32.
2. 16 U.S.C. §§ 757a–757g (2012).
3. 16 U.S.C. §§ 3601–3608 (2012).
4. 16 U.S.C. §§ 5152–5154, 5156, 5158 (2012).
5. 16 U.S.C. §§ 971–971k (2012).
6. 16 U.S.C. §§ 972, 972a–972h (2012).
7. 16 U.S.C. §§ 1151–1159, 1161–1169, 1171–1175 (2012).
8. 16 U.S.C. §§ 952, 953, 962, 1362, 1371, 1374, 1378, 1380, 1385, 1411–1418 (2012).
9. 16 U.S.C. §§ 772–772j, 773–773k (2012).
10. 16 U.S.C. §§ 973–973r (2012).
11. 16 U.S.C. §§ 781–785 (2012).
12. 16 U.S.C. §§ 916–916*l* (2012).
13. 16 U.S.C. §§ 1361–1421h (2012).
14. 16 U.S.C. §§ 1531–1544 (2012).
15. *Ibid.*, §§ 1532(16), 1533(a).
16. U.S. Commission on Ocean Policy, *An Ocean Blueprint for the 21st Century* (Washington, DC: 2004), 55.
17. Ibid., 5.
18. 16 U.S.C. § 758e (2012).
19. 16 U.S.C. §§ 4001–4017 (2012).
20. 46 U.S.C. §§ 1274, 1279f, 1279g (2012).
21. 22 U.S.C. §§ 1971–1980 (2012).
22. 16 U.S.C. §§ 1801–1882 (2012).
23. Ibid., § 1801(a)(6).
24. Pub. L. No. 104-297, 110 Stat. 3559 (Oct. 11, 1996).
25. 16 U.S.C. § 1851(a)(1), (3) (2012).
26. Ibid., § 1802(21).
27. 16 U.S.C. §§ 1431–1445c.
28. Ibid., § 1431(a)(2).
29. Ibid., § 1433(a)(2)(C).

30. Robert J. Wilder et al., "Saving Marine Biodiversity," *Issues in Science & Technology Online* (Spring 1999): 2–3, http:// www.nap.edu/issues/15.3/wilder.htm/.
31. *Plymouth Herald*, "UK Fishermen Protest over Marine Reserves," *National Fisherman* (February 22, 2016), https://www.nationalfisherman.com/national-international/uk-fishermen-protest-over-marine-reserves/.
32. Bret Yager, "Fishermen Protest Marine Reserve at Kaupulehu," *West Hawaii Today* (June 5, 2016), https://www.westhawaiitoday.com/2016/06/05/hawaii-news/fishermen-protest-marine-reserve-at-kaupulehu/.
33. Robin Kundis Craig, "Taking Steps Toward Marine Wilderness Protection? Fishing and Coral Reef Marine Reserves in Florida and Hawaii," *McGeorge Law Review* 34 (2003): 155, 224–39.
34. Ibid., 238.
35. 54 U.S.C. § 320301(a) (formerly 16 U.S.C. § 431(a)).
36. Pres. Barack Obama, "Proclamation 9496—Northeast Canyons and Seamounts Marine National Monument" (Sept. 15, 2016).
37. 54 U.S.C. § 320301(a) (formerly 16 U.S.C. § 431(a)).
38. "Northeast Canyons and Seamounts Marine National Monument," NOAA Fisheries, updated December 19, 2010, https://www.fisheries.noaa.gov/new-england-mid-atlantic/habitat-conservation/northeast-canyons-and-seamounts-marine-national/.
39. United Nations Convention on the Law of the Sea (December 10, 1982, in force November 16, 1994), art. 57.
40. 43 U.S.C. §§ 1301(a)(1), 1311(b) (2012).
41. Ibid., § 1302.
42. Pres. Obama, "Proclamation 9496," 1.
43. Ibid.
44. Ibid.
45. "Northeast Canyons and Seamounts Marine National Monument," NOAA Fisheries, updated December 19, 2018, https://www.fisheries.noaa.gov/new-england-mid-atlantic/habitat-conservation/northeast-canyons-and-seamounts-marine-national/.
46. President Obama, "Proclamation 9496," 1.
47. Ibid., 3.
48. Ibid., 4.
49. Ibid., 5–6.
50. Ibid., 5.
51. Ibid.
52. Ahjond Garmestani, J. B. Ruhl, Brian C. Chaffin, Robin K. Craig, Helena F. M. W. van Rijswick, David G. Angeler, Carl Folke, Lance Gunderson, Dirac Twidwell, and Craig R. Allen, "Untapped capacity for resilience in environmental law," *PNAS* 116:40 (Oct. 1, 2019): 19,899–19,904 (arguing that environmental law has sufficient "wiggle room" to allow for enhanced adaptive and transformative capacity in the Anthropocene).

53. 54 U.S.C. § 320301(a) (formerly 16 U.S.C. § 431(a)) (emphasis added).
54. 16 U.S.C. §§ 1431–1445c (2012).
55. "Mallows Bay–Potomac River National Marine Sanctuary," NOAA, https://sanctuaries.noaa.gov/mallows-potomac/.
56. Pres. Proclamation No. 1733 (Feb. 26, 1925); 16 U.S.C. § 410hh-1(1) (2012).
57. Alaska v. United States, 545 U.S. 75, 78–79 (2005).
58. Ibid., 101 (citations and internal references omitted).
59. Pres. Proclamation No. 2112 (Jan. 4, 1935).
60. 16 U.S.C. § 410xx (2012).
61. Pres. Proclamation No. 2281, 53 Stat. 1541 (1938).
62. Pres. Proclamation No. 2825, 63 Stat. 1258 (1949).
63. United States v. California, 332 U.S. 19, 38–39 (1947); *see also* United States v. California, 332 U.S. 804, 805 (1947) (affirming the Court's holding that the United States was "possessed of paramount rights in, and full dominion and power over, the lands, minerals and other things underlying the Pacific Ocean lying seaward of the ordinary low-water mark on the coast of California, and outside of the inland waters, extending seaward three nautical miles").
64. United States v. California, 436 U.S. 32, 36 (1978).
65. 43 U.S.C. §§ 1301–1315 (2012).
66. 43 U.S.C. §§ 1331–1356b (2012). President Truman proclaimed federal control over the continental shelf in 1945, Policy of the United States with Respect to the Natural Resources of the Subsoil and Seabed on the Continental Shelf, Proclamation No. 2667 (H. Truman, Sept. 28, 1945), and under international law the Outer Continental Shelf now extends two hundred nautical miles out to sea. United Nations Convention on the Law of the Sea, art. 76.1.
67. United States v. California, 436 U.S. 32, 41 (1978).
68. Pres. Proclamation No. 3443, 76 Stat. 1441, 1441 (Dec. 28, 1961), 27 Fed. Reg. 31 (Jan. 4, 1962).
69. Ibid.; Pres. Proclamation No. 4346, 40 Fed. Reg. 5127 (Feb. 4, 1975); Pres. Proclamation No. 7392, 66 Fed. Reg. 7335 (Jan. 17, 2001).
70. "Buck Island Reef: Natural Features and Ecosystems," National Park Service, https://www.nps.gov/buis/learn/nature/naturalfeaturesandecosystems.htm/.
71. Pres. Proclamation No. 4614, 43 Fed. Reg. 57,025 (Dec. 5, 1978).
72. Proclamation No. 4620, 43 Fed. Reg. 57,067 (Dec. 5, 1978).
73. Pres. Proclamation No. 4623, 43 Fed. Reg. 57,087 (Dec. 5, 1978).
74. Pres. Proclamation No. 4614, 43 Fed. Reg. 57,025, 57,025 (Dec. 5, 1978).
75. Ibid., 57.026.
76. Pub. L. No. 96-487, § 201(2), 94 Stat. 2371 (Dec. 2, 1980).
77. Pres. Proclamation No. 4620, 43 Fed. Reg. 57,067, 57,067 (Dec. 5, 1978).
78. Ibid.
79. Pub. L. No. 96-487, § 201(5), 94 Stat. 2371 (Dec. 2, 1980).
80. Pres. Proclamation No. 4623, 43 Fed. Reg. 57,087, 57,087 (Dec. 5, 1978).
81. Ibid., 57,087–57,088.

82. Pub. L. No. 96-487, § 503, 94 Stat. 2371 (Dec. 2, 1980).
83. Pres. Proclamation No. 7264, 65 Fed. Reg. 2821, 2822 (Jan. 11, 2000).
84. Pres. Proclamation No. 8947, 78 Fed. Reg. 18,789 (March 25, 2013).
85. Pres. Proclamation No. 9089, 79 Fed. Reg. 14,603, 14,604 (March 14, 2014).
86. Pres. Proclamation No. 9563, 82 Fed. Reg. 6131, 6131 (Jan. 18, 2017).
87. Pres. Proclamation No. 7399, 66 Fed. Reg. 7364, 7364 (Jan. 17, 2001).
88. Ibid., 7365.
89. Pres. Proclamation No. 8031, 71 Fed. Reg, 36,443 (June 15, 2006).
90. Pres. Proclamation No. 8335, 74 Fed. Reg. 1557 (Jan. 6, 2009).
91. Pres. Proclamation No. 8336, 74 Fed. Reg. 1565 (Jan. 6, 2009).
92. Pres. Proclamation No. 8337, 74 Fed. Reg. 1577 (Jan. 6, 2009).
93. Pres. Proclamation No. 8327, 73 Fed. Reg. 75,293 (Dec. 5, 2008).
94. Pres. Proclamation No. 8031, 71 Fed. Reg, 36,443. 36,443 (June 15, 2006).
95. Pres. Proclamation No. 9478, 81 Fed. Reg. 60,227, 60,227–30 (Aug. 26, 2016).
96. "About Papahānaumokuākea," Papahānaumokuākea Marine National Monument, revised August 7, 2019, https://www.papahanaumokuakea.gov/new-about/.
97. John Lynham, Anton Nikolaev, Jennifer Raynor, Thaís Vilela, and Juan Carlos Villaseñor-Derbez, "Impact of two of the world's largest protected areas on longline fishery catch rates," *Nature Communications* 11: art. 979 (2020), https://doi.org/10.1038/s41467-020-14588-3/.
98. Massachusetts Lobstermen's Association v. Ross, Complaint, 3 ¶ 8.
99. Ibid., 3–4 ¶ 10.
100. Ibid., 4 ¶ 11.
101. Ibid., 5 ¶ 13.
102. Massachusetts Lobstermen's Association v. Ross, 349 F. Supp. 3d 48, 54–55 (D.D.C. 2018), *aff'd*, 945 F.3d 535, 540, 545 (D.C. Cir. 2019).
103. Ibid., 349 F. Supp. 3d at 56–58, *aff'd*, 945 F.3d at 540–41.
104. Ibid., 349 F. Supp. 3d at 59, *aff'd*, 945 F.3d at 541–42.
105. Ibid., 349 F. Supp. 3d at 59–66, *aff'd*, 945 F.3d at 542–44.
106. Ibid., 349 F. Supp. 3d at 67–68, *aff'd*, 945 F.3d at 544.
107. Massachusetts Lobsterman's Association v. Raimondo, --- U.S. ----, 141 S. Ct. 979, 979 (2021).
108. Ibid., at 980.
109. "Fishing activity before closure, during closure, and after reopening of the Northeast Canyons and Seamounts Marine National Monument," *Scientific Reports* 12: art. 917 (2022), https://doi.org/10.1038/s41598-021-03394-6/.
110. President Donald J. Trump, "Proclamation on Modifying the Northeast Canyons and Seamounts Marine National Monument," June 5, 2020, https://www.whitehouse.gov/presidential-actions/proclamation-modifying-northeast-canyons-seamounts-marine-national-monument/.
111. Ibid.
112. Ibid.

113. Pres. Joseph Biden, A Proclamation on Northeast Canyons and Seamounts Marine National Monument (Oct. 8, 2021), https://www.whitehouse.gov/briefing-room/presidential-actions/2021/10/08/a-proclamation-on-northeast-canyons-and-seamounts-marine-national-monument/.
114. Fehily v. Biden, Case No. 3:22-cv-02120 (D.N.J. filed Apr. 12, 2022). The complaint is available at: https://pacificlegal.org/wp-content/uploads/2022/04/2022.04.12-Fehily-et-al.-v.-Biden-et-al.-PLF-Complaint.pdf/.
115. Case No. 1:20-cv-01589-JEB (D.D.C. 2020).
116. *See* NRDC v. Trump, Case No. Case 1:17-cv-02606 (D.D.C. filed Dec. 7, 2017).

CHAPTER 9

One Step Forward, Two Steps Back

Reforming the National Ocean Policy for the Twenty-First Century

NATHANIEL E. BROADHURST

ON JUNE 19, 2018, President Donald J. Trump signed Executive Order 13840, entitled "Ocean Policy to Advance the Economic, Security, and Environmental Interests of the United States"[1] (Trump EO). The Trump White House proclaimed numerous regulatory and management improvements that would be affected by the new Ocean Policy, including a "more streamlined process for federal coordination," "eliminating duplicative federal bureaucracy," and "promoting a strong ocean economy."[2] Despite its purported advancement of environmental interests, the Trump EO removed most of the management framework and environmental safeguards that were put in place by President Barack H. Obama's National Ocean Policy executive order (Obama EO).[3] Instead, the Trump EO represented a return to many of the same policy guidelines established by President George W. Bush in the first National Ocean Policy executive order (Bush EO), issued in 2004.[4]

The promulgation of a National Ocean Policy unilaterally via executive order may seem unorthodox, yet this has been the modus operandi since the George W. Bush (Bush 43) Administration. The Bush EO was developed in response to the recommendations from two groundbreaking reports issued by the U.S. Commission on Ocean Policy[5] and the Pew Oceans Commission.[6] Both of these reports, in turn, were derived from congressional guidance in the Oceans Act of 2000.[7] All of these developments in the early 2000s represented a dramatic shift in federal ocean and coastal law, marking the first time efforts to conjunctively manage ocean resources across the federal government resulted in an interagency policy-coordinating body.

Historically, federal ocean management has occurred piecemeal through hundreds of disjointed statutes, disparately divided among dozens of agencies.[8] The National Oceanic and Atmospheric Administration (NOAA) was established in 1970 via executive order to administer several federal oceanic programs.[9] Over time, NOAA's responsibilities have grown pursuant to statutes, executive orders, and administrative restructuring, yet the agency lacks coherent direction. A national, comprehensive ocean policy, adopted via statute, has been proposed again and again since the late 1960s to address the enduring administrative and legal fragmentation and consequent resource management problems.[10]

The Trump National Ocean Policy returned to the Bush 43 model of using an Ocean Policy Committee as the appropriate interagency coordinating body, rather than the Ocean Council created by President Obama.[11] Furthermore, unlike the Obama EO, which mandated council representation by state, local, and tribal governments, the Trump EO merely directs agencies to coordinate and consult with other levels of government "as appropriate."[12]

Although the Biden Administration has taken several executive actions to protect the ocean, President Joseph R. Biden has neither adopted his own National Ocean Policy nor has he explicitly revoked the Trump EO. As a result, the policy shift from President Obama to President Trump remains an important case study of how policy regarding the ocean can change. The policy shifts have facial management implications, making it more difficult to address current jurisdictional and spatial management deficiencies. While there are several structural and substantive drawbacks of the Trump Administration's changes to the Obama EO's directives, some promising changes were also made. For instance, the Trump EO directs the Ocean Policy Committee to "coordinate the timely release of unclassified data" related to the ocean.[13] Although new intergovernmental data-sharing portals were implemented pursuant to the National Ocean Policy under Obama, neither the Bush EO nor the Obama EO placed as great an emphasis on regional data sharing and public access to information.[14] Additionally, although the Trump EO does not mention climate change specifically, Section 2 does provide that "[i]t shall be the policy of the United States" to coordinate agency actions so as "to provide economic, security, and environmental benefits for present and future generations of Americans."[15]

This chapter evaluates the efficacy of the bureaucratic restructuring instituted under the three National Ocean Policies to date and issues recommendations for reform. Section II presents a historical and current overview of federal ocean governance, demonstrating the legal and administrative fragmentation that plagues marine resource management in the United States. Section III examines why the United States needs a comprehensive ocean policy to thrive in the Anthropocene, explores the advent of a National Ocean Policy by executive order, and analyzes three most recent executive orders. Section IV discusses possible solutions to the lingering governance and management deficiencies with the current legal framework. The chapter concludes in Section V by recommending that the Biden Administration, or a future administration, should apply its political capital toward passing an organic act for NOAA, as that is the most politically attainable of the potential solutions. A NOAA organic act would also serve as a critical first step toward future omnibus ocean management legislation, and stakeholders across the board are likely to support the enhanced clarity and structure that a NOAA organic act would bring to the federal marine management paradigm.

The Problem: Deficiencies in Governance

Who Controls What? A Brief Overview of Ocean Jurisdiction

International Law

Traditionally, the world's oceans were largely considered to be a commons, with the exception of waters in close proximity to the coasts.[16] These coastal waters became known as the "territorial seas," with coastal nations' jurisdiction over territorial waters having essentially the same scope as their jurisdiction on dry land.[17] The extent of the territorial seas was considered to be the length of a cannon shot from the shoreline.[18] Over time, the "cannon shot rule" resulted in the territorial seas extending three nautical miles from the shoreline.[19]

The jurisdictional paradigm began to shift in the mid-twentieth century with what became known as the "enclosure of the oceans."[20] The movement involved asserting the exclusive rights of coastal nations over marine waters beyond their traditionally recognized territorial seas.[21] One of the most vocal proponents of this shift in governance was the United States, motivated at least in part by the nascent expansion of offshore oil and gas drilling into deeper waters.[22] This geopolitical movement culminated with the adoption of four United Nations conventions on the Law of the Sea in Geneva in 1958.[23] However, the 1958 UN conventions failed to adequately define the parameters of jurisdiction over natural resources development.[24] For instance, the jurisdictional continental shelf was defined exclusively by the ability of a given country to develop resources on the ocean floor.[25] This definition bred uncertainty as technological advancements during the second half of the twentieth century led to resource recovery on the continental shelf in ways that were not anticipated by the drafters of the 1958 conventions.[26]

Further international consensus was reached in 1982 with the landmark United Nations Convention on the Law of the Sea (UNCLOS).[27] UNCLOS altered the international paradigms established by the 1958 conventions—although it did not supplant them entirely—and it largely defines the parameters of oceanic jurisdiction today.[28] UNCLOS essentially serves as "an international constitution for the oceans."[29] Although the United States never actually ratified UNCLOS, it mostly abides by the jurisdictional boundaries that UNCLOS established as customary international law.[30] Under UNCLOS, the baseline is the dividing line that jurisdictionally separates the ocean from land and is typically the "low-water line along the coast."[31] UNCLOS extended the territorial sea from the baseline to twelve nautical miles out to sea.[32] Each nation has sovereignty over its territorial seas, exercising exclusive jurisdiction over the airspace above the territorial sea, the water column, seafloor, and subsoil,[33] subject to some limited rights of innocent passage for foreign vessels.[34] Under UNCLOS, the area twelve to twenty-four nautical miles from the baseline is called the contiguous zone.[35] Countries have jurisdiction to enforce laws related to customs, immigration, and sanitation within their contiguous zones.[36] UNCLOS also recognized the ability for each nation to establish an exclusive economic zone (EEZ) measuring twelve to two hundred nautical miles from the baseline.[37] Within the EEZ, countries can assert "sovereign rights" over the exploration, development, and management of resources.[38] These exclusive resource rights extend to living and nonliving resources found in the water column, seafloor, and subsoil.[39]

Finally, UNCLOS outlined a legal definition for the continental shelf, which confers on countries the ability to develop resources in the seabed and subsoil within their continental shelf.[40] Each nation's continental shelf automatically extends two hundred nautical miles from the baseline, but UNCLOS also gave each signatory nation the ability to assert a claim to the full geological extent of its adjacent continental shelf if it shows that the shelf physically extends farther than two hundred nautical miles.[41] However, because the United States is not a signatory to UNCLOS, it technically does not have the ability to assert a jurisdictional continental shelf beyond two hundred nautical miles.[42]

Federal and State Jurisdiction in the United States

Jurisdictional issues also impact ocean and coastal management at the domestic level. Within the United States' ocean waters, the coastal states have jurisdiction over the water and ocean floor from the baseline to three nautical miles seaward—the area traditionally considered the territorial seas.[43] The states were granted this authority by the Submerged Lands Act of 1953,[44] which gave title to the states over the ocean floor within the territorial seas, but retained federal jurisdiction to regulate commerce, navigation, and national security interests in the states' waters.[45] Thus, while states have primary authority over resource development within their territorial seas, the regulatory authority that the federal government retains means that state territorial waters are a critical part of federal management strategies.

The Coastal Zone Management Act[46] (CZMA) further developed the shared jurisdiction of federal and state governments over the "coastal zone," which it defined as including all of the states' waters, "transitional and intertidal areas, salt marshes, wetlands, and beaches," that "extends inland from the shorelines only to the extent necessary to control shorelands, the uses of which have a direct and significant impact on the coastal waters."[47] Characterized as a "marriage of federal activism and states' rights," the CZMA created a voluntary program to incentivize states to develop land use management plans for their coastal areas.[48] The CZMA established a federal grant system to aid states with the development and implementation of their coastal zone management plans, as long as the plans are consistent with the national interest in conservation of natural resources.[49] The CZMA also obligates the federal government to ensure that any federal actions affecting a state's coastal zone comply with that state's management plan.[50] Today, thirty-four of the thirty-five coastal states (including the Great Lakes states) participate in the voluntary coastal zone management program.[51]

In sum, the federal government cooperatively manages coastal land and wetlands with the states and shares authority with the states over management of the territorial seas between the baseline and three miles into the ocean—including submerged lands.[52] Beyond three miles, the federal government exercises sovereign jurisdiction over a territorial sea to twelve nautical miles,[53] enforces laws throughout the contiguous zone to twenty-four nautical miles,[54] and exercises exclusive authority over resources located in the EEZ—from twelve to two hundred nautical miles offshore.[55] The federal government also controls all uses of the continental shelf beyond three miles from the coast.[56]

Administrative Fragmentation

The Bush, Obama, and Trump National Ocean Policy EOs are reflective of—and responses to—decades of marine regulatory fragmentation in the United States. Federal management of ocean and coastal resources is disjointed, and understanding the history of that disjunction is necessary to understanding the presidential EOs. Accordingly, this section provides a brief survey of the last fifty years of marine regulation in the United States.

U.S. ocean policy has traditionally consisted of a "hodgepodge" of laws, often driven by reaction to crises rather than proactivity.[57] Roughly 140 federal laws affect ocean and coastal management, yet these laws are often insular and approach management from an issue- or agency-specific standpoint.[58] Many of these statutes were enacted during the wave of omnibus federal environmental laws in the 1970s, which made several dramatic changes to ocean and coastal law. Passage of the National Environmental Policy Act[59] (NEPA) initiated this movement by adopting a "look before you leap" approach, requiring agencies to conduct analyses of environmental impacts before any proposed "major federal action."[60] Thus, any major action authorized or carried out by a federal agency in the EEZ that potentially affects the environment will be subject to some level of NEPA analysis.[61]

NEPA also established the Council on Environmental Quality (CEQ), which serves as an advisory group to counsel the president on national environmental policy goals.[62] Among other duties, NEPA charges the CEQ with "develop[ing] and recommend[ing] to the President national policies to foster and promote the improvement of environmental quality to meet the conservation, social, economic, [and] health . . . goals of the Nation."[63] Given these responsibilities, the CEQ has historically played a vital role in implementing and evaluating federal ocean and coastal policies.[64] Despite its institutional importance, CEQ's focus is both too narrow (viewing marine management exclusively through an environmental lens) and too broad (high-level, national policies) to adequately serve as an interagency facilitator for a national ocean policy.[65]

The National Oceanic and Atmospheric Administration (NOAA) was created by executive order[66] in 1970 in response to recommendations from the Stratton Commission calling for the creation of a new agency to administer the United States' marine programs.[67] At first, President Nixon's advisors suggested placing NOAA within the Department of Interior (DOI) as part of a broader initiative to restructure and replace the DOI with a new Department of Natural Resources.[68] However, the decision was ultimately made to house NOAA within the Department of Commerce because Commerce staff would comprise roughly two-thirds of the nascent agency's employees.[69] Ultimately, personnel and programs from the Environmental Sciences Services Administration, parts of the Bureau of Commercial Fisheries and Bureaus of Sport Fisheries and Wildlife, the Sea Grant program, parts of the U.S. Army's Lake Survey, the U.S. Navy's National Oceanographic Data and Instrumentation Centers, the marine Minerals Technology Center, and the national Data Buoy Project were reorganized into NOAA.[70]

Congress enacted the CZMA in 1972, extending the planning and stewardship ethos of NEPA into ocean and coastal areas specifically.[71] The CZMA charged NOAA with implementing the cooperative planning programs established under the act.[72] The same year, Congress passed the Ocean Dumping Act[73] in order to study and regulate the practice of intentionally dumping waste into the ocean.[74] The Ocean Dumping Act delegates primary authority to the Environmental Protection Agency (EPA) to regulate waste disposal from vessels and other facilities at sea.[75] The U.S. Army Corps of Engineers (Corps) regulates disposal of dredged material into the ocean.[76] Similarly, the EPA and the Corps work together to enforce provisions of the Clean Water Act,[77] which regulates discharges of pollutants into navigable waters (including the ocean) from land-based point sources.[78]

The Marine Mammal Protection Act[79] (MMPA), also enacted in 1972, prohibits the "taking" of any marine mammal in United States waters.[80] The MMPA defines "take" as "hunting, killing, capture, and/or harassment of any marine mammal, or the attempt of such."[81] The MMPA also bans the importation and sale of marine mammal parts or products, except in some limited circumstances.[82] The MMPA was passed in response to public pressure over whaling industry practices and the amount of marine mammals killed incidentally by fishing operations.[83]

The Endangered Species Act[84] (ESA), enacted in 1973, has a similar structure to the MMPA, prohibiting the taking of a listed species and the importation and sale of listed species' parts.[85] Under the ESA, "take" is defined as "harass, harm, pursue, hunt, shoot, wound, kill, trap, capture, or collect, or to attempt to engage in any such conduct."[86] The key difference between the ESA and MMPA is the listing process; whereas the MMPA automatically protects all marine mammals, a species must be deliberately listed in order to enjoy protection under the ESA. NOAA is responsible for the listing of marine and anadromous species.[87]

In 1976, Congress enacted what became known as the Magnuson-Stevens Act[88] to address the presence of foreign fishing vessels operating in waters traditionally dominated by American fishermen.[89] The Magnuson-Stevens Act expanded exclusive use of fisheries for American vessels from twelve nautical miles into the ocean to two hundred nautical miles, which was an unprecedented expansion of jurisdiction in maritime law prior to the 1982 adoption of UNCLOS.[90] In addition to promoting domestic fisheries use, the Magnuson-Stevens Act also placed important conservation goals at the forefront: regional fisheries management, rebuilding depleted stocks, and preventing future overfishing.[91] While the Coast Guard is tasked with enforcing the exclusivity of American fisheries within the EEZ, NOAA is tasked with approving regional fishery management plans.[92] The plans are developed and submitted to NOAA by Regional Fishery Management Councils (multistakeholder bodies established by the Magnuson-Stevens Act)[93] for particular species, with the intent to use local expertise and data.[94]

In 1977, four new offices were created within NOAA.[95] The Office of Fisheries was tasked with conserving and responsibly managing fish within two hundred miles of the American coast, as well as generally supporting the wellbeing of the American fishing industry.[96] The Office of Coastal Zone Management was established to fulfill NOAA's role under the CZMA and aid the states' implementation of their coastal zone management plans.[97] This department was also tasked with supporting the creation of estuarine or marine sanctuaries.[98] The Office of Oceanic and Atmospheric Services was charged with creating navigational charts, managing NOAA's fleet of vessels, conducting weather forecasts and issuing warnings, and operating NOAA's databases.[99] The Office of Research and Development manages NOAA's laboratories, supports academic research relating to NOAA's mission and needs, and administers the Sea Grant program.[100]

Bucking the generally pro-environment and pro-conservation trends of the 1970s, Congress amended the Outer Continental Shelf Lands Act (OCSLA) in 1978,[101] leading to a dramatic expansion of oil and gas leasing on the United States' outer continental shelf.[102] OCSLA was originally passed in 1953[103] in the wake of the Submerged Lands Act to assert federal jurisdiction over oil and gas resources under the submerged lands of the continental shelf.[104] The 1978 amendments sought to promote American energy development in response to the 1973 oil embargo, but ultimately included compromise provisions to assuage environmental concerns.[105] These additional safeguards included planning requirements, the adoption of environmental and safety regulations, and consultation with state and local governments.[106] To administer federal offshore oil and gas leases, a new agency was created within the Department of Interior: the Minerals Management

Service.[107] In 2010, after the *Deepwater Horizon* oil spill catastrophe, this agency was renamed the Bureau of Ocean Energy Management, Regulation and Enforcement and split into two subagencies: the Bureau of Ocean Energy Management and the Bureau of Safety and Environmental Enforcement (BSEE).[108]

Despite the sweeping legal changes over the second half of the twentieth century, federal agencies still "lacked an overarching vision," and NOAA was ill-equipped to set priorities among the other marine-related agencies.[109] Thus, federal governance of the United States' ocean and coastal domain remained fragmented, with agencies operating mostly within their individual legal and administrative silos.

The National Ocean Policy: A Solution to Fill the Legal Vacuum

Humble Beginnings: The Stratton Commission

In addition to the substantive environmental protections put in place during the 1960s and 1970s,[110] milestones were also reached in the governance and management arena. The Marine Resources and Engineering Development Act of 1966 established the Commission on Marine Science, Engineering and Resources to advise the president[111]—colloquially known as the Stratton Commission.[112] The Stratton Commission was tasked with completing a "comprehensive investigation and study of all aspects of marine science in order to recommend an overall plan for an adequate national oceanographic program that will meet the present and future national needs."[113]

After nineteen meetings, the Stratton Commission released a report that centered around three broad policy problems: the ocean as "the new frontier," environmental protection and pollution prevention, and the need to reorganize federal ocean and coastal initiatives.[114] The Stratton Commission's report was lauded as the first truly comprehensive study of federal marine management, and the Commission's recommendations led (directly or indirectly) to most of the significant legislative and administrative developments in ocean and coastal law during the latter half of the twentieth century.[115]

The Stratton Commission also made the important contribution of proposing a new independent and centralized agency to implement the United States' ocean and atmospheric programs.[116] NOAA was officially formed in 1970 to fulfill this recommendation, and, as noted above, several existing programs and departments were administratively reorganized into the new agency.[117]

The creation of NOAA succeeded in achieving a degree of administrative streamlining and cohesiveness over marine resources management by housing several of the federal marine programs within one agency.[118] Nevertheless, although NOAA is the agency with the most influence over ocean and coastal resources management, numerous other agencies continued to administer programs related to the ocean.[119] Indeed, since the creation of NOAA and the passage of the Magnuson-Stevens Act, "governmental attitudes toward ocean policy . . . remained largely unaffected."[120] Thus, a need for further collaboration and direction among agencies, state and local governments, and other stakeholders remained.

Oceans Act of 2000

The first legislative codification of the principles underlying a comprehensive national ocean policy was in the Oceans Act of 2000.[121] The Oceans Act established the U.S. Commission on Ocean Policy[122] and called for the Commission to make recommendations that would promote the Act's listed purposes and objectives.[123] Although these purposes and objectives included various resource development goals,[124] Congress also emphasized a clear balance between resource use and sustainable management. For instance, while working toward the "enhancement of marine-related commerce and transportation," the Commission was also to engage "the private sector in innovative approaches for *sustainable* use of living marine resources and *responsible* use of non-living marine resources."[125] Additionally, while calling for investment in "technologies designed to promote national energy use and food security," the Oceans Act also called for further study into "the role of the oceans in climate and global environmental change."[126] This section also listed "responsible stewardship, including use, of fishery and other ocean and coastal resources" and "the protection of the marine environment and prevention of marine pollution" as objectives.[127]

Most importantly, the Oceans Act of 2000 directed the nascent Commission to promote "close cooperation among all government agencies and departments" in order to effectuate a truly national ocean policy.[128] Congress elaborated that this interagency cooperation was to incorporate coordinated management of ocean and coastal activities, regulatory consistency, efficient distribution of agency funding and resources, and the development of partnerships with state and local governments and the private sector.[129]

The Ocean Commissions and Their Reports

The U.S. Commission on Ocean Policy

Pursuant to the mandates in the Oceans Act of 2000, the U.S. Commission on Ocean Policy (COP) began meeting in September 2001 to fulfill its mission to develop policy recommendations.[130] The COP consisted of sixteen commissioners representing diverse perspectives, including NOAA, academia, state port authorities, finance, and interest groups.[131] The final report, issued in September 2004, pronounced that there was a "historic opportunity" to implement a "comprehensive and coordinated national ocean policy."[132]

In its opening letters to President Bush, the Senate, and the House, the COP called for "immediate action" on three main recommendations: the establishment of a national ocean policy (including the important corollary of "strengthening support for territorial, tribal and local efforts to resolve issues at the regional level"); basing all ocean and coastal resources decisions on unbiased and best-available science; and educating the public to promote a new "stewardship ethic" of the ocean.[133]

In response to the COP Report, President Bush established the Committee on Ocean Policy via executive order, establishing the first embodiment of a national ocean policy put into action.[134]

The Pew Oceans Commission

At the same time the COP was getting underway, the Pew Charitable Trusts convened the Pew Oceans Commission (Pew Commission), "a bipartisan, independent group of American leaders," consisting of scientists, philanthropists, fishermen, public officials, researchers, industry leaders, and conservationists.[135] The Pew Commission identified four core issues that it would analyze: "governance, fishing, pollution, and coastal development."[136] Before issuing its report on June 3, 2003, the Commission met in fifteen regional meetings over two years and conducted public hearings, focus groups, and workshops in coastal communities.[137]

Unlike the COP, the Pew Report focused on the United States' living marine resources.[138] The report articulated that there was a basic failure in governance and conception of the United States' ocean domain and opined that the oceans should be managed in accordance with their importance to the economy, culture, and the land.[139] To address these failures, the Pew Report advocated for creating healthier ecosystems, which would in turn lead to a more economically productive and resilient ocean.[140]

The Pew Commission emphasized the need for a paradigmatic shift in order to achieve the management changes.[141] Specifically, it called for the nation to adopt "an ethic of stewardship" and manage the oceans as a public trust.[142]

In line with these broader societal changes, the Pew Commission also recommended several concrete governance reforms: enacting a National Ocean Policy Act; implementing comprehensive and coordinated governance institutions at the regional level; expanding marine protected reserves and creating a national system to manage them; establishing an independent national marine management agency; and establishing a "permanent federal interagency oceans council."[143]

The Pew Report also took a different stance than the COP on the utility of maintaining NOAA as the centralized administrative entity for federal marine management. It advocated instead for the establishment of an independent agency that would absorb NOAA's responsibilities and "as many federal ocean programs as is practical."[144] Nonetheless, the Pew Commission also recognized the importance of an organic act.[145] In the context of federal public lands, the Pew Report noted the benefits of agencies receiving congressional guidance and exercising delegated authority when making management decisions, including codification of the "purposes of the system . . . [and] a framework within which the cumulative effects of all uses of public lands can be assessed, coordinated, and managed."[146] Accordingly, the organic acts for federal land management agencies could serve as a model for crafting the guidance language necessary to formulate an effective marine resources management organic act.[147]

The National Ocean Policy Under President Bush

Committee on Ocean Policy

On December 17, 2004, the Bush 43 Administration issued its response to the COP recommendations: the U.S. Ocean Action Plan.[148] The Ocean Action Plan noted that the primary component to the Administration's response was President Bush's creation of the Cabinet-level Committee on Ocean Policy (Committee) by executive order.[149] Under the Bush EO, the Chairman of the President's Council on Environmental Quality (CEQ) would also serve as the chairman of the new Committee.[150] The Committee's membership also included numerous Cabinet members, as well as Bush's assistants for national security, homeland security, domestic policy, and economic policy.[151] Finally, the Committee was to include one employee designated by the Vice President and potentially any other federal employees that the Chairman might designate.[152]

In creating the new Committee, the Bush EO recognized that the new policy of the United States would be to "coordinate the activities of executive departments and agencies regarding ocean-related matters . . . to advance the *environmental*, economic, and security interests of *present and future generations* of Americans."[153] This broad policy directive appears to have implicitly recognized the reality of climate change and that marine resources should be managed with future generations in mind.

Intergovernmental Planning Framework

Additionally, the Bush EO articulated a policy of "facilitat[ing], as appropriate, coordination and consultation regarding ocean-related matters among Federal, State, tribal, local governments, the private sector, foreign governments, and international organizations."[154] In meeting this directive, the nascent Committee was to collect information and guidance from state, local, and tribal governments and private parties on the development and implementation of marine policies.[155] The Bush EO clarified that the Committee was only to consult the parties for "their individual advice," which would "not involve collective judgment or consensus advice or deliberation."[156]

At the same time, however, the Bush Action Plan acknowledged that the Bush 43 Administration "strongly value[d] the importance of local involvement" and that its National Ocean Policy would create robust partnerships between the federal government, states, tribes, local governments, the private sector, and/or international bodies.[157] The Action Plan noted two specific examples of how this strategy was already being implemented in the Great Lakes area: the Great Lakes Regional Collaboration, which brought together governors of the Great Lakes states, the states' congressional delegation, several mayors, tribal leaders, and Canadian officials; and a collaborative effort between federal agencies, states, and tribes to address harmful algal blooms in the Great Lakes.[158]

The Bush EO also directed the Committee to review any ocean-related policy action and provide recommendations upon request from any federal department or agency.[159] The review process included providing general information, facilitating dissemination and exchange of information, "develop[ing] and implement[ing] common principles for the conduct of governmental activities," promoting regionalized practices, and using science to drive policy decisions.[160]

Past Efforts to Pass a NOAA Organic Act

The Bush 43 Administration also recognized the importance of enacting an organic act to codify NOAA's structure and responsibilities under the Department of Commerce.[161] The Administration drafted a NOAA organic act pursuant to COP Recommendation 7-1,[162] and the bill was introduced during the 108th Congress.[163] Subcommittee hearings were held, but no further action was taken.[164] The Action Plan outlined the Administration's intentions to pass the organic act in the 109th Congress,[165] and two bills were introduced in 2005.[166] The second bill was passed by the House and referred to the Senate, but no further actions were taken.[167] An organic act for NOAA still has yet to gain any real traction in Congress.[168]

The National Ocean Policy under President Obama

On July 22, 2010, President Obama signed an executive order implementing his own National Ocean Policy, establishing a brand new entity: the National Ocean Council (Council).[169] The Council largely consisted of the same governing membership as the Committee established under the Bush EO.[170] Unlike the Bush Committee, however, the Obama EO endowed the Council with a full-time staff, including a director and deputy director who were directed to assist the co-chairs with implementing the National Ocean Policy.[171] At the close of the Obama Administration, the Council had established twenty committees, subcommittees, and working groups relating to various marine management initiatives.[172]

Among its broader policy goals, the Obama EO largely adopted many of the Pew Report's recommendations relating to ecosystem-based management, encouraging coastal resiliency, setting up regional management councils, and managing to protect biodiversity.[173] President Obama's National Ocean Policy was largely considered a success because of its significant progress made in federal ocean management.[174] Specifically, new management measures included the implementation of bottom-up participation through marine spatial planning; instituting adaptive, ecosystem-based management; and a "re-commitment to strengthening the value of our oceans."[175]

Advent of Marine Spatial Planning

The "most advanced supported and discussed" part of the National Ocean Policy under President Obama was the implementation of marine spatial planning in the United States.[176] Marine spatial planning is "a process for analyzing and allocating the spatial and temporal distribution of human activities in coastal and marine areas to achieve ecological, economic, and social objectives."[177] Most importantly, it is a planning tool that uses the best available science and sets up the ability to form Regional Planning Bodies, so that federal actions can be tailored to local needs where appropriate.[178] Critics liken marine spatial planning to "ocean zoning," and bemoan the new regional institutions as creating additional bureaucracy.[179] Yet this criticism is misplaced, at least regarding the planning structure implemented under President Obama. The Obama EO simply directed existing federal, state, local, and tribal authorities to work together and with other stakeholders to develop their own regional frameworks.[180] The Obama Administration thus defined marine spatial planning as a "public policy process," incorporating adaptive management and "ecosystem-based" approaches "to reduce conflicts among uses, reduce environmental impacts, facilitate compatible uses, and preserve critical ecosystem services."[181]

Grading the Obama National Ocean Policy

Shortly after the Obama EO was adopted, an organization called the Joint Ocean Commission (JOC) issued "Report Cards" evaluating the United States' progress in marine governance following the implementation of President Obama's National Ocean Policy.[182] The JOC gave the Obama EO a "C" in national support and leadership for insufficient communication networks, citing concerns about the Council's adopted "precautionary approach," which critics feared would lead to the marginalization of some industries.[183]

The Obama EO received its highest grade, an "A-," for its "regional, state, and local leadership and implementation efforts.[184] The JOC cited the progress made through the Regional Planning Bodies, which yielded increased local knowledge and better management of marine resources.[185] The JOC gave President Obama's National Ocean Policy a "C" on "research, science, and education" because of funding and program cuts.[186] The JOC did note, however, that the establishment of a new publicly accessible data portal kept the grade from being even lower.[187] The lowest grade was for funding, which received a "D-."[188] According to the JOC, even during the Obama Administration, federal marine programs were "chronically underfunded."[189]

The National Ocean Policy under President Trump

The Trump EO explicitly revoked the Obama EO and reinstated the committee structure, signaling a return to aspects of the Bush EO.[190] Under the

Trump EO, the Committee's governing membership was roughly the same as the Obama Council and Bush Committee, with the notable inclusion of the Commandant of the Coast Guard, who had not been explicitly included in the Bush and Obama EOs.[191] However, the director and permanent, full-time staff established under President Obama were no longer provided for under the Trump EO, reflecting the Trump Administration's view that the Council structure was "overly bureaucratic."[192]

Improving Transparency & Public Access to Data

One of the few features of the Trump EO that received praise was the order's emphasis on improving access to data.[193] The Obama Administration also received praise for its efforts to make government ocean data more accessible,[194] although the Obama EO did not explicitly identify this as an objective. The Trump National Ocean Policy, which has not been revoked by President Biden as of June 2022, specifically lists "improved public access to marine data and information" as one of its principal policy objectives.[195] Indeed, the Trump EO directed the Committee to "coordinate the timely public release of unclassified data and other information related to the ocean, coasts, and Great Lakes that agencies collect, and support the common information management systems ... that organize and disseminate this information."[196]

Energy Dominance: How "America First" Interacts with the National Ocean Policy

Despite the ostensible benefits of better data and transparency, critics noted that a focus on data and science, without also including stewardship principles, suggests that the true intent behind the Trump EO's expanded data collection and availability was to increase resource development at the expense of conservation and ecological considerations.[197] Such concerns are bolstered by considering the Trump EO in the context of the Trump Administration's broader "America First" and "Energy Dominance" initiatives.[198]

When examining other provisions of the Trump EO itself, the emphasis on energy development becomes more apparent. For instance, within its "Purpose" section, the Trump EO opined that "[d]omestic energy production from Federal waters strengthens the Nation's security and reduces reliance on imported energy."[199] One of the policy directives was for the federal government to "facilitate the economic growth of coastal communities and promote ocean industries, which ... enhance America's energy security."[200] Further, the EO directed agencies to "modernize the acquisition, distribution, and use of the best available ocean-related science and knowledge, in partnership with marine industries . . . to inform decisions and enhance entrepreneurial opportunities."[201] Both of these directives, without clarifying sustainability or conservation language, intimate an emphasis on energy development—without several of the structural and planning limitations imposed by previous National Ocean Policies.

Similar themes can be found in other Trump executive orders, including the order that reopened areas of the outer continental shelf that were previously withdrawn from oil and gas development.[202] For instance, the "Findings" section of the Offshore Energy EO stated:

> America must put the energy needs of American families and businesses first and continue implementing a plan that ensures energy security and economic vitality for decades to come. The energy and minerals produced from lands and waters under Federal management are important to a vibrant economy and to our national security.[203]

Reading this language together with the broader Trump EO objectives promoting domestic energy development and other extractive industries, it becomes clear that the Trump Administration was intending to shift the marine resource paradigm from balancing development with sustainable and responsible use to prioritizing shortsightedness and aggressive development.[204]

The Offshore Energy EO also directed the Secretary of the Interior to "develop and implement, in coordination with the Secretary of Commerce and to the maximum extent permitted by law, a streamlined permitting approach for privately funded seismic data research and collection aimed at expeditiously determining the offshore energy resource potential."[205] This language suggests that the Trump Administration's intentions regarding enhanced data availability in the Trump EO may have been ultimately aimed at streamlining the offshore energy development process.

Although a portion of the Offshore Drilling EO, which attempted to revoke President Obama's drilling withdrawals, was recently invalidated for violating the OCSLA in *League of Conservation Voters v. Trump* in March 2019,[206] the DOI took other actions to implement the expanded resource development pursuant to the Offshore Drilling EO.[207] For example, the DOI evaluated permit applications to use seismic surveys to locate oil and gas deposits below the Atlantic seabed.[208] Critics and biologists pointed out that the seismic surveys use earsplitting sounds that can injure whales.[209]

Yet, the permitting process went forward despite the fact that development of the five-year DOI leasing plan was postponed while the Trump Administration appealed the *League of Conservation Voters* decision.[210] Further, President Trump issued a memorandum on November 2, 2019, directing the Committee to "coordinate the development of a national strategy for mapping, exploring, and characterizing the U.S. EEZ, and for enhancing opportunities for collaboration ... with respect to those activities."[211] The Administration posited that the strategy "is necessary for a systematic and efficient approach to understanding our resources," potentially identifying "new sources of critical minerals, biopharmaceuticals, energy, and other resources."[212] These developments further support the claim that energy interests were the intended beneficiaries of the Trump EO's data accessibility provisions.

The Trump Administration's energy-first philosophy was further evidenced by BSEE's 2019 proposed rule amendment[213] to revise regulations relating to well control and blowout preventer systems, seeking to roll back the changes that were adopted by the Obama Administration in the wake of the *Deepwater Horizon* spill.[214] While BSEE asserted in the proposed rule that the changes would ensure that "safety and environmental protection" were maintained, the primary purpose of changes was to identify "provisions that can be amended, revised, or removed to reduce significant burdens on oil and natural gas operators" and "fortify the [Trump] Administration's position towards facilitating energy dominance leading to increased domestic oil and gas production."[215] Indeed, BSEE highlighted how the rule was estimated to save the industry over $900 million over the next ten years by removing numerous former regulatory obligations.[216] For instance, the final rule removed a requirement that safety and pollution prevention technology be inspected by a BSEE-approved third party and reinstated a pre-*Deepwater Horizon* practice that merely encourages operators to adhere to industry-recommended guidelines.[217] In short, many of the new rules governing blowout prevention systems became "filled with 'should' instead of 'must.'"[218]

Thus, throughout his term, President Trump and his Administration made their intentions toward the ocean clear: the replacement of a measured, conservative approach to marine energy development with full-speed-ahead "energy dominance."[219]

Planning and Management Deficiencies

As noted, one of the touchstones of the Obama EO was mandatory representation and consultation with local and tribal governments.[220] The

Trump EO removed this requirement, instead making these intergovernmental relations optional by providing that the national policy was to "facilitate, *as appropriate*, coordination, consultation, and collaboration regarding ocean-related matters . . . among Federal, State, tribal, and local governments, marine industries, the ocean science and technology community, other ocean stakeholders, and foreign governments and international organizations."[221]

The Trump EO also removed the requirements for agencies to engage in marine spatial planning, instead directing the Committee to "engage and collaborate, under existing laws and regulations, with stakeholders, including regional ocean partnerships, to address ocean-related matters that may require interagency or intergovernmental solutions."[222] While this provision technically allowed the Regional Planning Bodies established during the Obama Administration to continue operating, the Trump National Ocean Policy no longer provided support for these institutions and rendered the two regional plans approved by the Obama Administration no longer controlling.[223]

Subsequent guidance from the Trump Administration expressed a preference for regional collaboration to be initiated by states and characterized the previous Regional Planning Bodies as "duplicative, Federally-driven," and "unnecessary."[224] Federal agencies were permitted to continue executing the regional plans under the Trump EO, but any continued implementation had to be consistent with the new state-led partnerships.[225]

The decision to maintain an emphasis on regional planning was an important retention of a positive development from the Obama EO, placing all of the emphasis on the states established a policy with negative consequences. First, it placed the onus on the states to organize new partnerships and invite agencies and other stakeholders to participate.[226] The Trump EO also removed influence from local and tribal governments by not mandating their representation in the process, focusing instead entirely on state governments.[227] The level of indigenous representation on the Regional Planning Bodies when they were meeting regularly was "unprecedented," and tribal representatives were justifiably concerned that they would be left without a seat at the table under the Trump National Ocean Policy.[228]

Ocean Policy Developments under President Biden

As this volume goes to press, President Biden has yet to issue his own National Ocean Policy, and the Trump EO remains on the books.[229] But even without an overarching policy, the Biden Administration has made several policy decisions during its first two years that indicate a significant change in direction from the prior administration. On his inauguration day, President Biden issued an executive order addressing multiple environmental issues, including within its provisions a repeal of President Trump's Offshore Energy EO and concomitant reinstatement of the Obama-era offshore drilling withdrawals in the Arctic Ocean and Bering Sea.[230]

A few days later, he issued another sweeping climate-related executive order (Biden Climate EO) that affects United States ocean governance in several ways.[231] First, the Biden Climate EO creates a National Climate Task Force, which includes the Secretary of Commerce among its membership, and directs the task force to facilitate federal actions that will, in conjunction with state, local, and tribal governments, lead to ocean and biodiversity conservation.[232] The Biden Climate EO directs the Secretary of the Interior, in consultation with the NOAA Administrator, to facilitate an increase in renewable energy production in offshore waters "with the goal of doubling offshore wind by 2030."[233] It also institutes a "pause" on new federal oil and gas leases in offshore waters.[234] Finally, the Biden Climate EO directs the Secretary of the Interior, again in consultation with the Secretary of Commerce, to "recommend[] steps that the United States should take, working with State, local, Tribal, and territorial governments, . . . fishermen, and other key stakeholders, to achieve the goal of conserving at least 30 percent of our lands and waters by 2030"[235]—a goal colloquially referred to as President Biden's "30x30" initiative.[236]

Under President Biden, June has been declared National Ocean Month by presidential proclamations issued in 2021 and 2022.[237] The 2021 proclamation recognizes that the ocean is "critical to the success of our Nation and, indeed, to life on Earth" because it "powers our economy, provides food for billions of people, supplies 50 percent of the world's oxygen, offers recreational opportunities for us to enjoy, and regulates weather patterns and our global climate system."[238] It states that the Biden Administration is "dedicated to improving our Nation's public health by supporting resilient ocean habitats, wildlife, and resources" through implementing the 30x30 initiative, expanding the United States' offshore wind capability, and addressing climate change.[239] The 2022 Ocean Month Proclamation not only reinforces many of the same themes, but also reflects some other important commitments, such as "recogniz[ing] and elevat[ing] Indigenous and local knowledge" to work toward equity and climate justice.[240] The 2022 Proclamation also highlights ocean-related accomplishments made possible by the 2021 Bipartisan Infrastructure Law, such as improved coastal mapping and forecasting.[241]

On June 8, 2022 (World Ocean Day), President Biden further committed to several more actions to protect the ocean, including: the establishment of a new national marine sanctuary to protect the Hudson Canyon in the Atlantic Ocean; reducing the use of single-use plastics on lands managed by the Department of the Interior; nation-to-nation stewardship with resident tribes of the North Bering Sea Climate Resilience Area; the drafting of a whole-of-government National Ocean Climate Plan; and strengthening the United States' leadership in the International Clean Seas Campaign.[242] In addition, Biden Administration officials took an important step by convening the first meeting of President Biden's National Ocean Committee in October 2021. Committee membership largely mirrors that of the department and agency officials comprising the Committee under the Trump EO. One notable early development in the October 2021 meeting was initiating the development of the nascent Ocean-Climate Action Plan and ways in which agencies can focus their "actions on ocean-based mitigation of climate change effects."[243]

While all of these steps by the current administration are certainly encouraging—indeed the National Ocean Month proclamations adopt language intimating a rhetorical revaluation of the ocean—they stop short of the sort of permanent, tangible, and comprehensive changes needed to truly revalue the ocean from a policy standpoint.

Recommendations

The Problem with National Policymaking by Executive Order

Considering the strategic, economic, social, cultural, and environmental importance of the ocean, it is curious that the National Ocean Policies have been adopted via executive order rather than through omnibus legislation. However, executive orders have been used to implement central administrative decisions since the advent of the presidency and have evolved to take on an ever-expanding role in federal law.[244]

Generally, executive orders serve one of two purposes: as written statements of presidential decisions, or a means by which the president issues directives to executive branch personnel (including most federal agencies).[245] The power to issue executive orders derives from the Constitution or a statutory delegation from Congress.[246] This authority has historically been almost uncontroverted; between 1789 and 2010, courts overturned only two executive orders.[247] Indeed, executive orders often become "legislative" in character and effect, because they can fill in the gaps left by Congress when addressing policy problems.[248]

Given the importance of executive orders in modern federal law and agency governance, it may actually be fitting that the National Ocean Policies have been adopted through executive orders thus far. Comprehensive management of the oceans likely fits within the defined parameters of executive power, particularly because the general guidance for the policies can be traced to congressional goals set forth in the Oceans Act of 2000.[249] Furthermore, given the historically disjointed nature of ocean management in the United States, with responsibilities spread across twenty agencies,[250] the chief executive is particularly well equipped to develop and implement new governance strategies.

On the other hand, executive orders receive mixed reviews as far as solutions to policy problems because they do not provide the same degree of permanence as legislation. Although executive orders have the force of law, they are "unilateral in nature" and do not require any additional review before being implemented by executive agencies.[251] As a result, new policies that live by the executive order often die by executive order,[252] meaning that existing executive orders are just as seamlessly repealed when a new administration steps in. Both in general, and with marine-related issues in particular, executive orders that are supported by explicit congressional authorization or codification are more enduring.[253]

For example, prior executive orders placing a moratorium on offshore drilling were subsequently lifted with the stroke of a pen during both the Bush 43 and Trump Administrations.[254] Congressional offshore drilling moratoria, on the other hand, are far more difficult to revoke.[255] Even though congressional moratoria typically include sunset clauses, or may have to be renewed during the appropriations process, a codifying act of Congress imparts both structural legitimacy and stability.[256] For instance, the Trump Administration attempted to open up large portions of the ocean that were previously withdrawn from oil and gas leasing by revoking an Obama executive order.[257] Although the ability to completely lift a previous president's withdrawal rather than tweaking it slightly is in serious doubt following the holding in *League of Conservation Voters v. Trump*,[258] Trump's revocation demonstrates the ease with which broad national policies adopted via executive order can be quickly and radically changed. [259] Contrast the lack of legal permanence for offshore drilling moratoria with the opening of the Arctic National Wildlife Refuge to oil and gas development, which has been an ongoing controversy for decades.[260] Although Congress ultimately opened the coastal plain of the Refuge to drilling in 2017,[261] it took thirty years to lift the ban from when drilling in the Refuge was first proposed because of public pressure and the relative inefficiency of the legislative process.[262]

Additionally, executive orders without statutory or constitutional authorization may have no legal effect if challenged in court.[263] Thus far, there have not been legal challenges to any of the National Ocean Policy executive orders, presumably because the Bush EO was adopted pursuant to congressional direction in the Oceans Act. However, certain stakeholders raised the threat of potential lawsuits in the wake of both the Obama and Trump EOs.[264] Although the specific grounds upon which such suits would be brought is unclear, opponents of the Obama National Ocean Policy in a 2017 Senate hearing cited its "questionable statutory authority" and "conflicts with existing laws" as creating risks for litigation.[265]

Codifying a National Ocean Policy

Given the numerous issues with national policymaking via executive order, the time is ripe for Congress to step in and adopt a truly comprehensive National Ocean Policy.[266] In addition to longevity concerns, codifying the National Ocean Policy would help secure permanent funding for NOAA and other federal marine programs, which remain drastically underfunded.[267] Codifying a National Ocean Policy also presents the opportunity to go beyond the Obama EO and include important common law principles such as the public trust doctrine[268] and the relevant statutory provisions from the CZMA, NEPA, and other federal environmental laws that apply to the ocean.[269] All of these legal changes would play a critical role in re-envisioning the United States' relationship with the ocean to meet twenty-first-century challenges.

While pundits have noted that an omnibus National Ocean Policy law may not be practically possible given the prevailing political winds of the day,[270] one such bill was advanced in June 2018 in the wake of the Trump EO.[271] The National Ocean Policy Act, sponsored and introduced by Congressman Jimmy Panetta (D-Calif.), sought simply to reinstate and codify provisions of the Obama EO, giving the Obama National Ocean Policy "the full force and effect of law."[272] Congressman Panetta is the son of Leon Panetta—former congressman, Secretary of Defense, and the Chairman of the Pew Oceans Commission[273]—and has committed to carrying on his father's legacy by promoting the sustainable management of ocean and coastal resources.[274] Congressman Panetta's bill also would have explicitly given the Trump EO "no force or effect."[275] Although the bill did not make it out of subcommittee, it evinces at least some support among members of Congress to adopt a National Ocean Policy by statute.

NOAA Needs an Organic Act

Although an omnibus marine management law is almost certainly out of reach in the current political climate, enacting an organic act for NOAA is a more feasible legislative goal. It would also provide a critical first step toward adopting more comprehensive statutory reforms in the future. While the COP and the Pew Commission advocated for slightly different structural reforms to NOAA's fragmentation and institutional limitations, both of their reports emphasized the general need for a legislative solution.[276] The COP Report called for an organic act that would strengthen NOAA and expand its responsibilities, as well as budgetary increases commensurate with the agency's critical role.[277] The COP's ideal organic act would also legislatively establish NOAA as the lead ocean federal agency and reorganize it to improve collaboration with other federal agencies.[278] The Pew Commission, on the other hand, recommended that Congress establish a completely new and independent agency and consolidate under one roof "as many federal ocean programs as is practical."[279]

Forming a brand "new independent ocean agency may be an unattainable goal due to lack of political will."[280] An organic act for NOAA, on the other hand, is a more realistic goal, given that the agency is already "the nation's lead civilian ocean agency."[281] An organic act would solidify this role.[282] There have been at least twelve attempts to reorganize NOAA under a new department or to restructure it as an independent agency, but none have been successful thus far.[283] As noted, only one of these bills has passed in either house of Congress—H.R. 5450 in 2006—but it was met with inaction in the Senate.[284]

NOAA's structural weaknesses have been attributed to its lacking "cabinet status, independence, a congressional charter, and control over many federal maritime activities."[285] The agency currently operates pursuant to approximately two hundred distinct legislative authorities.[286] As a result of these challenges, NOAA's leadership is consistently unable to "establish and implement cross-cutting priorities."[287] Agency personnel have noted that an organic act would enable them to more effectively manage agency operations and improve their understanding of its mission and role in the federal government.[288]

A recent report laid out the coalition-building that would be needed to secure passage of a NOAA organic act: it "would require a multi-year effort by agency leaders, executive and legislative branch advocates and outside influencers."[289] Given the recent increase in severity of natural disasters and the cumulative impacts on ocean and coastal resources, NOAA's programs are increasingly attracting the attention of federal decision makers.[290] For instance, Congress adopted the Weather Research and Forecasting Innovation Act of 2017 to address the insufficiency of the United States' weather and climactic modeling.[291] Framing this need for reform as a national security and economic issue may generate additional supporters as well.[292]

Although challenging, these conditions are not insurmountable. If the movement is "driven by a small, close-knit cadre of committed leaders and influencers compelled by a common purpose and able to develop a mutual understanding and trust among all stakeholders," enactment could be achieved after two to three years of coalition-building within the executive and legislative branches.[293]

Reinstituting Marine Spatial Planning

The benefits of marine spatial planning are numerous: instituting ecosystem-based management, reducing cumulative impacts by viewing regional uses systematically, reducing conflicts among competing stakeholders, and increasing certainty by adopting regional plans.[294] As noted, when the Obama Administration implemented marine spatial planning, the institutions received recognition internationally and regionally for the voice they gave to local stakeholders.[295] While it remains to be seen what the new national Ocean Climate Action Plan, announced June 2022, will entail, the Biden Administration has not committed to comprehensive marine spatial planning as this book goes to press. As such, the Biden Administration should strongly consider revoking the Trump EO in its entirety and reinstituting the marine spatial planning framework established under the Obama EO. At the very least, the Biden Administration could also amend the Trump EO to reinstitute a watered-down form of marine spatial planning. Although the role of states is paramount in the Trump EO as written, local governments, tribal governments, and nongovernmental stakeholders have an equally important role to play and must be given a seat at the regional planning table. Thus, marine spatial planning has a critical role to play in helping humanity to re-envision and re-evaluate its complex relationship with the oceans. Any future omnibus ocean legislation that might be proposed must therefore include robust marine spatial planning provisions.

Conclusion

Re-envisioning humankind's relationship with the ocean to meet twenty-first-century challenges begins with re-envisioning existing management and policy frameworks. Oceanic governance in the United States has always been and remains fragmented. A National Ocean Policy attempts to solve the United States' marine management fragmentation by instituting a comprehensive ocean and coastal governance framework. Although this panacea was conceptualized as early as 1969, the first National Ocean Policy was not instituted until President George W. Bush adopted it via executive order in 2004. Despite the last three presidents' efforts to streamline management by instituting their own national policy visions and integrated planning frameworks, a permanent and comprehensive solution remains elusive. The three National Ocean Policies promulgated to date have been inadequate for several reasons, but principally because they were adopted as executive orders. Executive orders are an inauspicious source of law for instituting comprehensive, lasting reform over a resource as dynamic and important as the ocean. They lack judicial enforceability and are easily—and often drastically—changed from one president to the next.

In order to effectively and comprehensively manage the United States' marine resources for the twenty-first century, Congress should enact a National Ocean Policy Act. Ideally, the statute would incorporate the best aspects of each of the National Ocean Policy executive orders promulgated to date, as well as the Pew and COP Reports' key recommendations. Such a statute should, as a start, institute a mandatory marine spatial planning framework, include organic provisions to clarify and bolster NOAA's role as the primary agency to administer federal marine programs, and identify permanent and stable funding sources for the agency.

Considering the contemporary political climate in the United States, the outlook is bleak for enacting an omnibus marine management statute in the near future, particularly given conservative policymakers' and the oil and gas industry's dismissive opinions of marine spatial planning.[296] However, enacting an organic act for NOAA is significantly more attainable[297]—although admittedly still difficult. The Biden Administration should direct Commerce Secretary Gina Raimondo to start by building a coalition of NOAA personnel, federal legislators, and other stakeholders that NOAA frequently works with, such as fishermen, researchers, meteorologists, and state and local officials from coastal areas. Given the multitude of statutory responsibilities placed on NOAA and the dramatic challenges the agency and associated stakeholders will face in an increasingly volatile climate, an untapped opportunity exists to build support for a NOAA organic act.[298] The Biden Administration should begin expeditiously laying the groundwork for developing, drafting, and eventually passing an organic act for NOAA. Such an accomplishment would represent a generational governance shift and serve as an important first step in reforming how the United States values the ocean.

Notes

1. Ocean Policy to Advance the Economic, Security, and Environmental Interests of the United States, Exec. Order No. 13,840, 83 Fed. Reg. 29,431 (June 19, 2018) (hereinafter Trump EO).
2. The White House, "President Donald J. Trump Is Promoting America's Ocean Economy," June 19, 2018, https://www.whitehouse.gov/briefings-statements/president-donald-j-trump-promoting-americas-ocean-economy/.
3. See Stewardship of the Ocean, Our Coasts, and the Great Lakes, Exec. Order No. 13,547, 75 Fed. Reg. 43,021 (July 19, 2010) (hereinafter Obama EO).
4. See Committee on Ocean Policy, Exec. Order No. 13,366, 69 Fed. Reg. 76,589 (Dec. 17, 2004) (hereinafter Bush EO).
5. U.S. Commission on Ocean Policy, *An Ocean Blueprint for the 21st Century*, July 2004, https://govinfo.library.unt.edu/oceancommission/documents/full_color_rpt/000_ocean_full_report.pdf/ (hereinafter *COP Report*).
6. Pew Oceans Commission, *America's Living Oceans: Charting a Course for Sea Change*, May 2003, https://www.pewtrusts.org/en/research-and-analysis/reports/2003/06/02/americas-living-oceans-charting-a-course-for-sea-change/ (hereinafter *Pew Report*).
7. Pub. L. No. 106-256, 114 Stat. 644 (2000).
8. See discussion in Section II.B, *infra*.
9. See discussion in Section II.B, *infra*.
10. See discussion in Section III.A–III.C, *infra*.
11. See discussion Section III.D–III.F, *infra*.
12. Trump EO, § 2(g).
13. Trump EO, § 5(c).
14. EELP Staff, "National Ocean Policy Executive Order," *Harvard Law School Environmental & Energy Law Program*, September 20, 2018, https://eelp.law.harvard.edu/2018/09/national-ocean-policy-executive-order/.

15. Trump EO, § 2(a).
16. *COP Report*, 49.
17. Bowen L. Florsheim, "Territorial Seas—3000-Year-Old Question," *Journal of Air Law and Commerce* 36 (1970): 74.
18. Florsheim, "Territorial Seas," 79.
19. Ibid. A nautical mile is equal to one minute of latitude, or approximately 1.15 miles. "What Is the Difference Between a Nautical Mile and a Knot?," NOAA, https://oceanservice.noaa.gov/facts/nauticalmile_knot.html/.
20. *COP Report*, 49.
21. Ibid.
22. Ibid. For instance, in 1945, President Harry Truman proclaimed jurisdiction over natural resources located under the continental shelf under the high seas adjoining the United States' territorial seas.
23. See United Nations Convention on the Territorial Sea and the Contiguous Zone, Apr. 29, 1958, 516 U.N.T.S. 7477; United Nations Convention on the High Seas, Apr. 29, 1958, 450 U.N.T.S. 6465; United Nations Convention on Fishing and Conservation of the Living Resources of the High Seas, Apr. 29, 1958, 559 U.N.T.S. 8164; United Nations Convention on the Continental Shelf, Apr. 29, 1958, 499 U.N.T.S. 7302.
24. *COP Report*, 72–73.
25. Ibid., 73.
26. Ibid.
27. See United Nations Convention on the Law of the Sea, Dec. 10, 1982, 1833 U.N.T.S. 3 (hereinafter UNCLOS).
28. *COP Report*, 444.
29. Ibid.
30. Ibid.
31. UNCLOS, art. 5, 1833 U.N.T.S. 27. In the United States, the baseline has been further defined in the courts for international and domestic purposes. See discussion in Section II.A.2. *infra*.
32. UNCLOS, art. 3–4, 1833 U.N.T.S. 27.
33. UNCLOS, art. 2, 1833 U.N.T.S. 27.
34. UNCLOS, art. 17, 1833 U.N.T.S. 30.
35. UNCLOS, art. 33, 1833 U.N.T.S. 35.
36. UNCLOS, art. 33, 1833 U.N.T.S. 35.
37. UNCLOS, art. 57, 1833 U.N.T.S. 44.
38. UNCLOS, art. 56, 1833 U.N.T.S. 43.
39. UNCLOS, art. 56, 1833 U.N.T.S. 43.
40. UNCLOS, art. 77, 1833 U.N.T.S. 54.
41. UNCLOS, art. 76, 1833 U.N.T.S. 53–54.
42. *COP Report*, 73. This was one of the principal reasons cited by the USCOP for recommending that the United States ratify UNCLOS. *COP Report*, 444–45 (Recommendation 29-1). The Pew Oceans Commission also recommended acceding to UNCLOS. See *Pew Report*, 80–81.

43. *COP Report*, 49. The territorial sea extends to nine nautical miles for Texas, the west coast of Florida, and Puerto Rico. *COP Report*, 71.
44. Submerged Lands Act of 1953, Pub. L. No. 83-31, 67 Stat. 29 (codified at 43 U.S.C. §§ 1301–1315 [2019]). The Submerged Lands Act was passed in response to the Supreme Court's decision in *United States v. California*, 332 U.S. 804 (1947), which held that the federal government had jurisdiction over all ocean resources from the high tidemark seaward. 332 U.S. at 805.
45. Submerged Lands Act of 1953 § 3. See *COP Report*, 71.
46. Coastal Zone Management Act of 1972, Pub. L. No. 92-583, 86 Stat. 1280 (codified at 16 U.S.C. §§ 1451–1465 [2019]).
47. Coastal Zone Management Act § 304(1).
48. *COP Report*, 51.
49. *Pew Report*, 27.
50. Ibid.
51. "Coastal Zone Management Programs," NOAA, https://coast.noaa.gov/czm/mystate/. Alaska withdrew from the program in 2011. See ibid. for additional information.
52. Or nine miles from the coasts of the jurisdictions listed in note 43, *supra*.
53. See Territorial Sea of the United States of America, Proclamation No. 5928, 54 Fed. Reg. 777 (Dec. 27, 1988).
54. See Contiguous Zone of the United States, Proclamation No. 7219, 64 Fed. Reg. 48,701 (Sept. 8, 1999).
55. See Exclusive Economic Zone of the United States of America, Proclamation No. 5030, 48 Fed. Reg. 10,601 (Mar. 10, 1983).
56. See Policy of the United States with Respect to the Natural Resources of the Subsoil and Sea Bed of the Continental Shelf, Proclamation No. 2667, 10 Fed. Reg. 12,305 (Sept. 28, 1945).
57. *COP Report*, 26.
58. Ibid.
59. National Environmental Policy Act of 1969, Pub. L. No. 91-190, 83 Stat. 852 (codified at 42 U.S.C. §§ 4321–4370 [2019]) (*hereinafter* NEPA).
60. EELP Staff, "NEPA Environmental Review Requirements," *Harvard Law School Energy & Environmental Law Program*, August 15, 2018, https://eelp.law.harvard.edu/2018/08/nepa-environmental-review-requirements/. See NEPA § 102(2)(C).
61. The application of NEPA to the entire EEZ has been upheld by United States District Courts. See, e.g., National Resources Defense Council v. U.S. Department of the Navy, CV-01-07781, 2002 U.S. Dist. LEXIS 26360, at *41 (C.D. Cal. 2002) ("Because the United States exercises substantial legislative control of the EEZ in the area of the environment stemming from its 'sovereign rights' for the purpose of conserving and managing natural resources . . . NEPA applies to federal actions which may affect the environment in the EEZ").
62. 42 U.S.C. § 4344.

63. 42 U.S.C. § 4344(4).
64. *COP Report*, 77.
65. Ibid.
66. Exec. Order No. 11,564, 35 Fed. Reg. 15,801 (Oct. 8, 1970).
67. Eileen L. Shea, *A History of NOAA*, ed. Skip Theberge (National Oceanic and Atmospheric Administration, 1999), 3, https://www.history.noaa.gov/legacy/noaahistory_1.html/. See Section III.A., *infra*, for background and discussion of the Stratton Commission.
68. Shea, *A History of NOAA*, 3.
69. Ibid.
70. Ibid.
71. *COP Report*, 51. See additional discussion of the CZMA in Section II.A.2, *supra*.
72. 16 U.S.C. § 1456-1.
73. Marine Protection, Research and Sanctuaries Act of 1972, Pub. L. No. 92-532, 86 Stat. 1052 (codified as amended at 16 U.S.C. §§ 1431–47f; 33 U.S.C. §§ 1401–45, 2801–05 [2019]).
74. *COP Report*, D12.
75. See 33 U.S.C. § 1412.
76. See 33 U.S.C. § 1413.
77. Federal Water Pollution Control Act Amendments of 1972, Pub. L. No. 92-500, 86 Stat. 816 (codified as amended at 33 U.S.C. 1251–1387 [2019]).
78. See 33 U.S.C. § 1311.
79. Marine Mammal Protection Act of 1972, Pub. L. No. 92-522, 86 Stat. 1027 (codified at 16 U.S.C. §§ 1361–62, 1371–89, 1401–07, 1411–18, 1421–21h, 1423–23h [2019]).
80. 16 U.S.C. § 1371.
81. 16 U.S.C. § 1362(13).
82. See 16 U.S.C. §§ 1371–89.
83. *Pew Report*, 27.
84. Endangered Species Act of 1973, Pub. L. No. 93-205, 87 Stat. 884 (codified at 16 U.S.C. 1531–1544 [2019]).
85. *Pew Report*, 27.
86. 16 U.S.C. § 1532(19).
87. *COP Report*, 309. The U.S. Fish & Wildlife Service is responsible for listing land-based species. Ibid.
88. Fishery Conservation and Management Act of 1976, Pub. L. No. 94-265, 90 Stat. 331 (codified at 16 U.S.C. §§ 1801–85 [2019]).
89. *COP Report*, 52.
90. Ibid., 52–53.
91. Ibid., 52.
92. Ibid.
93. See 16 U.S.C. § 1852.
94. *COP Report*, 52.

95. Shea, *A History of NOAA*, 13.
96. Ibid.
97. Ibid.
98. Ibid.
99. Ibid.
100. Ibid.
101. Outer Continental Shelf Leasing Act Amendments of 1978, Pub. L. No. 95-372, 92 Stat. 629 (codified at 43 U.S.C. §§ 1331–43 [2019]).
102. *COP Report*, 52.
103. Outer Continental Shelf Lands Act of 1953, Pub. L. No. 83-212, 67 Stat. 842 (codified as amended at 43 U.S.C. §§ 1331–1356b [2019]).
104. "OCS Lands Act History," Bureau of Ocean Energy Management, https://www.boem.gov/OCS-Lands-Act-History/.
105. *COP Report*, 52.
106. Ibid. See 43 U.S.C. §§ 1345–1348.
107. BOEM, "OCS Lands Act History."
108. Ibid.
109. *COP Report*, 52.
110. See Section II.B., *supra*.
111. Marine Resources and Engineering Development Act of 1966, Pub. L. No. 89-454, § 5, 80 Stat. 203, 205–7.
112. The name was derived from the chair of the Commission, Julius Adams Stratton. Angela T. Howe, "The U.S. National Ocean Policy: One Small Step for National Waters, But Will It Be the Giant Leap Needed for Our Blue Planet?," *Ocean and Coastal Law Journal* 17 (2011): 73.
113. Marine Resources and Engineering Development Act of 1966, § 5(b).
114. Howe, "The U.S. National Ocean Policy," 74.
115. *COP Report*, 50.
116. Shea, *A History of NOAA*, 3.
117. Ibid. See discussion regarding the history of NOAA in Section II.B., *supra*.
118. *COP Report*, at 51.
119. See *supra* Section II.B.
120. Helen V. Smith, "A Summary Analysis of the U.S. Commission on Ocean Policy's Recommendations for a Revised Federal Ocean Policy, and the Bush Administration's Response," *Southeastern Environmental Law Journal* 14 (2005): 134.
121. Oceans Act of 2000, Pub. L. No. 106-256, § 2, 114 Stat. 644 (2000).
122. Howe, "The U.S. National Ocean Policy," 75. See Oceans Act of 2000 § 3.
123. Oceans Act of 2000 § 2 ("The purpose of this Act is to establish a commission to make recommendations . . . that will promote . . . enhancement of partnerships with state and local governments with respect to . . . the management of ocean and coastal resources").
124. See, e.g., Oceans Act of 2000 §2(d).

125. Oceans Act of 2000 § 2(4) (emphasis added).
126. Oceans Act of 2000 §§ 2(6), 2(5).
127. Oceans Act of 2000 §§ 2(2), 2(3).
128. Oceans Act of 2000 § 2(7).
129. Oceans Act of 2000 § 2(7).
130. *U.S. Ocean Action Plan: The Bush Administration's Response to the U.S. Commission on Ocean Policy*, December 2004, https://data.nodc.noaa.gov/coris/library/NOAA/other/us_ocean_action_plan_2004.pdf/ (hereinafter *Bush Action Plan*).
131. See *COP Report*, xiii.
132. Ibid., vii.
133. Ibid., vii–xii.
134. Howe, "The U.S. National Ocean Policy," 75.
135. *Pew Report*, iii–iv, ix.
136. Ibid., ix.
137. Ibid.
138. Ibid.
139. Ibid., vii.
140. Ibid., ix–x.
141. Ibid., x.
142. Ibid.
143. Ibid., x–xi.
144. Ibid., 34.
145. Ibid., 30–32.
146. Ibid., 32.
147. Ibid.
148. See *Bush Action Plan*.
149. Ibid., 4.
150. Bush EO § 3(b)(i).
151. Bush EO § 3(b)(ii)–(iii).
152. Bush EO § 3(b)(iv)–(v).
153. Bush EO § 1(a) (emphasis added).
154. Bush EO § 1(b).
155. Bush EO § 4(b).
156. Bush EO § 4(b).
157. *Bush Action Plan*, 3.
158. Ibid., 10, 16.
159. Bush EO § 4(c).
160. Bush EO § 4(d)(i)–(iv).
161. *Bush Action Plan*, 6.
162. See *COP Report*, 111.
163. *Bush Action Plan*, 6. See National Oceanic and Atmospheric Administration Organic Act of 2004, H.R. 4607, 108th Cong. (2004).

164. Tim Hall and Mary Kicza, *An Organic Act for NOAA to Formalize Its Purpose and Authorities*, Aerospace, Spring 2018, 5.
165. *Bush Action Plan*, 6.
166. Hall and Kicza, *An Organic Act for NOAA*, 5. See National Oceanic and Atmospheric Act, H.R. 50, 109th Cong. (2005); National Oceanic and Atmospheric Act, H.R. 5450, 109th Cong. (2006).
167. Hall and Kicza, *An Organic Act for NOAA*, 5.
168. Howe, "The U.S. National Ocean Policy," 95.
169. Obama EO § 1.
170. See Obama EO § 4.
171. Obama EO § 4(g).
172. The White House, "President Donald J. Trump Is Promoting America's Ocean Economy."
173. Obama EO § 2.
174. Howe, "The U.S. National Ocean Policy," 102.
175. Ibid.
176. Ibid., 83.
177. Ibid.
178. "Marine Planning," *The Obama White House*, https://obamawhitehouse.archives.gov/administration/eop/oceans/marine-planning/.
179. Emily Migliaccio, "The National Ocean Policy: Can It Reduce Marine Pollution and Streamline Our Ocean Bureaucracy?," *Vermont Journal of Environmental Law* 15 (2014): 649.
180. Ibid., 650. See Obama EO § 3(b).
181. Obama EO § 3(b).
182. Migliaccio, "National Ocean Policy," 651 (citing "Joint Ocean Commission Initiative," Meridian Institute, www.merid.org/en/Content/Projects/Joint_Ocean_Commission_Initiative.aspx?view=news).
183. Ibid., 651–52.
184. Ibid., 653.
185. Ibid.
186. Ibid.
187. Ibid.
188. Ibid., 654.
189. Ibid.
190. Trump EO §§ 4, 7.
191. Trump EO § 4(a)(i)(2). Cf. Obama EO § 4(b)(ii); Bush EO § 4(b)(ii).
192. See The White House, "President Donald J. Trump Is Promoting America's Ocean Economy."
193. See, e.g., Amy Trice, "President Trump Rescinds the National Ocean Policy," *Ocean Conservancy: Ocean Currents Blog*, June 19, 2018, https://oceanconservancy.org/blog/2018/06/19/president-trump-rescinds-national-ocean-policy/ (noting that the "new policy . . . provide[s] ocean management

tools that have the potential to build on the work of previous administrations to improve ocean data and reduce conflicts").

194. See Section III.E.3, *supra*.
195. Trump EO § 1.
196. Trump EO § 5(c).
197. Sarah Winter Whelan, "So, Trump Repealed the National Ocean Policy and Replaced It with His Own Policy on the Ocean. Now What?," *Healthy Oceans Blog*, July 25, 2018, https://healthyoceanscoalition.org/blog/so-trump-repealed-the-national-ocean-policy-and-replaced-it-with-his-own-policy-on-the-ocean-now-what/.
198. Whelan, "So, Trump Repealed the National Ocean Policy."
199. Trump EO § 1.
200. Trump EO § 2(d).
201. Trump EO § 2(f).
202. See Implementing an America-First Offshore Energy Strategy, Exec. Order No. 13,795, 82 Fed. Reg. 20,815 (Apr. 28, 2017) (hereinafter Offshore Energy EO). Submerged lands in the Arctic and Atlantic Oceans were previously withdrawn from oil and gas development by President Obama in 2015 and 2016. See Memorandum on Withdrawal of Certain Areas of the United States Outer Continental Shelf Offshore Alaska from Leasing Disposition, DCPD201500059 (Jan. 27, 2015); Exec. Order 13,754, 81 Fed. Reg. 90,669, § 3 (Dec. 9, 2016); Memorandum on Withdrawal of Certain Portions of the United States Arctic Outer Continental Shelf from Mineral Leasing, DCPD201600860 (Dec. 20, 2016); Memorandum on Withdrawal of Certain Areas off the Atlantic Coast on the Outer Continental Shelf from Mineral Leasing, DCPD201600861 (Dec. 20, 2016). President Biden revoked the Offshore Energy EO by executive order in January 2021. See Protecting Public Health and the Environment and Restoring Science to Tackle the Climate Crisis, Exec. Order No. 13,990, 86 Fed. Reg. 7037, § 7(a) (Jan. 20, 2021).
203. Offshore Energy EO § 1.
204. David Malakoff, "Trump's New Ocean Policy Washes Away Obama's Emphasis on Conservation and Climate," *Science*, June 19, 2018, https://www.sciencemag.org/news/2018/06/trump-s-new-oceans-policy-washes-away-obama-s-emphasis-conservation-and-climate/.
205. Offshore Energy EO § 3(c).
206. League of Conservation Voters v. Trump, 363 F. Supp. 3d 1013, 1020–1030 (D. Alaska 2019).
207. Darryl Fears, "Trump Rolls Back Safety Rules Meant to Prevent Another *Deepwater Horizon* Spill," *Washington Post*, May 2, 2019, https://www.washingtonpost.com/climate-environment/2019/05/02/trump-rolls-back-safety-rules-meant-prevent-another-deepwater-horizon-spill/?utm_term=.6b98a4ff9960/.

208. Fears, "Trump Rolls Back Safety Rules." See also Darryl Fears, "Sound and Fury: Trump Administration Pushes Forward on Seismic Mapping in the Atlantic," *Washington Post*, May 1, 2019, https://www.washingtonpost.com/climate-environment/2019/05/01/sound-fury-trump-administration-pushes-forward-seismic-mapping-atlantic/?utm_term=.b1056ec3fa4d/.
209. Fears, "Trump Rolls Back Safety Rules."
210. Fears, "Sound and Fury." Ultimately, the trial court decision in this case was vacated as moot on appeal because by that point President Biden had signed an executive order rescinding President Trump's Offshore Energy EO and reinstating President Obama's withdrawals. See League of Conservation Voters v. Biden, 843 Fed. App'x 937, 938–39 (9th Cir. 2021).
211. Memorandum of November 19, 2019: Ocean Mapping of the United States Exclusive Economic Zone and the Shoreline and Nearshore of Alaska, 84 Fed. Reg. 64,699 (Nov. 22, 2019).
212. Memorandum of November 19, 2019, 84 Fed. Reg. at 64,700.
213. Oil and Gas and Sulfur Operations in the Outer Continental Shelf—Blowout Preventer Systems and Well Control Revisions, 84 Fed. Reg. 21,908 (May 15, 2019) (codified at 30 CFR pt. 250).
214. Fears, "Trump Rolls Back Safety Rules."
215. Oil and Gas and Sulfur Operations in the Outer Continental Shelf—Blowout Preventer Systems and Well Control Revisions, 83 Fed. Reg. 22,128–29 (proposed May 11, 2018) (to be codified at 30 CFR pt. 250).
216. 83 Fed. Reg. at 22,144.
217. Fears, "Trump Rolls Back Safety Rules." See 84 Fed. Reg. at 21,910.
218. Fears, "Trump Rolls Back Safety Rules" (quoting Nancy Leveson).
219. Coral Davenport, "Trump's Order to Open Arctic Waters to Oil Drilling was Unlawful, Federal Judge Finds," *New York Times*, Mar. 30, 2019, https://www.nytimes.com/2019/03/30/climate/trump-oil-drilling-arctic.html?module=inline/.
220. See Section III.E.3, *supra*.
221. Trump EO § 2(g) (emphasis added).
222. Trump EO § 5(c).
223. EELP Staff, "NEPA Environmental Review Requirements."
224. The White House, "President Donald J. Trump Is Promoting America's Ocean Economy."
225. EELP Staff, "NEPA Environmental Review Requirements."
226. Maya Wei-Haas, "Trump Just Remade Ocean Policy—Here's What That Means," *National Geographic*, July 13, 2018, https://www.nationalgeographic.com/environment/2018/07/news-ocean-policy-indigenous-sustainability-fisheries-industry-economy-marine/.
227. Ibid.
228. Ibid.

229. See "2018 Donald Trump Executive Orders," *Federal Register*, 2021, https://www.federalregister.gov/presidential-documents/executive-orders/donald-trump/2018/ (collating executive orders issued in 2018 and noting whether they have been amended or revoked).
230. Protecting Public Health and the Environment and Restoring Science to Tackle the Climate Crisis, Exec. Order No. 13,990, 86 Fed. Reg. 7037, §§ 4(b), 7(a) (Jan. 20, 2021).
231. See Tackling the Climate Crisis at Home and Abroad, Exec. Order No. 14,008, 86 Fed. Reg. 7619 (Jan. 27, 2021) (hereinafter Biden Climate EO).
232. Biden Climate EO § 203.
233. Biden Climate EO § 207.
234. Biden Climate EO § 208.
235. Biden Climate EO § 216.
236. Rick Steiner, "Scientists Urge Biden to Go Big on Ocean Protection," *Seattle Times*, Jan. 14, 2021, https://www.seattletimes.com/opinion/scientists-urge-biden-to-go-big-on-ocean-protection/.
237. National Ocean Month, Proclamation No. 10,266, 86 Fed. Reg. 30,143 (June 7, 2021) (hereinafter 2021 Ocean Month Proclamation). National Ocean Month, Proclamation No. 10,413, 87 Fed. Reg. 33,613 (June 3, 2022) (hereinafter 2022 Ocean Month Proclamation).
238. 2021 Ocean Month Proclamation.
239. Ibid.
240. 2022 Ocean Month Proclamation.
241. Ibid.
242. Fact Sheet: Biden-Harris Administration Celebrates World Ocean Day with Actions to Conserve America's Deepest Atlantic Canyon, Cut Plastic Pollution, and Create America's First-Ever Ocean Climate Action Plan," *The White House* (June 8, 2022), https://www.whitehouse.gov/briefing-room/statements-releases/2022/06/08/fact-sheet-biden-harris-administration-celebrates-world-ocean-day-with-actions-to-conserve-americas-deepest-atlantic-canyon-cut-plastic-pollution-and-create-americas-first-ever-o/.
243. The White House, "Readout of the First Ocean Policy Committee Meeting," Oct. 29, 2021, https://www.whitehouse.gov/ostp/news-updates/2021/10/29/readout-of-the-first-ocean-policy-committee-meeting/.
244. John C. Duncan Jr., "A Critical Consideration of Executive Orders: Glimmerings of Autopoiesis in the Executive Role," *Vermont Law Review* 35 (2010): 333, 338–39.
245. Ibid., 335.
246. Ibid., 337.
247. Ibid.
248. Ibid., 410.
249. See discussion in Section III.B, *supra*.
250. *COP Report*, 55.

251. Duncan, "A Critical Consideration of Executive Orders," 335.
252. Or, as often put by critics of presidential policymaking via executive order, "If you live by the pen, you die by the pen." See, e.g., James Hohman, "If you live by the pen, you die by the pen," *The Washington Post*, June 10, 2015, https://www.washingtonpost.com/news/powerpost/wp/2015/06/10/if-you-live-by-the-pen-you-die-by-the-pen/?noredirect=on&utm_term=.48b8731ae8e8/.
253. Howe, "The U.S. National Ocean Policy," 80.
254. Ibid., 81–82. However, the Trump Administration's action revoking the previous administration's withdrawal was judicially overturned. See discussion in Section III.F.3, *supra*.
255. Howe, "The U.S. National Ocean Policy," 81.
256. Ibid., 81–82.
257. Exec. Order No. 13,795, 82 Fed. Reg. 20,815 (Apr. 28, 2017).
258. 363 F. Supp. 3d at 1030. See discussion in Section III.F.3, *supra*.
259. Nathan Rott and Merrit Kennedy, "Trump Signs Executive Order on Offshore Drilling and Marine Sanctuaries," *National Public Radio*, Apr. 27, 2017, https://www.npr.org/sections/thetwo-way/2017/04/27/525959808/trump-to-sign-executive-order-on-offshore-drilling-and-marine-sanctuaries/.
260. Steve Mufson, "Trump Administration Takes Another Step Toward Oil Drilling in Arctic National Wildlife Refuge," *Washington Post*, Dec. 20, 2018, https://www.washingtonpost.com/national/health-science/trump-administration-takes-another-step-toward-oil-drilling-in-arctic-national-wildlife-refuge/2018/12/20/5fb93f40-0469-11e9-b5df-5d3874f1ac36_story.html?utm_term=.0d0c2922209e/.
261. The drilling ban was lifted by a rider attached to the Tax Cuts and Jobs Act. See Tax Cuts and Jobs Act of 2017, Pub. L. No. 115-97, § 20001–03, 131 Stat. 2054, 2235–38.
262. Brad Plumer and Henry Fountain, "Trump Administration Finalizes Plan to Open Arctic Refuge to Drilling," *New York Times*, Aug. 17, 2020, https://www.nytimes.com/2020/08/17/climate/alaska-oil-drilling-anwr.html/.
263. Duncan, "A Critical Consideration of Executive Orders," 348.
264. See, e.g., Rep. Richard Norman Hastings, "Obama's National Ocean Policy Threatens Jobs and Economic Activities Onshore and Off," *Fox News*, June 19, 2012, https://www.foxnews.com/opinion/obamas-national-ocean-policy-threatens-jobs-and-economic-activities-onshore-and-off/ (opining that the Obama EO's "vague and undefined objectives" would result in "costly frivolous lawsuits"); Karina Brown, "Environmentalists Sue EPA for Pollution that Acidifies Ocean," *Courthouse News Service*, Nov. 27, 2018, https://www.courthousenews.com/environmentalists-sue-epa-for-pollution-that-acidifies-ocean/ (reporting that Trump revocation of Obama EO could be fodder for potential future lawsuits).
265. National Ocean Policy Coalition, "National Ocean Policy Coalition U.S. Senate Hearing Recap," *Saving Seafood*, Dec. 14, 2017, https://www

.savingseafood.org/news/council-actions/national-ocean-policy-coalition-u-s-senate-hearing-recap/.

266. Howe, "The U.S. National Ocean Policy," 95.
267. Ibid., 95–96. See Section III.E.3, *supra*.
268. The public trust doctrine "protects public rights in communal water-related resources, including public use of the shore and the bed of waterways, the rights of fishing and public access, and perhaps even the right to the water itself for certain communal uses." Mark Squillace, "Restoring the Public Interest in Western Water Law," *Utah Law Review* (2020): 644–45.
269. Howe, "The U.S. National Ocean Policy," 95. See Section II.B., *supra*.
270. Ibid., 80.
271. Congressman Jimmy Panetta, "Congressman Panetta Introduces National Ocean Policy Act to Protect Ocean Resilience and Economy," June 29, 2018, https://panetta.house.gov/media/press-releases/congressman-panetta-introduces-national-ocean-policy-act-protect-ocean/.
272. National Ocean Policy Act of 2018, H.R. 6300, 115th Cong. § 2(a) (2018).
273. See *Pew Report*, i, iii.
274. Jeremy P. Jacobs, "Campaign 2016: Famous Name Has Two Legacies to Live Up To," *E&E News*, June 6, 2016, https://www.eenews.net/stories/1060038320/.
275. H.R. 6300 §2 (b).
276. Katherine J. Mengerink, "The Pew Oceans Commission Report: Navigating a Route to Sustainable Seas," *Ecology Law Quarterly* 31 (2004): 710–11. See discussions of the two Commissions' recommendations relating to NOAA in Section II.C.1 and Section II.C.2, *supra*.
277. *COP Report*, 112–13.
278. Ibid.
279. *Pew Report*, 34.
280. Mengerink, "The Pew Oceans Commission Report," 711.
281. *COP Report*, 110.
282. Ibid.
283. Mengerink, "The Pew Oceans Commission Report," *supra* note 251, at 711.
284. Hall and Kicza, *An Organic Act for NOAA*, 2.
285. *COP Report*, 51.
286. Hall and Kicza, *An Organic Act for NOAA*, 1.
287. Ibid.
288. Ibid.
289. Ibid.
290. Ibid., 2.
291. Ibid.
292. Ibid.
293. Ibid., 3.
294. Howe, "The U.S. National Ocean Policy," 83.

295. See discussion in Section III.E.3, *infra.*
296. See, e.g., The White House, "President Donald J. Trump Is Promoting America's Ocean Economy" (describing the regional planning bodies as "duplicative" and "unnecessary").
297. See Section IV.B, *infra.*
298. See Hall and Kicza, *An Organic Act for NOAA*, 4.

CHAPTER 10

Rights of Nature

The Answer to Our Oceanic Issues?

ABIGAIL BENESH

THE LEGAL SYSTEM of the United States discriminates against nature. Artificial "persons" such as ships, corporations, partnerships, and trusts are permitted to bring suit on their own behalf. Nature, however, cannot. For example, the United States Court of Appeals for the Ninth Circuit explored the question of whether the Endangered Species Act permits animals to protect themselves in court. Ultimately, the Ninth Circuit determined that "[t]he statute is set up to authorize 'persons' to sue to protect animals whenever those animals are 'endangered' or 'threatened.' Animals are not authorized to sue in their own names to protect themselves."[1] Why are living, breathing creatures denied the same rights that are given to lifeless entities, such as corporations? The answer lies in U.S. culture and what we as humans value.

The law is inherently anthropogenic. It views humans as the only subjects or legal persons. This leaves animals and the environment unprotected under our current system. These subjects may be unlawfully possessed or even destroyed by humans—the sole legal person—but they have no means of becoming whole again. It is the owner of the damaged land that can receive legal benefit, not the land itself. If an entity, such as a polluted river, has no legal owner and thus no human has a specific interest in that river, the waterway has no independent right to continue to exist, let alone thrive. The anthropocentric approach to law seems unavoidable in most contexts, but it is leaving large gaps in the legal protection of the environment and natural systems.

While rights of nature begin to take off on land, a huge hole is left in environmental protection. We tend to focus on protecting land resources, and even cleaning up the ocean, but very rarely emphasize the need to protect the ocean as a whole. The ocean covers 72 percent of our planet and supplies over half our oxygen,[2] and it is in trouble. In order to best defend the ocean, the world desperately needs to make a shift from anthropocentric views, legislation, policies, and regulations to a systematic approach to ocean restoration and management through a "rights of nature" approach. Rights of nature essentially grant different entities—the ocean, for example—the same legal rights that are typically assigned to humans or that corporations enjoy, including legal standing. This chapter looks at current threats to the ocean, like climate change and pollution, and examines the different policies in place to protect it, describes rights-of-nature models in different countries, explores the possibility of a rights-of-nature approach being a viable approach for ocean protection in the United States, and ultimately concludes that marine protected areas located in the United States' exclusive economic zone (EEZ) are an appropriate entity to receive legal rights in order to best protect the ocean and its resources and restore it back to optimal health.

Why We Need a New Vision

Current Threats

The ocean is one of the world's most valuable resources, if not the most valuable. It controls the weather, cleans our air, provides for recreation, and is home to the most biodiversity on the planet.[3] Even so, the ocean is suffering momentously from the effects of human life. The main issues plaguing the ocean today include overfishing, pollution, invasive species, and climate change.[4] Although there are currently laws in place to protect against these problems, they are inflexible and do not cover the broad range of issues our changing ocean is currently facing.

Present-Day Solutions

There are currently several laws and regulations in place to protect the ocean. The main issue with these rules is that they are completely human-centric and do not consider the ocean as an independent entity that deserves to exist in peace in order to heal and flourish.

For example, in 1972, Congress determined that "certain species and population stocks of marine mammals are, or may be, in danger of extinction or depletion as a result of man's activity" and enacted the Marine Mammal Protection Act.[5] Although Congress admitted that it is humanity's touch that is harming marine mammals, it reasoned that because marine mammals have proven themselves to be valuable for the economy, they are entitled to protection.[6] The Marine Mammal Protection Act is an example of human-centric legislation because it affords protections to the ocean purely for the benefit of man. It fails to recognize the importance of marine mammals separate from the economy.

That same year, Congress enacted the Coastal Zone Management Act (CZMA) to address the need to regulate development in the coastal zone.[7] Congress found a "national interest in the effective management, beneficial use, protection, and development of the coastal zone."[8] This language suggests that Congress is encouraging the country to prioritize protection of the coastal zone for purposes of future development. Similar to the Marine Mammal Protection Act, the CZMA has human-centric language throughout. For example, it notes that:

> Because of their proximity to and reliance upon the ocean and its resources, the coastal states have substantial and significant interests in the protection, management, and development of the resources of the exclusive economic zone that can only be served by the active participation of coastal states in all Federal programs affecting such resources and, wherever appropriate, by the development of state ocean resource plans as part of their federally approved coastal zone management programs.[9]

This language signifies that purely because coastal states have an interest in preserving coastal resources, they should actively participate in these types of federal programs.

The National Marine Sanctuaries Act (NMSA) became effective in 1972. The Secretary of Commerce is in charge of designating and managing certain areas of the marine environment that he or she feels are significant and require federal management. Unfortunately, the NMSA does not prohibit any use within the designated sanctuaries; rather, it is up to the secretary to designate, through public process, what activities will be permissible.[10] The Secretary of Commerce is likely to encourage recreation, development, and other activities that will bring in revenue for the area and support the economy, rather than supporting preservation of ecosystem. The NMSA is flawed for that very reason and is inherently human-centric.

Enacted in 1976, the Magnuson-Stevens Act (MSA) serves as the chief federal law governing marine fishery management in the United States. The MSA seeks to promote long-term "biological and economic sustainability" of the country's fisheries.[11] In enacting the MSA, Congress found that:

> Commercial and recreational fishing constitutes a major source of employment and contributes significantly to the economy of the Nation. Many coastal areas are dependent upon fishing and related activities, and their economies have been badly damaged by the over-fishing of fishery resources at an ever-increasing rate over the past decade. The activities of massive foreign fishing fleets in waters adjacent to such coastal areas have contributed to such damage, interfered with domestic fishing efforts, and caused destruction of the fishing gear of United States fishermen.[12]

This finding demonstrates that Congress's concern about maintaining fisheries focuses on job creation and the economy, not on marine biodiversity per se. There is no mention of the devastating effects that overfishing has on marine ecosystems. Congress is mainly concerned about how fisheries benefit humans.

The anthropocentric perspective also pervades international law. In 1999, the International Union for the Conservation of Nature (IUCN) released guidelines for marine protected areas. The guidelines' purpose is to help establish marine protected areas in an effort to better manage coastal marine areas.[13] According to the IUCN, marine protected areas are key to marine sustainability because they "maintain essential ecological processes and life support systems, preserve genetic diversity, and ensure the sustainable utilization of species and ecosystems."[14] The issue with this language, however, is that the framework supports and conserves ecological systems for the benefit of humans.[15]

The principal and purpose that animates the vast majority of these laws is economic sustainability and human benefit. The way we currently view the ocean as property and a resource creates a huge barrier in our ability to protect the ocean. What we need is a new perspective, a shift in the way we regard the ocean. We need to stop asking "how can the ocean better serve us" and start asking "how can we allow the ocean to heal itself." This shift honors the ocean as a life-giving entity and allows humans to develop a new sense of responsibility and respect for this force.

What Are Rights of Nature?

The policies currently in place are inflexible and do not sufficiently protect the ocean from present-day stressors. As a result, we need a new approach to ocean security and restoration. One such approach is to assign legal personhood to a particular natural feature found in the ocean, as has been done for other natural resources in Ecuador, Bolivia, Australia, New Zealand, and India. This approach is commonly known as the rights-of-nature movement, which this part explores in more depth.

Rights of Nature

Several countries have successfully assigned legal personality to different entities in nature, such as rivers, producing a suite of rights-of-nature models. Each country has a unique approach to granting this legal personhood. Some countries have done so through legislation, some through the courts, and others have created a provision in the national constitution. Nevertheless, regardless of what entity in nature these rights belong to and how they came about, these countries have one thing in common: they are developing a progressive approach to protecting and restoring natural resources.

The ocean has an inherent right to thrive. The assignment of legal rights is important because policy is traditionally centered around how the ocean can better serve us. Once policies are put into place that protect the ocean as an entity and not just as a common resource, we will develop a new sense of responsibility and accountability for maintaining a healthy ocean.

Basic Definition

Rights of nature embodies the idea that nature does not exist for humans; rather humans exist because of nature. Things like trees and rivers receive legal personhood, meaning that they have the right to bring suit in court as an injured party in addition to having a basic right to flourish and experience restoration.[16] The legal personality assigned to nature has three main elements: the right to sue and be sued (legal standing); the right to enter and enforce legal contracts; and the right to own property.[17]

In 1972, Christopher Stone wrote a radical article, "Should Trees Have Standing?," that laid a foundation for the rights of nature.[18] Stone acknowledged how shocking it is to suggest a previously right-less entity receive rights—but that was also true for women and African Americans.[19] History demonstrates that humans are always changing and evolving and constantly shifting their values.

It was the U.S. Supreme Court's decision in *Sierra Club v. Morton* that truly gave birth to "citizen-driven environmental enforcement" efforts, and Justice William O. Douglas's dissent specifically brought national attention to the rights-of-nature approach.[20] Essentially, the Sierra Club lacked standing, but the court effectively gave permission to adversely affected individuals to bring suit.[21] Justice Douglas, fueled by Stone's argument, said in his dissent:

> Inanimate objects are sometimes parties in litigation. A ship has a legal personality, a fiction found useful for maritime purposes. The corporation sole—a creature of ecclesiastical law—is an acceptable adversary, and large fortunes ride on its cases. The ordinary corporation is a "person" for purposes of the adjudicatory processes, whether it represents proprietary, spiritual, aesthetic, or charitable causes. So it should be as respects valleys, alpine meadows, rivers, lakes, estuaries, beaches, ridges, groves of trees, swampland, or even air that feels the destructive pressures of modern technology and modern life.[22]

While Justice Douglas's view did not prevail in the United States, rights of nature are now flourishing in other countries and becoming a very real approach to environmental conservation.

Examples from Other Countries

In slightly different ways, several governments have adopted rights of nature. These countries include Ecuador, Bolivia, Australia, New Zealand, Mexico, India, and Colombia, as well as individual cities within the United States. This section will compare and contrast the way in which a country has adopted rights of nature into its nation's policy and discuss the advantages and disadvantages of each method.

Ecuador

In 2008, Ecuador became the first nation to incorporate rights of nature into the country's constitution.[23] The constitution states that Pachamama, or nature, has the "right to exist and to maintain and regenerate its vital cycles, structure, functions, and evolutionary processes," as well as the right to restoration in the case of ecological destruction.[24] The constitution goes on to say:

> Nature, or Pacha Mama, where life is reproduced and occurs, has the right to integral respect for its existence and for the maintenance and regeneration of its life cycles, structure, functions and evolutionary processes. All persons, communities, peoples and nations can call upon public authorities to enforce the right of nature. To enforce and interpret these rights, the principles set forth in the Constitution shall be observed, as appropriate.[25]

The country rejoiced in this "historic victory" in late September 2008.

However, enforcement has proven to be quite an obstacle for Ecuador. The president is reluctant to implement the rights-of-nature amendments. Additionally, plaintiffs who bring suit under the new amendments are facing major obstacles, such as lack of standing and judicial corruption.

It goes without saying that, as a trailblazer in this area, Ecuador will run into issues with the implementation and enforcement of the rights of nature amendments. Nevertheless, Ecuador's constitution is among the most progressive in the world because it has officially made the leap from anthropocentric environmental regulations to an ecocentric approach.[26]

Bolivia

In 2010, Bolivia passed the Universal Declaration of the Rights of Mother Earth. The declaration is part of a reform of the country's legal system subsequent to a constitutional change in 2009. The change is greatly influenced by indigenous peoples.[27] The purpose of the Law of Mother Earth is to recognize the "rights of Mother Earth, and the obligations and duties of the Multinational State and society to ensure respect of these rights."[28] The law focuses on six main guiding principles: harmony; the collective good; a guarantee of the regeneration of Mother Earth; respect and defense of the rights of Mother Earth; no commercialism; and multiculturalism.[29] To support the new law, Bolivia created two new institutions: the Mother Earth Ombudsman's office and the Plurinational Mother Earth Authority.[30]

Like Ecuador, Bolivia is also running into several issues regarding the enforcement and implementation of its rights-of-nature legislation. For example, some believe that the legislation does not adequately set quantifiable targets, making it difficult to assess the success of implementation.[31] On the other hand, officials have noted that environmental awareness amongst constituents has increased since the law was passed, as made evident by the efforts of authorities to facilitate water-saving habits among the population.[32] The Bolivian government launched the "My Water Programme" in 2011 in an effort to preserve water and is also targeting sources of pollution and cleaning these areas up. Additionally, a bill is being pushed to reforest areas previously used for mining in order to improve air quality. These actions demonstrate a shift in the mindset of government officials, a shift towards environmental activism.

Australia

In the state of Victoria, Australia, the Victorian Environmental Water Reserve (EWR) controls all water assigned for environmental uses. The EWR ensures that the necessary water flow is available to support rivers, estuaries, and wetlands throughout Victoria. Until 2010, ownership of water for environmental purposes belonged to the Minister for Environment. In 2010, however, the Victorian Environmental Water Holder (VEWH) became the owner of water for environmental purposes. The VEWH received legal personhood and was given the authority to hold water rights, to make decisions about water allocation, and to buy and sell water. The VEWH is a statutory corporation, although it is not subject to Australia's governing corporation law; rather, it is a public entity. The VEWH has a committee consisting of three commissioners and a group of public service employees, and a levy placed on all Victorian water users provides sufficient funding. Basically, the VEWH acts as a guardian for water sources in Victoria.

As a result of the VEWH's creation, the rivers of Victoria possess the three key elements of legal personality: legal standing, the right to enter contracts, and the right to own property. While Victoria law does not explicitly delegate rights to the rivers, the purpose of the VEWH was to create a single voice independent from the government for decisions regarding Victoria's water for environmental uses.

The creation of the VEWH and its legal personhood did not intend to "revolutionize water resources management, but rather to provide a much-needed circuit-breaker in the political arguments about how much water should be recovered for the environment." The VEWH essentially serves as a hybrid between environmental values and market participation by acting as both a buyer and seller in the water market.[33]

New Zealand

In 2017, the Whanganui River in New Zealand received legal personhood through the Te Awa Tupua Act (referred to hereafter as the Act).[34] The local Whanganui Maori pushed for the passage of the Act when they protested the ownership of the riverbeds by the New Zealand government and management by local authorities. The Act recognizes the river and its catchment as an independent legal entity. Te Awa Tupua is the personified river and the Te Pou Tupua serves as its guardian. Te Pou Tupua consists of two people—the government assigns one and the Whanganui Iwi the other—and they work together as a unit to "act and speak on behalf of the river's health and well-being."

The Te Pou Tupua, although vested with the power to represent the river, has two groups that serve as advisories. The Te Karewao and the Te Kopuka na Te Awa Tupua. These groups will consist of a larger number of people from a range of backgrounds and thus will represent a number of different interests. Funding for this regulation comes from payments from the government.

Similar to the framework of the rights of nature in Australia, New Zealand gives the Whanganui River legal standing, the right to enter contracts, and the right to own property. However, rights of nature in New Zealand essentially take the shape of a legal person with appointed guardians. Although the entity itself is separate from the government, the funding comes from the Crown, which poses a potential risk for too much government involvement. Fortunately, however, the Te Awa Tupua Act provides for decentralized decision-making by creating rules at the local level.[35]

United States

A handful of states and cities in the United States have adopted rights of nature. For example, Santa Monica, California, passed the Sustainability Rights Ordinance in 2013 that establishes rights of nature for "natural communities in Santa Monica." An eight-mile easement of land received legal personhood in Hawaii in 2017. The owner of the easement, Joan Porter, noted that "I established the easement in hopes that other landowners and governments will also understand the need to change the status of nature from property to bearing rights."[36] The effectiveness of smaller-scaled rights-of-nature movements around the United States indicates the possibility of the success of larger action.

India

Compared to the models discussed above, India has a unique approach to rights of nature. In March 2017, the High Court of Uttarakhand declared rights of nature for the Ganges and Yamuna Rivers. Specifically, the court stated that "the Rivers Ganga and Yamuna, all their tributaries, streams, every natural water flowing with flow continuously or intermittently of these rivers, are declared as juristic/legal persons/living entities having the status of a legal person with all corresponding rights, duties and liabilities of a living person." The basis for the decision was the court's determination that the rivers are important for religious purposes and are essential to the life of India's citizens, present and future. India assigned guardians by deeming the rivers minors under the law. Several individuals are in charge of the management of the rivers, including the Director NAMAMI Ganges, the Chief Secretary of the State, and the Advocate General of the State.

This setup has already proven problematic for India. In July 2017, the Supreme Court heard an appeal regarding the framework on the basis that the guardians' responsibilities were ambiguous because the rivers in question extend beyond the borders of Uttarakhand.[37] This appeal froze the implementation of the rights of nature.

Colombia

Since 2017, the Atrato River in Colombia has been a legal person. Similar to India, it was the courts that made the decision to grant the river legal personhood. The High Court judgment reads:

> [I]t is the human populations that are interdependent of the natural world—and not the opposite—and that they must assume the consequences of their actions and omissions with the nature. It is a question of understanding this new sociopolitical reality with the aim of achieving a respectful transformation with the natural world and its environment, as has happened before with civil and political rights. . . . Now is the time to begin taking the first steps to effectively protect the planet and its resources before it is too late."

Further, the court made the decision to assign a partnership to the river as guardians: a member of the government and an indigenous person from the affected community.[38] This duo creates a balance between power and authority and local concerns and allows both voices to be heard.

Dealing with the Arguments Against Rights of Nature

Much like any new idea, there is initial pushback to the rights-of-nature movement. One argument is that humans cannot possibly know what nature is "saying" or what it really needs and therefore assigning a guardian to represent a voiceless entity—a tree, for example—isn't a feasible option for protection. While it is true that humans cannot possibly know exactly what is best for nature, the needs of nature are simple: to exist and continue to exist and restore itself when it is possible. A second debate is how rights of nature can possibly exist if humans are giving nature these rights: does that not make rights of nature anthropocentric just like the regulations this movement is trying to fix? This argument is unsuccessful, however, because we must realize that humans are powerful creatures that have the ability to write laws that will influence (and require) other humans to follow those laws. Realistically, humans must initially create these rights for nature. However, once nature receives these rights, they will belong exclusively to nature and no human intervention will be permissible, with the exception of the times the entity's guardian will need to step in to provide a voice for the otherwise voiceless.[39]

Moving Rights of Nature to the Ocean

When deciding the best method of assigning rights to the ocean, there are several things we must consider. For example, to what do we assign legal personality and what will those rights look like? This section explores both of these questions in depth.

To What Do We Assign Legal Personality?

There are three viable options for granting legal personality to the ocean: rights could be granted to a species, an ecosystem, or an already protected area. This section looks at how legal personhood would work for each potential candidate.

A Species

One idea is to give legal rights to threatened oceanic species. Each nation would establish legal standing for that particular species, say, a type of shark. A shark would be very specific and easily identifiable, and trackable. However, the species-specific approach is probably insufficient. For example, the Galapagos Islands have recently expressed interest in establishing additional protections for endangered sharks in the area. The islands of Darwin and Wolf have the highest concentration of sharks in the world. To make these conservation goals a reality, the Galapagos Islands has opted to create several no-take zones within marine reserves, thus creating a shark sanctuary. It is these marine reserves that received legal personality in order to better safeguard the sharks.

The Galapagos example is instructive. First, although the country was seeking to protect a specific species, it did not assign any protections to the sharks themselves. Instead, it created a protected area for the sharks. Second, policymakers believed that establishing no-take zones within a marine protected area was the best option to reach their objective of shark conservation. This decision suggests that it may be simplest to use an area that is preestablished and just make small changes to the rules that govern that area. Third, policymakers did not stop at just the creation of no-take zones; they are working to assign legal personality to this area, as well.[40] This example shows that although no-take zones within marine protected areas are the best option, there is more we can do to protect the ocean and the life within it. It also suggests that assigning legal personality to a single species is, in fact, too small an idea and would not be a sufficient option to maximize oceanic protection and restoration.

An Ecosystem

A second way to create rights of nature in the ocean is by establishing legal rights for a specific ecosystem, for instance, a river like New Zealand and India have done. Similar to a species, an ecosystem would be capable of possessing legal standing as well as the right to enforce contracts and own property. Giving an entire ecosystem legal personality affords protection to a greater number of organisms than just one species, and thus benefits a larger part of the ocean.

Nevertheless, ecosystems pose some practical problems. Boundary lines need to be established around previously unprotected areas of the ocean. Establishing boundaries can be time-consuming, costly, and confusing for people who had previously used the area for fishing, recreation, or other activities. In addition, marine ecosystems overlap, making clear definition of this legal person difficult. Though the concept of protecting a whole ecosystem seems like the most effective option, the logistics will prove difficult.

Giving legal personality to an entire ecosystem is a better idea than giving those rights to a specific species. The ecosystem will include these species that would otherwise be the only organism receiving protection. Things get complicated, however, when deciding where to draw boundary lines around the ecosystem we wish to give legal rights to. The ideal situation would be to protect an entire ecosystem that is already protected so as to work with pre-existing boundary lines and areas that humans already recognize as special and protected.

A Previously Protected Area

Marine protected areas (MPAs) are some of the most well established forms of ecosystem protection in the ocean. MPAs essentially limit human activity in a designated area. The United States defines these areas as "any area of the marine environment that has been reserved by federal, state, tribal, territorial, or local laws or regulations to provide lasting protection for part or all of the natural and cultural resources therein."[41] President Bill Clinton issued an executive order in 2000 that included the following provision:

> This Executive Order will help protect the significant natural and cultural resources within the marine environment for the benefit of present and future generations by strengthening and expanding the Nation's system of marine protected areas (MPAs). An expanded and strengthened comprehensive system of marine protected areas throughout the marine environment would enhance the conservation of our Nation's natural and cultural marine heritage and the ecologically and economically sustainable use of the marine environment for future generations.[42]

Perhaps the most attractive reason for assigning legal rights to MPAs rather than to a species or an ecosystem is that boundary lines already exist. If we were to assign legal personality to an ecosystem, for example, experts would need to conduct extensive research and studies to determine what part of the ecosystem is in danger and to what extent and then determine appropriate boundary lines. The advantage to using a previously protected area like an MPA is that people are aware that the area is already protected as well as cutting out the long process of research. Establishing rights of nature for MPAs is the simplest and most efficient way to protect the ocean.

The Oceanic Rights of Nature Model

Like Pachamama in Ecuador's constitution, MPAs in the United States' EEZ should have "the right to exist and to maintain and regenerate [their] vital cycles, structure, functions, and evolutionary processes," as well as the right to restoration in the case of ecological destruction.[43] Even so, multiple different characteristics are vital for the success of rights of nature. These characteristics include legal standing, the right to enter contracts, the right to own property, a specific legal form, the explicit creation of legal rights, giving rights of nature to a particular entity in nature, a specific aim of creation, a certain method of creation, assigned legal representatives, independence from the government, and financial support.[44] The following sections will discuss the structural aspects of creating legal rights for MPAs.

Legal Standing

Christopher Stone believed that the park at issue in *Sierra Club v. Morton* was a legal person for the purpose of establishing legal standing. According to Stone, the first prong of Article III's standing requirement is met because the park was suffering a direct, concrete, nonhypothetical personal injury.[45] Stone's argument was ultimately unsuccessful because the U.S. Supreme Court, in a 4–3 decision, determined that the plaintiff organization had to show that its members had suffered direct, immediate injury.[46] Stone suggested that courts should grant nature legal standing through the appointment of a guardian. He reasoned that it is possible to appoint guardians for humans who are unable to competently represent themselves, like minors or mentally incompetent persons. Additionally, Stone noted potential conflicts of interest, for example with a federally appointed guardian to represent a river that flows from state to state, but it was not enough to sway him from the opinion that rights of nature are a viable option to environmental protection and that the appointment of a legal guardian to an entity in nature is the key to establishing standing in courts for such entities.[47]

Following Stone's model, it is critical for an MPA, in order to establish legal standing, to have a well-educated committee of guardians who are ready to represent the ocean's best interests. The purpose of a committee, rather than one or two individuals, is to eliminate or, at the very least, decrease the possibility of a conflict of interest arising. The committee should include a diverse group of people that all hold different values and perspectives. For example, on the committee there should be a marine biologist, a lawyer, and an environmentalist, and possibly even a fisherman or a representative from an indigenous group.

The downfall of having a committee of guardians is the same as the positive of having a committee: the more minds, the more opinions. A healthy ocean to a fisherman may mean an ocean that has bountiful amounts of fish, to an activist it may be healthy coral reefs, and to others it may be an area of vast natural resources available for exploitation, like oil. It is thus crucial that this committee work as a cohesive unit.

As such, a concrete definition of a "healthy ocean" is necessary in order to create an objective or vision for the committee to work towards. According to the Earth Law Center, marine ecosystems are considered "healthy" when they are able to maintain their structure, organization, and function "over time in the face of external stress." Marine ecosystems will also receive the right to flourish and restore themselves. Given these rights, it makes sense for the definition of a "healthy ocean" to be something along the lines of an ocean that is able to sustain its structure and function and also have the ability to restore itself. Most importantly, the definition of a healthy ocean should consider the ocean's own wellbeing and natural state and not potential benefits for humans.[48]

With a clear understanding of what a "healthy" ocean looks like, the committee will be better able to work together towards a common goal, ultimately resulting in the better protection of MPAs and the ocean as a whole. When someone trespasses upon the MPA or it suffers damage from pollution or a fisherman violates the rules of a no-take zone, the guardians will bring suit in court against the responsible party for damages. If the MPA is successful, the award will be used for things like maintenance and enforcement.

Rights to Enter Contracts

Freedom to contract gives people the right to legally bind themselves in a contract absent external regulation or pressure.[49] Traditionally, in the EEZ, the United States has "sovereign rights for the purpose of exploring, exploiting, conserving and managing natural resources, whether living and nonliving, of the seabed and subsoil and the superjacent waters and with regard to other activities for the economic exploitation and exploration of the zone, such as the production of energy from the water, currents and winds [and] jurisdiction as provided for in international and domestic laws with regard to the establishment and use of artificial islands, installations, and structures, marine scientific research, and the protection and preservation of the marine environment."[50] Giving MPAs the right to enter contracts would strip the government of its jurisdiction over these areas and put the MPA in charge of its own management. However, the guardians would have to determine whether it would be in the ocean's best interest to enter into a contract to allow the government to continue to use the area for things like exploration, resource management, and scientific research.

Effectively, giving MPAs the right to enter contracts will rearrange decision-making authority, shifting it from the government to the MPA. After that, it is within the power of the MPA to decide if it would like to contract with the government to allow the government to continue to manage any of the things it previously had the authority to administer. The right to contract, however, is not limited to the government. MPAs will also have the option to contract with public or private businesses.

Right to Own Property

As is true in Australia, MPAs would have the right to "hold and dispose of real and personal property on behalf of the environment."[51] There are two different categories of property of interest for MPAs to purchase: property for protection and property for raising capital.

MPAs should be interested in purchasing property that will eliminate the possibility of that property causing harm to the MPA. For example, if there is an island located close to the MPA, it may be in the best interest of the MPA to purchase this area so there is less travel out to the island in general as well as eradicating the risk that someone may buy the island to live on or populate.

Additionally, an MPA may be interested in owning property along the coast in order to raise capital for its own maintenance and for staffing to support enforcement efforts independent of government funding. The MPA could own things like marinas and boat slips, gift shops, restaurants, and other tourist destinations. If the MPA permits recreational activities like kayaking, the MPA should consider owning a kayak rental booth and people would be able to tour through the area. This would also be a fun way to educate citizens about MPAs and their role in protecting the environment.

Legal Form

Rights can take form in multiple different ways. Australia, New Zealand, and India all use different methods of creating legal rights for the rivers in their country. Australia uses a corporate body, which is a hybrid of a corporation and a public entity, New Zealand uses a legal entity with the status of a legal person, and India makes the rivers a legal entity with the status of a legal minor, with the same rights and liabilities of a human.

In Australia, the legal form of the VEWH allows it to become a distinct organizational institution that is free of governmental interference. The VEWH, practically speaking, has "retained bipartisan political support in

Victoria, and has established itself as an independent decision maker for Victoria's rivers," although it is subject to Victoria's Financial Management Act. New Zealand has allowed the Te Awa Tupua Act to create an entirely new arrangement, one that takes the idea of rights of nature and the granting of legal rights to nature and blends it with existing legal frameworks. Authority comes from the national level, which allows for "decentralized decision making by creating a series of actors and rules to operationalize legal personhood at the local level and encourage multistakeholder and community participation."

Both Australia and New Zealand are having more success than the rights-of-nature movement in India because they opted to assign the rights legislatively rather than judicially. First, India will be unable to enforce the rights of the rivers across state and national borders; this is an issue because both the Ganges and the Yamuna Rivers are transboundary rivers. Second, the guardians of the rivers hold government positions and thus create a serious risk for a conflict of interest. Lastly, the guardians are not receiving any form of payment for the work they are doing on behalf of the river, which in turn reduces their motivation and commitment. As a result of all these issues, the Supreme Court of India is hearing an appeal lodged by the government, which is arguing that the rights of the rivers are unclear, thus halting the entire implementation process.[52]

New Zealand's approach is the most practicable for establishing legal rights for MPAs. There are already existing policies and legislation regarding MPAs in the United States, and making the MPA a legal person with human rights will allow us to blend the two frameworks into one that will cater to our ultimate goal of a healthy ocean. Additionally, MPAs will receive legal rights legislatively rather than judicially in order to prevent any petitioners that would halt the implementation process.

Potential Issues

New Zealand has had some issues to work out because it was a pioneer by incorporating rights of nature into its constitution. The beauty of not being the first country to implement rights of nature is that we are able to learn from other countries' models about what works and what does not. This section focuses on the mistakes India made in assigning rights to nature and evaluates what the United States can do differently to prevent similar problems, as well as a few enforcement issues every country has run into in creating rights of nature.

India's Problems

As discussed, India has run into a few issues in its attempts to implement rights of nature. Some of these problems have arisen from the fact that India created the rights through a judicial decision, and some have occurred because India made fatal errors in assigning representation to the rivers. These errors include the appointment of guardians who also hold a government position and not compensating the guardians for their work.[53]

To avoid these issues, assigning legal guardians to the MPAs must be done with care. The MPA's guardians would be a group of individuals with different backgrounds and values. The group constituted to act on behalf of the MPA would not include a member of the government to avoid potential conflicts of interest. Additionally, part of the MPAs funding (which is discussed in the following section) is for compensating the guardian committee. By doing these things, MPAs in the United States' EEZ should be able to avoid the same issues troubling India.

Enforcement Issues

There are several features that will determine the success of adequately enforcing the legal rights of the MPAs. They include the nature of the entity holding the rights, independence from government, and the source of funding and organizational support. Once again, Australia, New Zealand, and India all have different approaches to these characteristics.

New Zealand and India are most relevant when discussing the natural elements being protected; Australia merely protects water rights to provide for a specific environmental flow, while New Zealand and India give rights to an actual body of water. New Zealand has given rights to the Whanganui River "bed from the mountains to the sea, and its catchment" and India gave rights to the Ganges and Yamuna Rivers, but the extent of these rights is unclear. Again, New Zealand provides the better example because it is giving rights to the entire river, like India, but has made those rights more specific. Regarding the MPAs, rights belong to the specific MPA in its entirety, making it very clear that the rights extend throughout the entirety of the MPA's boundaries.

Being independent of the government is also an important characteristic. Both Australia and New Zealand created independent guardians, while India has elected to appoint only government employees as guardians of the rivers. The framework for MPAs is designed with the purpose of being free from government interference. The jurisdiction traditionally assigned to the government within the EEZ is transferred to the MPA, and the legal guardian committee will not have a seat for a member of government. The MPA will be free to regulate itself for itself without the influence of the administration, unless the MPA enters into a contract with the government.

Funding is the third point of possible contention for enforcing rights of nature. Australia uses a levy or a tax, while New Zealand has allocated funding specifically for the Whanganui, but the rivers in India receive no funding.[54] It makes the most sense for the MPAs to receive consistent funding, as in New Zealand. The funding will support enforcement staff compensation and legal guardian committee compensation, as well as things like maintenance. The MPA could also can raise additional revenue by owning the coastal property, as discussed earlier.

Conclusion

Rights of nature is a powerful tool that is capable of affording superior protections to the ocean. Although there are many regulations and policies in place to conserve ocean resources, they have proven inadequate to ensure a healthy ocean. Rights of nature gives rights directly to the ocean, allowing them to develop alongside the ocean. Although it is a tall task to bring MPAs to life by giving them legal rights, we are fortunate to have many example models from other countries that have implemented the rights of nature prior to the United States. Using those models, we are better situated to create a successful framework for MPAs. Rights of nature will give MPAs the right to flourish and the ability to continue to flourish for generations to come, despite the new challenges the ocean is facing, like climate change.

Notes

1. *Cetacean Community v. Bush*, 986 F.3d 1169, 1178 (9th Cir. 2004).
2. National Geographic Society, Why the Ocean Matters (2012), https://www.nationalgeographic.org/media/why-ocean-matters/.
3. Melissa Denchak, Ocean Pollution: The Dirty Facts (2018), https://www.nrdc.org/stories/ocean-pollution-dirty-facts/.
4. Robin Kundis Craig, Marine Biodiversity, Climate Change, and Governance of the Oceans. Diversity 4 (May 19, 2012): 224–238.
5. 16 U.S.C. § 1361.
6. Ibid. ("[M]arine mammals have proven themselves to be resources of great international significance, esthetic and recreational as well as economic, and it is the sense of the Congress that they should be protected.").

7. NOAA Office for Coastal Management | About the Office, https://coast.noaa.gov/czm/act/.
8. 16 U.S.C. § 1451. Congressional findings (Section 302(a)).
9. 16 U.S.C. § 1451. Congressional findings (Section 302(m)).
10. The National Marine Sanctuaries Act, Marine Conservation Institute, https://marine-conservation.org/what-we-do/program-areas/mpas/national-marine-sanctuaries/legislative-history-national-marine-sanctuaries-act/.
11. NOAA, Laws & Policies NOAA Fisheries, https://www.fisheries.noaa.gov/topic/laws-policies/.
12. 16 U.S.C. § 1801(a)(3).
13. Michelle Bender, "The Earth Law Framework for Marine Protected Areas, Earth Law Center," https://portals.iucn.org/library/sites/library/files/resrecrepattach/ELC%20MPA%20Framework%20Sept%204%20%282%29.pdf.
14. MPA News Staff, MPA News Poll: The Coming Challenges for MPAs, and How to Address Them, *MPA News* (2005), https://mpanews.openchannels.org/news/mpa-news/mpa-news-poll-coming-challenges-mpas-and-how-address-them/.
15. "The defined goal for creating a network of marine protected areas by the IUCN is 'to provide for protection, restoration, wise use, understanding and enjoyment of the marine heritage of the world in perpetuity . . . in accordance with the principles of the World Conservation Strategy of human activities that use or affect the marine environment.' Where wise use is defined as 'for the use of people on an ecologically sustainable basis.' This includes for the continued welfare of people affected by the creation of the marine protected area." Michelle Bender, "The Earth Law Framework for Marine Protected Areas."
16. Ibid.
17. Erin O'Donnell & Julia Talbot-Jones, Creating Legal Rights for Rivers: Lessons from Australia, New Zealand, and India (2018), https://www.ecologyandsociety.org/vol23/iss1/art7/.
18. Kurt Cobb, Living World: Should Natural Entities Be Treated As Legal Persons? (2017), https://www.resilience.org/stories/2017-04-09/living-world-natural-entities-treated-legal-persons/.
19. Hope M. Babcock, "A Brook with Legal Rights: The Rights of Nature in Court," *Ecology* 43, no. 1 (2016): 1–51, https://scholarship.law.georgetown.edu/facpub/1906.
20. Oliver A. Houck, Noah's Second Voyage: The Rights of Nature As Law, 31 Tul. Envtl. L.J. 1, 26 (2017).
21. *Sierra Club v. Morton*, 405 U.S. 727 (1972).
22. Ibid. at 742–43.
23. Kyle Pietari, Ecuador's Constitutional Rights of Nature: Implementation, Impacts, and Lessons Learned (2016), https://willamette.edu/law/resources/journals/welj/pdf/2016/2016-f-welj-pietari.pdf/.

24. Maria Akchurin, "Constructing the Rights of Nature: Constitutional Reform, Mobilization, and Environmental Protection in Ecuador," *Law & Social Inquiry* 40, no. 4 (Fall 2015): 937-968, https://doi.org/10.1111/lsi.12141.
25. Constitution of the Republic of Ecuador, Chapter 7, Article 71.
26. Mary Elizabeth Whittemore, The Problem with Enforcing Nature's Rights under Ecuador's Constitution: Why the 2008 Environmental Amendments Have No Bite (2011), https://digital.lib.washington.edu/dspace-law/bitstream/handle/1773.1/1032/20PacRimLPolyJ659.pdf?sequence=1/.
27. Bolivia—National Rights of Nature Legislation, Australian Earth Laws Alliance, https://www.earthlaws.org.au/what-is-earth-jurisprudence/rights-of-nature/bolivia/.
28. Law of Mother Earth, Chapter 1, Article 1.
29. Ibid., Article 2.
30. Bolivia—National Rights Of Nature Legislation, Australian Earth Laws Alliance, https://www.earthlaws.org.au/what-is-earth-jurisprudence/rights-of-nature/bolivia/.
31. Franz Chávez, Bolivia's Mother Earth Law Hard to Implement (2014), http://www.ipsnews.net/2014/05/bolivias-mother-earth-law-hard-implement/.
32. Ibid.
33. O'Donnell and Talbot-Jones, Creating legal rights for rivers.
34. Ibid.
35. Ibid.
36. Ibid.
37. Ibid.
38. Intercontinental.Cry, Colombia Constitutional Court Finds Atrato River Possesses Rights to "protection, conservation, maintenance and restoration" (2017), https://intercontinentalcry.org/colombia-constitutional-court-finds-atrato-river-possesses-rights-protection-conservation-maintenance-restoration/.
39. Houck, Noah's Second Voyage.
40. Shelia Hu & Michelle Bender, Rights of Nature to Save the Endangered Sharks of the Galapagos (2018), https://www.earthlawcenter.org/blog-entries/2018/7/rights-of-nature-to-save-the-endangered-sharks-of-the-galapagos/.
41. National Geographic Society, Marine Protected Areas (2011), https://www.nationalgeographic.org/news/marine-protected-areas/.
42. Executive Order 13158.
43. Akchurin, "Constructing the Rights of Nature," 942.
44. O'Donnell and Talbot-Jones, Creating Legal Rights for Rivers.
45. Babcock, "Brook with Legal Rights," 1–51.
46. *Sierra Club v. Morton*, 405 U.S. 727.
47. Babcock, A Brook with Legal Rights.
48. Bender, The Earth Law Framework.
49. Freedom of Contract Law and Legal Definition, https://definitions.uslegal.com/f/freedom-of-contract/.

50. What Is the EEZ?, NOAA's National Ocean Service (2018), https://oceanservice.noaa.gov/facts/eez.html/.
51. O'Donnell and Talbot-Jones, Creating Legal Rights for Rivers.
52. Ibid.
53. Ibid.
54. Ibid.

PART III

Re-envisioning Ocean Action

Editors' Introduction to Part III

Re-envisioning our connectivity to the ocean and its need for human protection naturally leads to rethinking how to pass these new lenses on the ocean to others so that they can share the reimagined visions of the marine world and humanity's relationship to it. To re-envision the Anthropocene ocean, in other words, is also to become a teacher and exhibitor of that new vision. Ideally, that sharing also encourages the audience to participate in shaping humanity's changing relationship to the changing ocean.

The essays in Part III provide different windows onto the process of this continual re-envisioning, both challenging readers to embrace the complexity of that process and providing tools to share with others. The essay writers are mainly scientists, but they are scientists who partake in broader realms of education and engagement, from politics to education to the fine arts. They all explore how new understandings of the ocean and its various planetary roles—including as human life-support system—lead people to new means of communicating these new visions, including the urgency of ocean protection and the many facets of adapting to a changing ocean.

Christopher Finlayson, for example, builds on the theme of creating more marine protected areas that Jackson and Craig explored in Part II, and complicates it by acknowledging, in the context of New Zealand's efforts to establish these areas, conflicts with Māori fishing rights and the process of settling those rights with the Crown and New Zealand Parliament. Like Swensen, therefore, Finlayson brings to the forefront the enduring legacy of European colonialism in marine resource management and coastal indigenous cultural preservation. In so doing, he reminds readers that re-envisioning the ocean for the Anthropocene is a complicated effort that must be attuned to many voices and must seek to correct multiple historical injustices as well as injuries to the ocean itself.

Kathryn K. Davies, Benjamin A. Davies, Paula Blackett, Paula Holland, and Nicholas Cradock-Henry then transition the reader to New Zealand's ongoing attempts to adapt to sea-level rise. They introduce the concept of serious gaming as a tool for re-envisioning the future of human coexistence with a changing ocean. Highlighting three such serious games—*Climate Adaptation Challenge*, *Maraeopoly*, and the especially new and online *Adaptive Futures* that they helped to develop—these researchers show how serious play can help all New Zealand communities (and, in the near future, coastal communities around the world) reimagine their coastal futures and priorities.

Like Finlayson, Brenda Bowen worries about ocean plastic pollution and strives to pass that concern along to her science students in a tangible way. Her essay shares her process for encouraging students not only to learn the science of ocean plastic pollution but also to create narratives about where that plastic comes from and what it means. She suggests that ocean science education for the Anthropocene works best when it embraces the humanities' techniques of storytelling and audience/learner engagement.

Marine biologist and National Geographic Explorer Tierney Thys similarly reaches across disciplines to create new visions of the marine realm, this time to dance. Her essay explores her role as chief science consultant to the Capacitor dance troupe, and their combined efforts to convey a vision of the Anthropocene ocean through the magic of movement, costumes, scenery, and lighting on stage. Like Bowen, she seeks to communicate science to a broader community. Like Finlayson, she emphasizes the importance of recognizing and respecting all members of the community.

In many ways, therefore, the essays in Part III circle back to the themes of Part I—the importance of community and of making connections, and the importance of the arts and humanities in creating new cultural visions of the world. At the same time, however, Part III emphasizes with new force the importance of an interdisciplinary re-envisioning of the Anthropocene ocean. Given that the quest for that interdisciplinary vision brought the authors in this volume together in the first place, these essays provide the fitting third and last act in that production.

CHAPTER 11

Plastic in the Pacific

How to Address an Environmental Catastrophe

CHRISTOPHER FINLAYSON

THE STORY OF THE MUTINY on the *Bounty* on April 28, 1789, is well known. Fletcher Christian and a group of sailors seized control of *HMS Bounty* and set adrift the ship's captain, William Bligh, and his supporters. The mutineers then voyaged to Tahiti and, together with their girlfriends, set sail again. Some time later they came across Pitcairn Island. The *Bounty* was set alight and Pitcairn became their new home. The story of the mutiny has been recounted in a novel, several films, and even a musical. Pitcairn Island is one of a number of islands that form Britain's last colony in the Pacific. The others are Henderson, Durie, and Oeno Islands. The governor of Pitcairn is also the high commissioner (or ambassador) of the United Kingdom in New Zealand.[1]

In recent years the colony has been in the headlines because of allegations of sex abuse.[2] Various trials were held on Pitcairn Island and the offenders were imprisoned there. Arranging trials was an extremely complex issue given the island's remoteness. Special legislation was passed by the New Zealand Parliament to enable some trials and appeals to be heard in New Zealand.[3]

One of the islands, Henderson, has been in the media in recent times albeit for very different reasons. This remote island, situated halfway between New Zealand and Panama, is littered with plastic. It is a UNESCO World Heritage Site.[4] Although it is only rarely visited by Pitcairners for their summer holiday, it is an environmental catastrophe.[5] The tragic state of this island has received a lot of international media exposure; in 2019 there was an expedition to the island to study the pollution and raise awareness of Henderson's marine environment.[6] Henderson Island is a metaphor for the Pacific. This mighty ocean is choking with plastic. Some have written of a gigantic floating plastic soup in the Pacific of up to fifteen million square kilometers—almost the size of Russia.[7]

Another major problem confronting the Pacific is overfishing. For example, the critical state of Pacific bluefin tuna stock. It is feared that the amount of this resource and the level of recovery remain at historically low levels.[8] There have, however, been some positive developments. In 2015, the Pacific Island Oceanic Fisheries Management Project was launched. This focused on building cooperation between states and developing skills and technology to track and manage tuna stock more effectively so that those who are dependent on the resource will be able to benefit from it for years to come. As a result of the program, fish stocks in the region have rebounded, leading to an increase in legal catches by almost 60 percent between 1997 and 2012.[9] That having been said, there is still more work to be done and continued international cooperation will be critical to preserving these hugely important fish stocks.

Farther south, one can see the consequences of countries failing to cooperate. Japanese "scientific" whaling in the Southern Ocean has been a major source of tension between Japan on the one hand, and Australia and New Zealand on the other. Some years ago, Australia took the issue to the International Court of Justice in the Hague (ICJ). New Zealand intervened and made a legal submission in support of the Australian case. Australia and New Zealand were successful.[10] A majority of the court held that Japan's whaling program in the Antarctic (JARPA II) was not authorized under the International Whaling Convention. In response to the judgment, the Japanese government abandoned JARPA II and a new program (NEWREP-A) took its place. In reality, the new program is not for the purposes of scientific research but New Zealand and Australia cannot go back to the ICJ because Japan has ruled out the jurisdiction of the court in "any dispute arising out of, concerning, or relating to research on, or conservation of, living resources of the sea."[11] In the meantime, warehouses in Tokyo are full of whale meat as younger Japanese have no interest in purchasing the product. It appears that only generational change in Japan can bring this idiocy to a halt.

This, then, is a very superficial overview of some of the serious problems facing the Pacific. The world is accustomed in this decade to hearing apocalyptic warnings of impending doom, but Henderson Island reminds us that the current situation simply cannot continue. So, what are some of the recent responses of Pacific countries to the challenges facing the Pacific Ocean?

The Response

In 2009, the United States announced the Pacific Remote Islands Marine National Monument.[12] This consists of seven areas totaling 1,058,848 square kilometers. Some areas have provisions for limited customary and recreational fishing. In March of 2015 the UK announced an 834,334-square-kilometer marine protected area around Pitcairn Island.[13] It was proposed that an area be set aside for customary fishing. In 2012 Australia announced the creation of the 989,836-square-kilometer Coral Sea Marine Reserve.[14] It is assigned IUCN category IV and is one of the world's largest protected areas. Around 51 percent of the reserve is zoned as a marine national park which is a "no take" zone. The remaining areas will have other levels of protection. Also in 2015, a 720,000-square-kilometer park around Easter Island in Chile was announced. This allows fishing by the local population.[15]

The New Zealand Response

And what about New Zealand? The situation is this country is complex and contentious and requires more detailed explanation.

Some years ago, the New Zealand government announced the establishment of the Kermadecs Sanctuary to be created in the Kermadec region of the South Pacific Ocean about 1,000 kilometers northeast of New Zealand. At 620,000 square kilometers, it would be one of the world's largest and most significant fully protected areas. It would be 35 times larger than the combined area of New Zealand's existing 44 marine reserves. The sanctuary would mean 15 percent of New Zealand's ocean environment will be fully protected. The sanctuary would cover an area of New Zealand's Exclusive Economic Zone (EEZ) from 12 to 200 nautical miles from the five Kermadec Islands of Raoul, Macauley, Cheeseman, Curtis, and L'Esperance which lie halfway between New Zealand and Tonga. It will be the first time an area of New Zealand's EEZ has been fully protected.[16]

The sanctuary follows the establishment in 1990 of the Kermadec Marine Reserve which consists of 7,500 square kilometers. This marine reserve extends 12 nautical miles, from the cliffs and boulder beaches of the various Kermadec Islands and rocks, out to the edge of the territorial sea. The Kermadec area is said to be one of the most pristine and unique places on the planet. It includes the world's longest chain of underwater volcanoes and its second deepest ocean trench at over 10 kilometers—deeper than Mount Everest is tall. Its waters are home to:

- over six million seabirds of 39 different species
- over 150 species of fish
- 35 species of whales and dolphins
- three species of sea turtles, all endangered
- many other marine species unique to this area such as corals, shellfish, and crabs.

Like other coastal states, New Zealand has sovereign rights in its territorial sea with very few limitations. Its rights and obligations in the EEZ are different but include the rights to manage fishing and mineral resources. These rights (e.g., over navigation and submarine cables) must be exercised with due regard for those of other states. Rights and limitations are:

- no fishing or mining applies to both the sanctuary and marine reserve;
- ships will be allowed to exchange ballast water in the sanctuary (subject to regulation) but not in the marine reserve;
- marine discharges from ships and yachts (subject to regulation) will be allowed in the sanctuary but not in the marine reserve; and
- submarine cables will be allowed in the sanctuary but are not permitted in the marine reserve.

All fishing and mining is prohibited in the marine reserve (the territorial sea out to 12 nautical miles around the Kermadec Islands). This is unchanged by the sanctuary.

Currently the 620,000-square-kilometer area where the sanctuary will be created is a benthic protection area (BPA). This was put in place in 2007 under the Fisheries Act of 1996 (NZ) and prohibits bottom trawling and dredging. The area is also subject to New Zealand's minerals legislation.

Under the International Union for Conservation of Nature (IUCN) Protected Areas classification system, the sanctuary would be classified as category I—strict nature reserve/wilderness area. This is the highest category of protection. Impacts of human activities and access will be strictly managed to protect the ecological integrity of the area.

The Kermadecs area is one of ten New Zealand fisheries management areas and is known as FMA10. The species caught around the Kermadecs are highly migratory and include swordfish, bigeye, and albacore tuna and blue shark. The quota for these highly migratory species is for New Zealand's entire EEZ and is not specific to FMA10. As the catch can be caught in other parts of New Zealand's EEZ, fishing interests will not be significantly impacted by the establishment of the sanctuary.

As all mining, exploration, and prospecting activities will be prohibited in the sanctuary, there will be an opportunity cost for New Zealand, but this is difficult to quantify. The logistics of mining in these very deep, remote waters is difficult and expensive.

The legislation to give effect to the sanctuary[17] is still stalled in the New Zealand Parliament because of objections of the indigenous people of New Zealand (The Maori) who say that the proposal will breach a historic settlement reached with them in 1992. In order to understand their contentions, it is necessary to provide some background.

Maori are the southernmost branch of the Polynesian people and the youngest grouping within that grouping of island peoples covering a truly enormous area of the Pacific. The ancestral roots of the Polynesian were in the South China Basin, but they came to the Pacific about 6,000 years ago. Maori became established in New Zealand only about 650 years ago. They are now formed into distinct tribal groups or *Iwi* and subtribal groups or *hapu*. Their distinctive customs and ethnologies have developed as a response to the wide variation of geographical physical environments in which they have found themselves. Although there are dialectical differences, they have a common language, Te Reo Maori.[18]

In the early nineteenth century, Europeans began visiting New Zealand on a regular basis and, in 1840, the British Crown signed a treaty with Maori. This treaty, known as the Treaty of Waitangi, was a very short document which:

- recognized British sovereignty over New Zealand;
- guaranteed Maori the full exclusive and undisturbed possession of their lands, forests, fisheries, and other properties which they wished to retain; and
- gave Maori all the rights and privileges of British subjects.

Maori complain on a regular basis that the treaty has not been honoured. Any objective observer would, however, have to admit that the Crown's promises were not kept. So, for example, in 1840 Maori owned most land but a hundred years later they owned virtually nothing. This was because of (1) confiscations sanctioned by Parliament; (2) dubious private and government land purchases; or (3) the work of the Native Land Court which wasn't a court so much as an agent of the Crown. So, for example, customary lands were converted to individual title and it wasn't unheard of for the cost of surveys to absorb the entire value of the land, so the owners were left with nothing.

When Europeans first came to New Zealand, local Maori took to capitalism and commerce very well. They exported fish in their own ships over to Sydney and Hobart in Australia, and trade with that colony was highly successful in other products like flax. Exports of other products, notably sealskins, went to China and Europe. In the north of New Zealand, Maori owned missionary-designed and missionary-built flour mills. For example, fifty were built in one region in the years following the signing of the Treaty of Waitangi. One of the great tragedies of New Zealand was that the actions of the Crown and settlers destroyed the commercial trade of the Maori and reduced them to penury. With the loss of land came the inevitable loss of population. At the time the treaty was signed it was estimated that there were around one hundred thousand Maori, but by 1900 that population had been reduced to around forty thousand. Life expectancy figures were appalling. Today there are around seven hundred thousand New Zealanders of Maori descent. Life expectancy figures are still behind Europeans (or Pakeha).

There was a dramatic surge in the Maori population in the years following World War II, and by the early 1970s there was a growing awareness that Maori aspirations and grievances could not be ignored. So, in 1975, Parliament established the Waitangi Tribunal which could inquire into contemporary grievances.[19] That jurisdiction was later expanded so that historical grievances going back to the 1840 signing of the Treaty of Waitangi were able to be investigated. Since the mid-1980s Maori tribes and individuals have made many claims to the tribunal.

After the first reports started to be released, the government had to determine how to respond to them. Since the mid-1990s the Crown has been settling these historical grievances. A deed of settlement is then given effect to by an Act of Parliament. Generally, these settlements have been supported across the political divide and by the community. There will always be those who claim that the settlements do not fairly compensate Maori for the losses they have suffered. On the other hand, there are those who contend that the treaty settlement work is an industry developed for the benefit of the Maori elite, lawyers, and accountants, and that the benefits do not flow to the ordinary tribal members.[20]

In 1986, the New Zealand Parliament passed the Fisheries Amendment Act 1986. This act substantially amended the Fisheries Act 1983 to bring into operation what is known as the Quota Management System (QMS). The 1986 act was a reaction against the former regime of open slather and government subsidy which had led to a massive expansion of the fishing industry from 1963. At the same time the inshore fishery dramatically declined as a result of overfishing. The 1986 amendment moved away from the older regulated system which contained no conservation incentives toward the creation of valuable and transferrable property rights in the resource.

Fundamental to the reforms was the development of the concept of a quota, a fraction of a particular "total allowable commercial catch" for a particular fish stock defined by a reference to species and particular quota management areas, these latter being divisions of the New Zealand territorial sea and the Exclusive Economic Zone. With one exception, quota is allocated in perpetuity, and the holders acquire a harvesting right, measured as a specific tonnage for a specific quota management area for a fixed time (one year). Quotas give rise to an "annual catch entitlement" in accordance with elaborate formulae set out in the Fisheries Act 1996.

The QMS was introduced on October 1, 1986. Maori went to court and persuaded the High Court of New Zealand that the government should not introduce fish stocks into the QMS until the issue of ownership had been resolved. The court confirmed that Maori customary fishing rights were controlled by "hapu and tribes" and that those customary rights contained *both* commercial and noncommercial elements. It acknowledged there was a strong case that before 1840 Maori had a highly developed and controlled fishery over the whole coast of New Zealand, at least where they were living. That was divided into zones under the control and authority of hapu and tribes of the district. Each of these hapu and tribes controlled those fisheries. The fisheries had a commercial element and were not purely recreational or ceremonial or merely for the sustenance of local dwellers.

To resolve claims and litigation involving fisheries, an interim settlement of fishing claims which acknowledged the full spectrum of Maori interests in fisheries was entered into between Maori and the Crown in 1989, and provided 10 percent of all fisheries then in the QMS, along with some funding for administration.[21]

The Fisheries Deed of Settlement, signed in 1992, was the final settlement of all Maori claims to customary fishing rights. Under the settlement, the Crown additionally:

- gave Maori funds to buy a 50 percent ownership stake in a major fisheries company;

- undertook to provide Maori with 20 percent of the quota for all new species brought within the QMS after that time;
- gave Maori positions on statutory fisheries management bodies;
- restructured the then Maori Fisheries Commission into the Treaty of Waitangi Fisheries Commission to enhance its accountability to Maori; and
- agreed to make regulations to allow self-management of Maori fishing for subsistence and cultural purposes (erroneously later labelled "customary fishing" in the regulations).

In return, Maori agreed:

- that the settlement settled all Maori commercial fishing rights and interests;
- to accept regulations for customary noncommercial fishing;
- to stop litigation relating to Maori commercial fisheries;
- to support legislation to give effect to the settlement; and
- to endorse the QMS.[22]

It was clearly understood by Maori that they were not only receiving existing rights but that there was also a right to development. So, for example, there could be an undeveloped fisheries management area that, with the passage of time and improved fishing techniques, could be worked up into a commercially valuable settlement. This right to development is very relevant in the context of the development of a marine reserve in the Kermadecs.

When Maori entered into the Treaty Fisheries Settlements, they accepted the QMS, which included defined QMAs, as the basis of a treaty settlement. It was a core condition on the Crown side agreed to by Maori. Any change requires Maori agreement. It should be noted that the Inshore Kermadec (twelve-mile) Zone currently has the highest possible international level of marine protection and this was imposed with the agreement of Maori.

The complaint of Maori about the Kermadecs is that if the Crown can unilaterally alter the system it entered into as a condition of the Fisheries Settlements of 1989 and 1992, it has the capacity to alter any treaty settlements on its own political whim. That could undermine the entire historical settlement framework.

Will the legislation be passed by the New Zealand Parliament? Applying a principled approach to the issue, the proposal in its current form should not proceed as it undermines the hard-fought rights of Maori which were gained as recently as 1992. In particular, a strong argument can be made that, because of the QMS, there is no need for a reserve. Conservation of fisheries is an essential ingredient of the system. To overlay the New Zealand Fishery System with a no-take marine reserve is to pile "Pelion on Ossa."

Moreover, there is no evidence of fish stack depletion in the Kermadecs region. Many say that the case for a sanctuary cannot on any evidence be made on any presently observable danger to biodiversity or ecology.

Where to from Here?

Some emphasis has been placed on New Zealand because it illustrates the important point that pure ideology is no substitute for science, and it also shows that what works in one part of the Pacific may not necessarily be successful in another region. It all depends on the circumstances and in particular the history, the expectations of the indigenous people, and the relevant legal framework. New Zealand has a fisheries regime built on the key concept of "evidence-based sustainability." That evidence can be based on abundance and environmental conditions. The system works without having to impose a no-take reserve. To upset the hugely important intergenerational work of resolving historical grievances is regarded by many as too high a price for New Zealand to pay.

In any event, there has to be a question about whether the reserves are enough. Will they protect the ocean from plastic? Of course not. These reserves are not hermetically sealed off so as to prevent the spread of plastic. A more comprehensive, indeed radical, approach is required. The time has come to consider the following:

First, a comprehensive treaty on transboundary pollution caused by plastic is required. It could be that the world will move in this direction but the gravity of the problem mandates urgency. The United Nations appointed Peter Thomson as UN Special Envoy for the Oceans[23] and, in 2017, sponsored an ocean conference to discuss the serious problems faced by the world's oceans.[24] Another UN conference, scheduled for summer 2022, is to take place in Lisbon, Portugal, cohosted by the governments of Kenya and Portugal. The overarching theme of the conference will be "scaling up ocean action based on science and innovation for the implementation of Goal 14: stocktaking, partnerships and solutions."[25] While it is excellent that the United Nations is leading work in this area, it is necessary to treat this issue as one of grave urgency, particularly given that it is generally accepted that plastic pollution has become "the new millennium's tragedy of the commons."[26] Evidence on how the issue can be addressed may be found in the excellent work done in recent years on transboundary air pollution. Since 1979 the Convention on Long-Range Transboundary Air Pollution has addressed some of the major environmental problems in Europe through scientific collaboration and policy negotiations.[27] The history of the convention dates back to the 1960s when scientists showed the interrelationship between sulphur emissions in continental Europe and acidification of Scandinavian lakes. Excellent work has also been done in North America to address transboundary flows on that continent. In 1991, for example, the United States and Canada agreed to address this issue. It led to a reduction in acid rain in the 1990s and was expanded in 2000 to reduce transboundary smog emissions.[28] The EPA has also done good work on how the marine environment can be better protected.[29] The world can successfully address complex environmental issues if there is a willingness to act.

Second, any convention will need to be supported by a clean-up fund contributed to by, in the case of the Pacific, all coastal states. Whether American politics allows the United States to join or not, their decision should not prevent other countries beginning the work. Such a fund should not be the sole responsibility of governments. Private enterprise has a major role to play. Imagine a multibillion-dollar lighting fund dedicated to the elimination of plastic from the Pacific. The world needs more Boyan Slats[30] whose efforts are heroic but just a start. Governments cannot do this job by themselves; private individuals and organisations also have a role.

Thirdly, nation-states have a responsibility to take immediate steps to limit the creation of plastic products and to try to stop the dumping of this material with the result that it escapes into waterways and thence to the oceans. Elimination of single-use plastic bags is an example of a worthy initiative. There are two answers to those who complain about such measures: first, there is no alternative, and second, learn to adapt.

Conclusion

Henderson Island is one of the saddest places on earth. Thousands of kilometres from any urban areas, it should be unspoilt. It isn't. Its current polluted state is an affront to humanity. Change is needed. Change is needed now. In this article, the case for marine reserves has been assessed. They can be a useful way of helping to preserve valuable at-risk fisheries. There is, however, no one-size-fits-all. They need to reflect, and be respectful of, local conditions. So, what may work in the waters around Rapa Nui or Easter Island may not work in New Zealand. An absolute no-take zone could cause problems in New Zealand which has constitutional responsibilities to its indigenous people and probably the world's most advanced system of historical dispute resolution. That cannot be put at risk.

Some who read this chapter may conclude that the author has spent far too much time talking about New Zealand. They could well be correct although one hopes they would appreciate that it is written by a former attorney general of New Zealand who, for nine years, was responsible for resolving historical grievances between the state and the indigenous people of that country. What the author has endeavoured to do is, first, highlight the issues confronting the Pacific; secondly, emphasise that they can be solved through individual action and international cooperation; and thirdly, that in designing solutions, one needs to be sensitive to the rights and aspirations of the indigenous peoples of the Pacific. Many years ago, one of Margaret Thatcher's advisers reminded her that politics is not about the imposition of dogma, but about the management of prejudice and the reconciliation of interests. Whether she followed that advice is debateable, but it is very good advice for all decisionmakers when dealing with complex environmental challenges in these troubled times.

Notes

1. My interest in these islands began when I acted for the governor many years ago in an employment dispute which raised interesting issues of international law. See *Sutton & Governor of Pitcairn* [1995], INZLR 426.
2. See Kathy Marks, *Lost Paradise: From Mutiny on the Bounty to a Modern-Day Legacy of Sexual Mayhem, the Dark Secrets of Pitcairn Island Revealed*, 2009, Free Press.
3. Pitcairn Trials Act 2002 (NZ), which enables Pitcairn courts to sit in New Zealand for the purpose of holding certain trials under Pitcairn law.
4. See whc.unesco.org.
5. See, for example, "Henderson Island—the World's highest density of plastic pollution," http://news.nationalgeographic.com/. See also "Henderson Island has nearly 40 million pieces of plastic waste—here's what it's like," http://m.youtube.com/.
6. Henderson Island Expedition June 2019, https://www.visitpitcairn.pn/resources/downloads/Henderson-Expedition-June-2019—Final-Flyer.pdf/.

7. There has been much written on this calamity. See, for example, "Plastic in the Ocean Facts," www.theworldcounts.com; "The Pacific Ocean—facts and information," https://www.nationalgeographic.com, March 4, 2019; and https://theoceancleanup.com/great-pacific-garbage-patch/.
8. See, for example, World Wildlife Fund Factsheet, *Current Situation of Pacific Bluefin Tuna*, July 2014.
9. See "*Sustainable Fishery in the Pacific saves global tuna stocks and small islanders' livelihoods*", www.gef.org, March 1, 2019.
10. I was the attorney general of New Zealand from 2008 to 2017 and presented the New Zealand case. Mark Dreyfus QC, then the Australian attorney general, presented the Australian case. See *Whaling in the Antarctic (Australian v. Japan: New Zealand intervening)*, www.icj-cij.org.
11. "Japan's new optional clause declaration at the ICJ," www.ejiltalk.org. The actual Japanese declaration may be found on the court's website, www.icj-cij.org.
12. Pacific Remote Islands Marine National Monument, www.fisheries.noaa.gov.
13. Pitcairn Islands Marine Reserve, www.visitpitcairn.pn.
14. See https://parksaustralia.gov.au/marine/.
15. Rapa Nui Marine Park, one of the world's largest MPAs. www.mpatlas.org. This reserve is roughly the size of the Chilean mainland. On the rights of the Rapa Nui people to fish, see "Ante Propuesta de Conservacion Marina Mesa del Mar, September 2015," www.scribd.com.
16. The proposed sanctuary is outlined on the website of the Ministry for the Environment: mfe.govt.nz/marine/Kermadec-ocean-sanctuary/about-sanctuary. A very interesting history of these islands of New Zealand is Steven Gentry, *Raoul and the Kermadecs: New Zealand's Northernmost Islands*, Roberts Publishers, 2013, ISBN 9781927242025.
17. Kermadec Ocean Sanctuary Bill. It is still awaiting its second reading while the current government tries to negotiate a solution with Maori interests.
18. The language is very close to Tahitian and has some similarities with Hawaiian.
19. Treaty of Waitangi Act 1975 (NZ).
20. At note 10 I referred to the fact that from 2008 to 2017, I was the attorney general of New Zealand. During that time, I was also minister for Treaty of Waitangi Negotiations which meant I was responsible for negotiating historical settlements with Maori tribes (or iwi). In my period as minister, I negotiated sixty settlements. These included the Whanganui River Settlement where a river was recognised as a legal person—see Te Awa Tupua (Whanganui River Claims Settlement) Act 2017 (NZ).
21. Maori Fisheries Act 1989 (NZ).
22. Treaty of Waitangi (Fisheries Claims) Settlement Act 1992 (NZ).
23. See oceanconference.un.org/special envoy. Thomson is a Fijian diplomat who was the seventy-first president of the United Nations General Assembly from September 2016 to September 2017.

24. See oceanconference.un.org/about.
25. See undocs.org/a/resolution/73/292. The reference to Goal 14 is to UN Sustainable Development Goal 14. See further UNGA Resolution 73/292.
26. See www.frontierism/org/articles, and in particular Vince and Hardesty, "Governance solutions to the Tragedy of the Commons that Marine Plastics have become." The two authors have written a very comprehensive article tracing the development of the global plastic problem in the context of governance and policy.
27. 1979 General Conventions on Long-range Transboundary Air Pollution, www.unce.org.
28. See epa.gov/airmarkets/us-canada-air-quality-agreement.
29. See epa.gov/international-cooperation/protecting-marine-environment.
30. Carolyn Kormann, "A Grand Plan to Clean the Great Pacific Garbage Patch," *The New Yorker*, January 28, 2019.

CHAPTER 12

Recrafting Narratives to Disrupt the Oceanic Plastic Plague

BRENDA B. BOWEN

In the end we will conserve only what we love; we will love only what we understand; and we will understand only what we are taught.
—Baba Dioum, 1968

TEACHING UNDERGRADUATE OCEAN SCIENCE in land-locked states can feel like an act of faith—the rhythms of the sea and the alien watery worlds we conjure are far removed from the lives of the inland college students. As an educator, I strive to empower developing scientists with observational and critical thinking skills needed to contribute to the process of discovery and becoming effective and informed consumers of science. The earth sciences are at the core of the most important issues facing society today as demands for energy, water, and natural resources collide with accumulation of pollutants and a warming climate. The possibility of a future where humanity thrives on this planet requires a geoscience-literate public, prepared and empowered to enact change. Yet, although students may be taught about complex resource and environmental issues, the growing scale of our unsustainable practices demonstrates that traditional disciplinary education is not enough. University scholars are in a privileged place and need to be willing to bridge their core disciplines with tools, perspectives, and techniques from other specialties if we are to uncover solutions to problems that are clearly beyond the scope of any single academic field. Science offers just one way of knowing the Earth. What might we learn from the humanities and the arts as we search for that something special that is needed to finally provoke change, to alter the trajectory of minds, hearts, and actions?

Students of the ocean who find their way into my classroom (be it within campus walls or virtually on our computer screens as the current times demand) explore all aspects of the Blue Marble. We tease apart the planetary impacts of being enshrouded in water vapor and saturated by biology.

We quantify the energy pulsing through H_2O phase changes, the forces driving the conveyer belt. We marvel at the bounty of physical and biological resources that have shaped our cultures, and worry about the temporally limited human lens of *now* as the only way of experiencing them. The ocean connects us, inspires us, and feeds us. But the ocean is changing. In response to our human-induced disruption of the carbon cycle, hydrocarbons long sequestered in the lithosphere now power our industries and escape into the atmosphere, changing the trajectory of energy budgets and ocean acidity. Abundant data from warming waters demonstrate the impacts of this disruption, and while we plot and analyze, our own culpability within this changing system can feel just out of reach. How can we touch it, see it, know it, and then change it?

Among the many rapid changes occurring on our planet in response to our unsustainable practices, the plastic plague—polyethylene and polypropylene products of our consumption inundating all aqueous ecosystems—presents alarming and disturbing images. Decaying albatross, loggerhead, and whale carcasses reveal innocent bodies of our oceanic cousins packed to the brim with ingested bottle caps, hair combs, grocery bags, lighters, and fishing line. Science measures it in the gyres, quantifies it in the rivers, characterizes it in our food, and models the conclusion that, yes, something drastic must be done.[1] The planetary scope and scale of the inundation can be difficult to fathom. Educating students about the sources, impacts, and possible solutions for this wicked plastic problem invites, and perhaps demands, new ways of integrating their personal experiences into the science. It is not enough just to know what is there. We must know *why* it is there and be able to imagine a world where it is not. In order to change the narrative, we must begin to articulate the story of the current system and see ourselves within the problem.

Why Study the Oceans?

Earth is defined by water, the only substance that exists in all three phase states—solid, liquid, and gas—on the surface of our Goldilocks planet. The majority of our water resides in the ocean, the critical connector of air and land, the keeper and mover of our planetary energy. Oceanic processes that govern our planet span vast spatial and temporal scales, from nanoscale instantaneous photochemistry to basin-scale ore formation processes that occur over millennia. The opening and closing of ocean basins over hundreds of millions of years defines the spaces where species and cultures bloom. Hydrocarbon energy resources we all depend on, no matter how reluctantly, are derived from ancient primary producers in the paleo surface oceans, photosynthesizing sunshine and capturing calories for us to combust now, millions and millions of years later. The ocean defines our planet's climate, absorbing heat and convecting it around the globe, feeding storms and droughts, in a constant complex dance with the atmosphere.

There is a fundamental need for citizens of Earth to understand these oceanic processes and connections, to appreciate the time and energy that has produced the abundance that we enjoy. As the ecological and environmental impacts of our consumption grow, the next generation of scholars needs not only to understand the interconnectedness of Earth systems, but also to grasp the complex connections between Earth and human systems. We need these scholars to be armed with critical thinking and reasoning skills that span the biophysical and social sciences and organically integrate the humanities and arts. Yes, we need hard science skills and analytical solutions, but we also need creative practice that will enable us to imagine a future where humans and ecosystems thrive.

Teaching undergraduate college-level interdisciplinary ocean science courses in land-locked states for the last fifteen years, I have had the opportunity to explore different classroom innovations that help to develop a new oceanic perspective on the world. My syllabus explains:

> This is an interdisciplinary course that includes origins, geology, climate, currents, circulation, chemistry, waves, biology, ecology, coastal processes, human interactions, and sustainability of the oceans. We unravel earth's processes over geologic time scales where humans only just entered the scene. We break down the biogeochemical cycles that link land, water, life, and climate. Developing an oceanic perspective of Earth warrants examination of the non-human biophysical processes of Earth systems, and also an appreciation of how the ocean influences societies and cultures, and how humans have affected and are now impacting the ocean.

It is an ongoing challenge to balance this oceanic interdisciplinary breadth with sufficient knowledge and critical thinking depth to facilitate understanding of complex ocean systems. Undergraduates spanning all levels from freshmen to seniors and from many different disciplines take my Oceans course. It functions as an elective for geoscience and environmental studies students and fulfills a science general education requirement for non-science majors. This mix of disciplinary backgrounds can present a challenge in delivering appropriate content, but also contributes to rich interactions and discussions within the classroom. Thus, while the challenge is real, this course provides an opportunity to give the next generation a common language and more complex vision for re-valuing the ocean. The course is now also identified with a "SUST" attribute, designating it as integrating fundamental concepts of sustainability including equity, economy, and environment throughout the course. I strive to help the students expand what they know, but also to recognize what they don't know and appreciate the limits of their biased perspectives: a humility and a willingness to occupy uncomfortable space where we recognize our own place within broken systems.

From day one, the problem of plastics rears its wicked head in our classroom discussion—the heaps in the gyres, the albatross and whales starving to death, the drifting microplastics outnumbering the phytoplankton, working its way up and out through disrupted food webs. We see the viral photo of the turtle with the straw through its nose, and all those six-pack holders choking the necks of our sea creatures. We've seen it for decades. A tragedy, we think as we watch on screens and drive in cars made from the same primordial ooze. Awkwardly, we disassociate this problem from the plastic pen we take notes with, the plastic lid on our coffee mug, the twist-on cap from my orange juice this morning. How do we handle the hypocrisy? How do we move from wide-eyed horror to motivation and change to stop the tragedies we see playing out in real time? It can feel paralyzing. At the end of our discussions on the first day of class, we talk though the stages of loss and grief, something I had the chance to talk about with David Orr, author of *Earth in Mind: On Education, Environment, and the Human Prospect*, when he visited our university a few years ago. Yes, he agreed, we need to recognize denial, anger, bargaining, depression, and acceptance as we learn about our environmental crises. The students quickly realize this is not just a geology class. It is a class about our intimate connections to our world. We can't just learn about it. We need to understand it and we need to intervene.

Plastic Plague

Increasingly, global news and social media present us with extreme and disturbing images depicting the dramatic problem of plastics in the global ocean. It is ubiquitous. We see white sand beaches on remote island atolls where the sand is barely visible beneath the anthropogenic debris awash on the shore. When researchers examined the contents of sea turtle stomachs (from turtles captured as accidental bi-catch and donated to science),[2] they are full of anthropogenic debris that had been ingested, making up the unnatural majority of the animal's diet. Plastics are now integrated into the geologic record,[3] providing a marker horizon to remind those seven generations from now of this moment in Earth history when humans showered the Earth's surface with the plastic remnant bits of our stuff. Small unidentifiable plastic bits ("confetti" it's called, not nearly as much fun as it sounds) create the framework grains of a new rock type on the south tip of the island of Hawai'i, plastic grains cemented together because that's what beach sediments do, to make a new type of beach rock, the plasticglomerate. Cores of Arctic sea ice include tiny microfibers of polystyrene, acrylic, polyethylene, polypropylene, nylon, polyester, and rayon. Did they travel on the wind or make it out with the laundry water?

Like oil and gas, monomers used to make plastic are primarily derived from fossil hydrocarbons. About 10 percent of the hydrocarbon we produce goes toward the production of plastics. The rise of plastic production and the establishment of our throw-away culture is credited with allowing women to spread their wings and enter the workforce, no longer burdened with the milkman's deliveries and the constant washing associated with re-use. The durability of plastic that makes it so valuable to humans is also what makes it a persistent petulance in the environment. By and large, plastic does not decompose but rather breaks down into smaller and smaller bits. Plastic production began in the 1950s and has accelerated from 30 million tons in 1970 to over 350 million tons in 2015.[4] The problem has transcended from something that impacts the world around us to something that impacts each of us as inhabitants of our polluted spaces. Particles of plastic are now found in tap water, commercial salt, and beer, at rates that suggest that an individual who consumes a single beer a day could be ingesting 500 to 1,800 plastic particles annually just from this product.[5] A study conducted by a scientific working group at the University of California–Santa Barbara's National Center for Ecological Analysis and Synthesis quantified the input of plastic waste from land into the ocean and concluded that every year, eight million metric tons of plastic end up in our oceans.[6] That's the equivalent of five grocery bags filled with plastic for every foot of coastline in the world. By 2025, the estimated annual input will be twice as great, or ten bags of plastic per foot of coastline. The largest market sector for plastic resins is packaging—that is, materials designed for immediate disposal. Researchers

Figure 12.1. Debris accumulated along the high-water line on the beach. Photo by Brenda B. Bowen.

estimate that roughly twenty million metric tons entered aquatic ecosystems in 2016 and anticipate that this may reach up to over fifty million metric tons per year by 2030 and that "extraordinary efforts to transform the global plastics economy are needed."[7]

The Coast

I have been very privileged to spend portions of my life along the Pacific Coast of California. I grew up in a bleached and tanned beach culture where nothing was better than a bright blue day splashing in the coastal surf and sprawling and digging in the sand with family and friends. I went to college in Santa Cruz where I could hear the sea lions barking from my bedroom window and had the chance to learn about the inner and outer workings of the planet, how to read the geologic record and find the stories presented to us by nature. I loved looking closely—my nose against the rocks—to see what the smallest grains and coatings might reveal about Earth's past ebbs

Figure 12.2. The beach where plastic debris was collected. Photo by Brenda B. Bowen.

and flows. As a young graduate student, I spent months at sea aboard a German research vessel mapping the seafloor of Walvis Ridge. I was so proud to cross the Atlantic Ocean by ship, collecting sonar depths over the legendary mid-Atlantic ridge. My days working and living near to the ocean frame how I know it, and surely how I teach it. Now, although I live a day's drive away from the sea, my extended family gathers annually to share a few weeks at the coast. It was during one of these annual sojourns to the coast a few years ago that it all began.

It was a precious sunny summer vacation morning in 2013 and I was out for a run along the sand. The beach was fairly narrow and backed by a cliff where the high tide of the night before had inundated the sand, leaving a high-water mark up against the cliff. A twinkle, a ribbon, and a flash of Barbie pink caught my eye. Suddenly, I saw it: abundant microplastic intimately mixed in with the pungent kelp lining the sand along the high-tide line. Once we've seen it, we cannot unsee it. I crouched down to the sand and began to carefully pick out bits of the anthropogenic debris, filling my sweatshirt pockets. I quickly ran out of personal storage space and returned to the beach later with my children and old grocery bags, easily filling several bags with thousands of tiny bits of trash from the sand. This ritual has now become a family tradition and we plan our travel knowing I will be bringing home several reusable bags of salvaged beach trash. It becomes an annual scavenger hunt for interesting pieces. Picking up each piece, one at a time, from my precious beach, I silently ask each one to tell me its story. Where did the partially melted hot pink Barbie brush begin and how did it find its way to this spot on the sand? I ponder a lone Mickey Mouse hand the size of a grape, his pudgy white fingers unmistakable. Small indentations across the surface suggest that maybe it wasn't left by a beach goer, but that it lost its Mickey long ago and far away. Has it traveled the high seas? Been tied up in sea ice for a season? We collect hundreds and hundreds of skinny green Starbucks stir sticks. Think of all the spills prevented. Was that worth it? Over the years I accumulate a motherload of tire air valve caps and wonder, was it like the rubber duckies, a container ship that spilled its load of caps, or are they there because of all the cars and bicyclists who cruise along the coast. Ironic plastic aquarium plants try to blend in with the real kelp. Deflated bits of birthday balloons. Most of it is plastic, but there is nonplastic debris as well—a cork, a sock, a hair clip, a bit of a crab trap.

I recognize that there is a size bias in the debris I collect—it has to be big enough for me to see and small enough to fit in my bag. My children and I also tend to collect debris in colors that stand out against the sand. Despite our sampling biases and limited time on the sand, after several years of collecting the trash, I began to notice patterns and to make basic forensic interpretations. There is an obvious size and weathering distinction between debris that the sea has washed and transported and the trash that beachgoers leave on the sand. There is less debris after El Niño seasons, with their big storms that bring erosion and not deposition to this beach. The strong south swells focus the debris on the north end of the beach, and I can't find a bit of it to the south. The framework of sediment transport cells becomes evident in the plastic bits and pieces. My beach trash has become a way that I know this coast. It is a part of the environment, and I have now interjected with my touch changing the trajectory. Something different will happen from here forward now that I have gathered each of these pieces. The realization is profound, and while my scale might be small compared to the global weight of the plastic plague, the possibility that we can actively intervene while we hear and retell the stories that our environment is presenting to us inspires me to share my trash with my students.

Touch It

Thus, since that year when I first saw the trash, I arrive to my fall Oceans class with a fresh collection of beach debris. A few weeks into the class, once we have gotten to know each other and have built some trust, I bring my beach trash into the classroom. I dump the bags of debris into giant piles on paper-lined tables. I offer gloves and invite them to touch it. Sort it. Examine it. Attempt to make sense of it. The debris is over 90 percent plastic, ranging in sizes and types, some recognizable products, but most of it has been broken into bits of bits. Students hand-sort, creating diverse classifications of debris. They order by size, composition, color, type, use, or evidence of transport (discoloration or physical breakage). They quantify size fractions of their types by weight and by piece. Single-use items, bits of food wrappers, and twist-off lids. So. Many. Lids. Dental floss holders, Chapstick tubes, Cigarillo tips, the torso of a toy soldier, lighters, strips of plastic bag, exploded bits of Styrofoam, pen and mechanical pencil carcasses, part of a shotgun shell, different sizes of plastic beads, a pill bottle lid, a plastic grape, a nail polish cap, a Lego wheel, a child's highly weathered plastic flower hair clip. They weigh and graph their data, turning my beach trash into science. The tactile experience changes the way we know this debris. The students touch, sort, and ponder each piece.

They shared:

> Even this small amount of trash on the table was a truly disturbing sight.
>
> The majority of the recognizable items were bottle lids.
>
> Mind boggling.
>
> Too much to fathom.
>
> Heartbreaking.
>
> Handling the debris in the bag that was given to us, made me feel gross and ashamed.
>
> Handling the plastic made me feel, above all, guilty.

Figure 12.3. Students sorting debris collected from the beach by Bowen. Photo by Brenda B. Bowen.

> Once I held actual plastic samples from the ocean, it really hit home. I could not image being an animal in the ocean swimming around foreign objects like plastic and maybe mistaking it for food.
>
> Our observations merge towards interpretation.
>
> The plastics range in size from several cm to less than 1mm. This high range of clast size represents a series of events that are likely going on to the various pieces separately so not a single explanation of transport can be used.
>
> Feeling the debris in my hand I kept thinking about how long each piece spent on the beach or in the ocean. Some of the pieces were very smooth, almost like they had been sanded down. They had rounded corners, and discoloration from the amount of time they spent in the water and the sun.... The trash felt gritty, scratchy, but at the same time, smooth with little or no sharp edges.

For the first few years we saw many plastic straws and observed the number decrease through time after California passed a bill in 2018 prohibiting restaurants from providing single-use plastic straws. How long beyond the policy change will we continue to see the remnants of past regulation in the beach debris?

Recrafting Our Narratives

Once the ocean debris is sorted, I invite students to pick a single piece and tell its story. Inspired by a single bit of debris, they craft short historical fiction narratives and tell the stories of the trash: its creation story, path toward existence, interactions with the world, and how it came to find itself here. Touching and knowing the debris has now changed the way we see it. It becomes personal as we wonder about the history of each individual piece. I ask them to describe possible stories of the virgin birth of the resource's origin (or was it created from recycled materials?) and what spurred the creation of the current iteration. What resources did it take to create it? Who made it, who paid for it, and who used it? Where did it travel as it found its way towards being deposited on my beach on the edge of the Pacific Ocean? The students scour the debris for clues about place of origin. Some find date stamps or cultural clues about place and time. We try to explore the full and complicated system that has led to this pollution, from creation through consumption, tracing its path from a hydrocarbon underground to a bit of trash that I picked up off the sand and put into their hands.

How long ago were the unidentifiable bits of colorful "confetti" usable products and how were they discarded? Did the people who made them know that they were creating something that would be with us for hundreds or thousands of years? Did they have a plan for what would become of it once they created it? Ethics of Frankenstein. Educating students about the sources of, impacts of, and possible solutions to this wicked problem invites new ways of integrating their personal experiences into the science. We weave together cogs of systems—commercial consumerism, environmental ecosystems, people with their messy and biased social, economic, and political contexts. We seek to identify the nodes of the plastic plague system and the flows that connect them, building feedback loops that show how it is all connected, and considering all the scales we know of space and time. We look for ourselves and our actions within the system and seek the intervention points where some action might have changed the fate of our beach debris. We consider the institutions we are a part of and their role in reinforcing the negative feedbacks that drive the plague, or how they might create a new kink in the cog that redirects the outcome.

Figure 12.4. Beach plastic sorted by color. Photo by Brenda B. Bowen.

The students' narratives reveal a wide range of imagined journeys and shed light into the way we excuse and rationalize by placing blame beyond ourselves. In their stories, the students discuss the petroleum products used to make plastic materials, the geopolitics of production sites, container ship ocean crossings, and energy associated with all aspects of the life cycle.

"All of this for another plastic comb"

They conjure factories in China, manufacturers in the Philippines, tsunamis capsizing container ships; outside forces driving abstract production. A ketchup packet invokes:

> [a] farm, where the tomatoes and corn (for the high fructose corn syrup) . . . they consumed water . . . fertilizers and pesticides . . . raw ingredients are then picked and transported by fossil fuel burning trucks to factories where they are made into corn syrup and then ketchup. Also, the packet must be made, which contains both plastics and metals in order to safely contain the acidic substance that ketchup is.[8]

Their stories give an individual human context to the vast and unfathomable scope and scale of system. They imagine time spent in the digestive systems of pelicans and sea lions, romantic encounters or celebrations on the beach.

Their stories illustrate an overwhelming framing of the plastics problem that places the blame for each piece on an individual in their narrative. A character in each story makes a poor decision or a lazy choice about how to discard the trash. Most of the student narratives include someone to blame based on a singular poor decision: "laziness and total lack of consideration and responsibility" and "hasty discarding" result in the rerouting of the debris to the beach. Students describe "lazy" people that "just simply do not care to throw away their garbage." Each piece of debris is associated with an anomalous exception that allows this one bit out into the environment.

> Everyone was so busy celebrating, eating catered food and drinking, they didn't think anything when the wind picked up a bit. The increased sea breeze rolled me off the table and I landed next to some beach grass in the sand. And there I stayed. I was forgotten about and left to lay on the sand near the beach grass. Days came and went. The tide ebbed and flowed. Seagulls flew overhead while small crabs tiptoed on the sand around me . . . until one day a very high tide came in. The ocean water flowed beneath me, picked me up, and retreated me back into the ocean.

> The trash bag had a hole in it and, therefore, several pieces along with the disposable flosser fell out of the bag on the way to the trash bin.

> The lonely lid that was jammed between the wires finally escaped the wires because as the truck turned, it loosened the wires allowing the wind to take the lid out of the truck, and blow it toward the coast. The lid finally landed on the coast in California in the nice warm sandy beach. The sand gently layered the plastic lid until the lid was not visible in the sand anymore.

> The lids were all used once, hypothetically thrown away into the garbage can. They were not. . . . They were thrown out of the car window, fell off of a table, fell out of a garbage can, etc. and ended up in the ocean.

> People tend to drop their soda caps where they can easily get buried and lost.
>
> Once she was done with her drink, instead of throwing away the trash or recycling it like a considerate human being that cares about our planet, she instead threw it out her window.
>
> Someone accidentally bumped into him and made Chang drop his plastic cap.
>
> In his momentary lapse in judgement, due to a strong bout of caffeine deprived road rage, he threw the cup and un-spilled coffee out the window.
>
> He was not in his right mind to go and find a recycle bin, so he tossed the bottle and cap to a spot on the beach.
>
> There is only one reason there is so much waste floating around the world today: laziness.

Although the students are very aware of scientific literature and data that show the global scale of the plastic plague and the corporate and political scale of the drivers, they made sense of the debris by placing responsibility on someone else who made a poor choice. One goal of the exercise, therefore, is to formulate a new narrative. Instead of viewing most people and producers as blameworthy polluters, we envision a well-intentioned populace reliant upon a civil sanitation system bursting at the seams, hemorrhaging in response to global consumerism and petroleum domination. We revise our stories to imagine the intervention points where a new path might be taken. We must change our relationship with the problem if we are to become a part of the solution. We begin to recognize the characters in our stories—the ones drinking from the straws, twisting off the caps, buying the plastic doo-dads for our whatnots. By dissecting the system and then retelling it within a context we know, we begin to see our place within it. We map out the connections between the nodes of the system, the feedbacks among the complex interactions, and work to find the points of intervention where a different and more desirable fate could have befallen our bit of debris. But what *is* that preferred fate?

The students' narratives point to a range of perceived solutions, and many of them advocate that "recycling plastics is the best way."

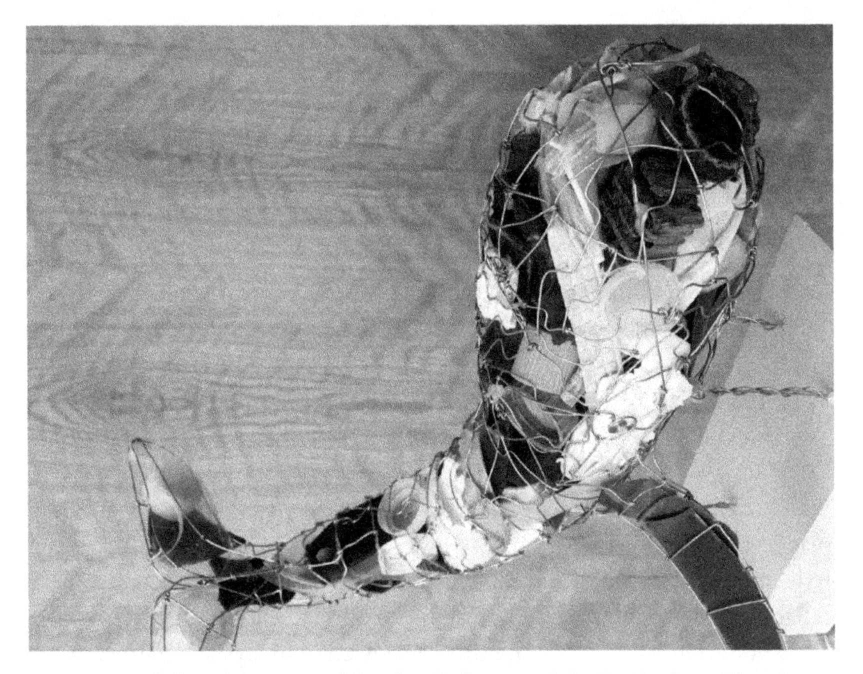

Figure 12.5. Whale sculpture created from beach plastic, made by Boe Erickson. Photo by Brenda B. Bowen.

With modern recycling techniques all that was needed was for the consumer to make the decision to put their trash into a recycling bin instead of wherever it was originally deposited before its journey to the ocean.

[Is there a need to simply supply] more garbage and recycle cans on the beach?

If people aren't allowed to bring food to the beach, plastic disposable bottles to the beach, and don't have the ability to buy food and toys to bring onto the beach, there would be less garbage on the beach.

Ultimately, the only way to keep this object off the beach would be to not bring it in in the first place.

Step one is simple: get your trash in the bin.

However, as China stopped importing the world's recycling two years ago, our eyes were opened to the extreme limits of what our current global manifestation of plastics recycling entails. Much of what had been sent to foreign shores ended up in landfills or in the sea.[9] It turns out that plastics are not made to be recycled. We have been fooled into thinking that we can consume and not be the ones to blame. If we put it in this certain bin, it disappears.

> I have been adding to the extensive problem of "throwing it away," because after all, where is "away"?

> It was surreal realizing that most of these pollutants are things that I, myself, consume all of the time.

We deduce what scholars have proposed—that mitigation of the plastic plague must begin with industry-led reduction in plastic production and reduced demand. Yes, the litter must be captured, and yes, we need more reliable and just global waste management and recycling systems. But even at its best, that won't solve the problem. A 2020 analysis suggests that 78 percent of the plastic pollution problem could be solved by 2040 through the "use of current knowledge and technologies" but that "substantial commitments to improving the global plastic system are required from businesses, governments, and the international community."[10] It's not just in the ocean, but also in our food.[11] Perhaps the only way to actually halt the flow of plastics to the sea is not to produce them in the first place, but how do we advocate for that change? It is nearly impossible to avoid plastic in our society. Try it. For decades, conservationists and climate change activists have championed the idea of keeping hydrocarbons in the ground. Is that what it will take, or can we innovate enough new industries to turn this into a green economy? I recognize our collective dependence on the industrial luxuries that allow some of us to thrive on a privileged personal level—tapping plastic keys on my keyboard, feeding my family from plastic casing. Within an arm's reach of where I sit: my Nalgene water bottle, my polyester fleece, plastic keychain, USB drive, markers, sunglasses, my son's eyeglasses, our shoes, the mouse to control the writing of these words. How do we break through the plastic realities of our daily consumption and move beyond it?

Breaking down that disassociation and productively connecting individual choices to the waste system is part of the goal of the course. Let's tell stories of transition to new materials, stewardship, and values. Cleanup projects are essential and admirable, but until the source of the plague is addressed,

Figure 12.6. Plastic debris in the beach ecosystem. Photo by Brenda B. Bowen.

the cleanups will not solve the problem. Many of my students have gone on to make spectacular art projects out of the beach debris. Sculptures and murals of the ocean's beauty juxtaposed with the debris that is killing it. It can be quite profound and moving, but I'm afraid we cannot art our way out of this plague. I challenge the students to come up with ideas for what I should do with my beach debris. If I put it in the recycling bin or the trash can, it might go through another cycle of environmental degradation until we pick it up again and contemplate its fate. For now, it piles up in my office, ready to inspire. As one of my students noted,

> As the tides go in and out and winds blow towards and then away from the shore, trash that may have been deposited in the morning can then be washed back out to sea yet again.

Notes

First image from New Yorker cover, "Waves" by J.J. Sempe, August 22, 2016. All other images are original photographs taken by B.B. Bowen.

1. Barnes, D.K.A., Galgani, F., Thompson, R.C., Barlaz, M., 2009, Accumulation and fragmentation of plastic debris in global environments: *Philosophical Transactions of the Royal Society* 364, 1985–1998. See also Cozar, A., Echevarria, F., Gonzalez-Gordillo, I.J., Irigoien, X., Ubeda, B., Hernandez-Leon, S., Palma, A.T., Navarro, S., Garcial-de-Lomas, J., Ruiz, A., Fernandez-de-Puelles, M., Duarte, C., 2014, Plastic debris in the open ocean: *Proceedings of the National Academy of Sciences of the United States of America* 111 (28), 10239–10244.
2. Wedemeyer-Strombel, K.R., Balazs, G.H., Johnson, J.B., Peterson, T.D., Wichsten, M.K., Plotkin, P.T., 2015, High frequency of occurrence of anthropogenic debris ingestion by sea turtles in the North Pacific Ocean: *Marine Biology* 162, 2079–2091.
3. Corcoran, P.L., Moore, C.J., Jazvac, K., 2014, An anthropogenic marker horizon in the future rock record: *GSA Today* 24 (6), 4–8.
4. Geyer, R., Jambeck, J.R., and Law, K.L., 2017, Production, use, and fate of all plastics ever made: *Science Advances*, v. 3, e1700782.
5. Kosuth, M., Mason, S.A., Wattenberg, E.V., 2018, Anthropogenic contamination of tap water, beer, and sea salt: *PLoS ONE* 13 (4): e0194970.
6. Jambeck, J.R., Geyer, R., Wilcox, C., Seigler, T.R., Perryman, M., Andrady, A., et al., 2015, Plastic waste inputs from land into the ocean: *Science*, v. 347, p. 768–771.
7. Borrelle, S.B., et al., 2020, Predicted growth in plastic waste exceeds efforts to mitigate plastic pollution: *Science* 369 (6510), 1515–1518.
8. "Class Notes, Ocean, 2017."
9. Sullivan, L., 2020, How big oil misled the public into believing plastic would be recycled: *NPR Investigations*, Morning Edition, September 11, 2020.
10. Lau, W.W. Y, et al., 2020, Evaluating scenarios toward zero plastic pollution: *Science* 369 (6510), 1455–1461.
11. Sharma, S., Chatterjee, S., Microplastic pollution, a threat to marine ecosystem and human health: a short review. *Environmental Science and Pollution Research Journal* (September 2017), 21530–21547.

CHAPTER 13

Adaptive and Interactive Futures

Developing "Serious Games" for Coastal Community Engagement and Decision-Making

KATHRYN K. DAVIES, BENJAMIN A. DAVIES, PAULA BLACKETT, PAULA HOLLAND, AND NICHOLAS CRADOCK-HENRY

CLIMATE CHANGE IS already affecting communities and livelihoods around the world through increased temperatures, prolonged droughts, violent storms, and other extremes. These impacts are expected to increase over time.[1] In Aotearoa, New Zealand (Aotearoa NZ), a country with more than 18,000 kilometres of coastline, planning for and adapting to the effects of sea-level rise are increasingly urgent concerns.[2] Today, there is a range of adaptation responses available to address climate change (e.g., hard structures, soft engineering options, relocation), but there is a need for community and provincial buy-in in order for specific adaptation options to actually be implemented. Ensuring buy-in requires that communities and interest groups understand the rationale behind options. However, there are multiple possible combinations of adaptation options to consider over time, often referred to in climate change literature as "pathways."[3] The opportunities to take many of these pathways are time-limited, and the choice of some adaption pathways in the near term may limit the ability to choose others in the future. Choosing from a range of pathways is likely to be complicated by strong values, vested interests, and unequal burdens, and therefore can lead to hotly contested adaptation discussions.[4]

Given the range of potential paths, their uncertain outcomes, and the social constraints on actions needed to initiate them, adapting to climate change presents as a "wicked problem"[5] with no clear, single solution and no easy means to test various options. Wicked problems are characterized by complexity, uncertainty, interdependence, and dispute, and are found in highly interconnected systems. In these systems, technical analysis alone is unlikely to lead to successful resolutions. Instead, stakeholder involvement in decision making is needed to ensure that multiple conflicting values are aired and negotiated, uncertainties are examined and understood by those likely to be affected, and increased risks are considered and managed.[6]

One way to enhance stakeholder involvement is to use serious games. Serious games, or games or simulations that are used for purposes beyond entertainment,[7] have been increasingly recognized for their potential to facilitate the exploration of value laden and contested decisions[8] and to support learning in diverse communities of stakeholders.[9] By using serious games to address climate change, players can learn about complexities of climate change and simultaneously develop skills necessary for adaptation.[10] For example, players may have to navigate multiple scales (time and/or space), individual and/or collective problems, and local and/or national issues. Serious games can also encourage players to consider alternative ways forward (pathways planning), and to trial innovative approaches to problem solving.[11] In this way, players challenge their own existing beliefs about strategies that will work, confronting their own mental models and removing potential barriers to adaptation action.[12] Game environments can provide opportunities for players to go through this process autonomously or collectively, and they can harness local knowledge or not, depending on their purpose and how they are designed.

Games encourage players to practice a range of different skill sets; for example, to be successful, players may have to follow rules, deploy strategies, make rapid decisions associated with trade-offs, take risks, or resolve conflicts. However, unlike performing these actions in real settings, where consequences could be detrimental to the wellbeing of the actor or others, the simulated environment offered by a game gives players the opportunity to learn, innovate, and experiment with these actions, with consequences that are real in the game world but pose no threat in reality.[13] The experimental mindset encouraged by this approach is particularly valuable for confronting wicked problems, where such safety is absent.[14]

While the study of serious games began more than forty years ago, the concept has only seen widespread use in addressing wicked or otherwise complex problems in the last two decades,[15] and there are still relatively few games designed for climate change adaptation planning.[16] This chapter describes the development and preliminary testing of novel serious games related to climate change adaptation in Aotearoa NZ. These games balance game complexity with social dynamics, exhibiting different strengths for different contexts. An online game called *Adaptive Futures* is highlighted to demonstrate some of these characteristics and how they are negotiated. Using compelling narratives based on real situations from coastal Aotearoa NZ, *Adaptive Futures* is tailored to engage a range of users in coastal adaptation planning (e.g., students, local and regional authorities, the wider public). The results of this game play are evaluated against a learning and outcomes framework based on Armitage et al.[17] Our preliminary findings indicate that through gaming, *Adaptive Futures* players learn about climate change impacts, implications, and adaptation options for coastal regions. The serious game also facilitates experimentation with robust management strategies applicable to a range of climate change scenarios. Access to the game is open to anyone with an internet connection and an interest in considering and reflecting on decision-making under uncertain and changing conditions.[18]

Serious Games from Aotearoa, New Zealand

While some problems and actions associated with climate change can occur at a local, individual level (for example, wildfire threats to a single-family home), many problems (for example, loss of markets as a result of resource depletion) and potential resolutions (for example, sustainable management practices and planning) occur across multiple temporal, spatial, and organizational scales.[19] Given this, and the complicating factors listed above, no single game can fully capture the complexity associated with climate change adaptation. Engaging stakeholders therefore requires developing games that use strategic trade-offs in how scenarios are represented and how gameplay targets different kinds of learning.

To help communities navigate some of the unique climate change adaptation challenges associated with Aotearoa NZ, a suite of serious games that enhance opportunities for learning, practice, and behaviour change have been developed with this specific context in mind. These games range in duration, complexity, and social interaction, and can be deployed in different settings and tailored for different audiences/interests.

Climate Adaptation Challenge

The NIWA-developed tabletop board game, *Climate Adaptation Challenge*[20] (Figure 13.1), allows players to explore different climate scenarios and pathways they could take to mitigate or realize potential opportunities

Figure 13.1. Aotearoa New Zealand's Climate Change Minister James Shaw plays *Climate Adaptation Challenge* with NIWA staff observing. Photo by Sarah Fraser/NIWA. Used by permission.

from a changing climate on their own farms. Using 3D-printed pieces and specially designed play money, players are invited to make decisions about their own farming practices in response to expected increases in extreme rainfall brought on by climate change. Each player rolls a twelve-sided die to simulate a climate scenario over ten years on the farm. Three sides of the die represent a "severe flood" outcome, four represent a "high flood" outcome, and five represent "no event." Players must then choose how to spend their money. The objective of the game is to avoid losing the farm's profitability when confronted with the effects of climate change over time.

Players use the "pathways approach" to explore, develop, and implement alternative strategies to address climate impacts, without compromising or shutting off other options. "Pathways thinking" is a strategic-planning approach, giving farmers a framework to consider many different options, how long these might be effective for, and when it might be time to change tack. The game has been deployed in a range of settings in Aotearoa NZ to encourage stakeholders with different interests to have conversations with

Figure 13.2. Playing *Maraeopoly* with the Tangoio community. Photo by Paula Blackett/NIWA. Used by permission.

one another about how climate change might affect their homes and businesses, hopefully in advance of a major event occurring. A similar version is under development for coastal communities.

Maraeopoly

The *Maraeopoly* game was developed by NIWA in partnership with members of Tangoio marae[21] to assist the Hawke's Bay[22] community to adapt to climate change. A marae is a sacred communal and collectively owned space that serves religious and social purposes in Māori society. The term is often used to include the complex of buildings around a central courtyard. Tangoio marae is central to the community, connecting family histories within the area and providing space for community activities and meetings. It is situated on a coastal flat within a low-lying catchment with high flood risk. The game was developed as part of a strategic, long-term plan to assist in mitigating the increasing threat of catastrophic flooding on the marae facilities while also supporting the development and maintenance of a vibrant community. *Maraeopoly* is in many ways the most tailored and complex game developed thus far with input from NIWA.[23]

Teams play the game in ten-year increments. At each turn, players must negotiate with their team to decide how to spend their money. Choices involve agreeing on issues such as: Do they want to invest in adaptation in the first place? If so, should they upgrade their embankment; retrofit the existing buildings; or buy new land and relocate the marae complex? Once players make their choices, they reflect on how they feel about these decisions. Computer software then simulates a series of events that will affect the investments. These may be small, medium, or large flood events that will affect the infrastructure in various ways, thus propelling the players to make another series of choices in the next round. This work has prompted important intergenerational discussions about what is important/valued, what outcomes are desirable, what options are reasonable, and what needs to be monitored to help clarify choices.[24] This is all occurring in advance of a devastating flood, rather than in response to one.

Path Diversity and Social Dynamics

Both *Climate Adaptation Challenge* and *Maraeopoly* share similar characteristics in their use of adaptation pathways to encourage thinking about climate change scenarios in local settings. However, these games differ in the diversity of potential paths towards climate change adaptation decision-making and role of social dynamics in gameplay. In *Climate Adaptation Challenge*, the game is short, the scenario is simple, and opportunities for decisions or chance events to change the course of the game are few, constraining the potential number of pathways through the game. At the same time, the decision-making is individualized; players do not need to consider anyone other than themselves as far as outcomes or responsibility. This combination of uncomplicated gameplay and individual decision-making helps to simplify the educational concepts, making it possible for players to grasp a few key concepts about climate change adaptation measures in a relatively short timeframe.

By contrast, in *Maraeopoly*, the opportunities to affect game outcomes are numerous, with a range of interventions that can be implemented in each turn. Additionally, players are required to engage with a group in order to make decisions, and group diversity, values, and power dynamics can have a strong influence over the pathways chosen. This combination of factors means that players are simultaneously exposed to complex concepts related to climate change adaptation measures, and also to the values and preferences of others. This approach improves social learning outcomes, where the participants learn collectively, rather than just improving individual learning outcomes.

Table 13.1. Game complexity typologies

Game	Type	Pathway diversity	Social dynamics
Climate Adaptation Challenge	Board game	Low	Individual
Adaptive Futures	Online game	Medium	Bounded
Maraeopoly	Board game	High	Collective

To address the range of challenges associated with climate change adaptation decision-making, it can be helpful to draw upon a spectrum of gaming options. The *Adaptive Futures* game provides a compromise between the pathway diversities and social dynamics extremes represented by *Climate Adaptation Challenge* and *Maraeopoly* (Table 13.1). By utilizing an open-access online platform, this serious game can also reach a much wider audience and can be adapted for alternative settings with relative ease.

Detailed Case Study: *Adaptive Futures*

Adaptive Futures is a serious game designed to introduce players to some of the complexities associated with community-level decision-making and climate change adaptation options. As such, a number of complex components of the game interact to create a plausible adaptation experience for players. The game targets learning and experimentation by engaging players in plausible climate change scenarios and encouraging them to make real world-type decisions in a "safe space."[25] The game can be deployed as a teaching tool either individually or in a facilitated group setting. It is programmed using the Twine platform,[26] which is an open-source tool for telling interactive, nonlinear stories. The format is similar to a Choose-Your-Own-Adventure book, in which the reader makes a decision on behalf of the character in the story. Twine can be used to rapidly prototype simple games, experiment with game features, and build in complex functionality as needed.[27]

The objective of the game is to protect the Seaview coastal community from the adverse effects of climate change. Players are positioned as leaders of a "Climate Committee" with the job of addressing climate change threats such as coastal inundation, which can harm local homes and businesses, and storm-induced coastal erosion, which can destroy the beach and harm local tourism-based jobs and businesses. To win the game, players must retain their seat on the committee without being ejected by disgruntled voters. The severity of climate change in the game is reflected in the rate of sea-level rise. This rate is randomly selected for the player based on different possible climate scenarios.[28] Options are also built into the game so that players can apply their own rates of change if they want to explore specific scenarios.

As with the *Climate Adaptation Challenge* and *Maraeopoly* games, all player decisions are made in ten-year blocks[29] and have physical, social, and economic consequences. The range of possible adaptation responses available to players is limited by factors such as committee finances and support from the diverse Seaview community. Achieving and maintaining support requires players to build trust with community members, who each experience the effects of climate change differently depending on their own values and location within the coastal space. Therefore, the game requires the player to balance the diverse wishes of the community with the need to mitigate the immediate and long-term effects of climate change.

At the beginning of each turn, players are presented with a summary of the condition of the town, as well as a dynamic map of Seaview that shows flood-affected areas (Figure 13.3). Different adaptation strategies can be selected to mitigate the effects of climate change on the community. Alternatively, players can choose to do nothing and "save" funds for more expensive options. Adaptation strategies include nourishing beaches, building seawalls, and relocating segments of the community. Each of the strategies is described in the game, including benefits, constraints, life span, and costs, so players can make informed decisions. The council must have sufficient funds to purchase an option, and selecting certain options may constrain the capacity to use other options later. Early development of the game considered a much wider set of strategies, including the pumping of flooded areas, upgrading of small beach walls to more structurally robust ones, and the maintenance of existing structures. However, these were eventually excluded, because it became apparent through trialling and testing that very similar dialogue over climate change and its implications could be achieved using a game with a select number of key adaptation options.

Community Components

The Seaview community is populated by nonplayer characters (NPCs) who represent a range of views typically expressed in coastal adaptation debates.[30] Winning the game is reflected in how well the NPCs consider the committee to have performed. Each NPC has an avatar image, location, back story, interests, preferences, values, and attitudes towards the Seaview

Figure 13.3. User interface for *Adaptive Futures*, illustrating the condition of the town in terms of a written description as well as a dynamic map of Seaview that shows flood-affected areas and community stakeholders.

Climate Committee. Players can learn about the NPCs through an interactive panel displaying the NPC avatars. When an avatar is clicked, the player can view information about the character's values and preferences along with a quote from the character indicating their attitude (the content of which is also determined by the character's position in the attitude matrix; see Figure 13.5).

The values expressed by each NPC are loosely associated with one of the four capitals (Figure 13.4) utilised by the NZ Treasury to consider intergenerational wellbeing in decision-making.[31] For example, Kim the Fish & Chips Shop Owner is primarily concerned about the stability of her business (financial capital), but she is aware of the important influence of the other three capitals (human, natural, and social) on her business and her wellbeing in general. By comparison, Fern, a working mother, is primarily concerned with ensuring the environmental integrity of the area although she is also aware of the need for other issues such as human, social, and financial security.

Throughout game play, NPCs shift their desires, responses, and attitudes to the Seaview Climate Committee, depending on the decisions of

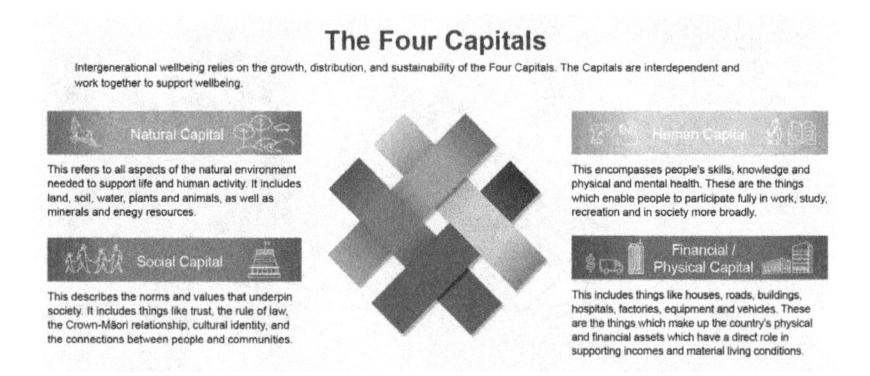

Figure 13.4. The four-capitals approach utilised by NZ Treasury to promote intergenerational wellbeing. Office of the Clerk/Parliamentary Service. Licensed by the NZ Treasury for re-use under the Creative Commons Attribution 4.0 International license. Full license available at https://creativecommons.org/licenses/by/4.0/.

the committee and any subsequent harm they experience from events. Their attitude toward the committee is positioned within a unique attitude matrix comprising "satisfaction vs trust" variables (Figure 13.5). The matrix functions as follows:

- Satisfaction: decreases when NPCs experience negative effects from climate change and increases when those effects are limited.
- Trust: decreases when the player pursues adaptation against the character's wishes and increases when player pursues adaptation options that align with the player's wishes.

As the game progresses, each NPC's position in the matrix will shift, pushing them between four states:

- Adaptive: The NPC is satisfied with their situation but trusts the player to make decisions.
- Reactive: The NPC is dissatisfied with their situation but trusts the player to make decisions.
- Lethargic: The NPC is satisfied with their situation but does not trust the player to make decisions.
- Outraged: The NPC is dissatisfied with their situation but does not trust the player to make decisions.

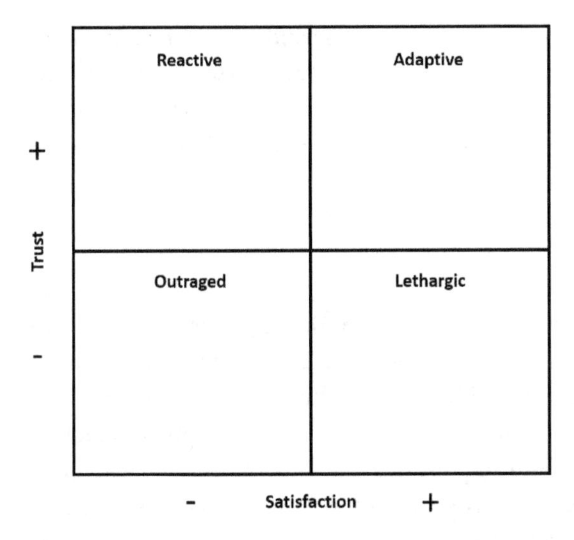

Figure 13.5. Attitude matrix that demonstrates how NPC attitudes towards the Seaview Climate Committee change.

Each of these states corresponds with a point value, with "adaptive" being the highest and "outraged" being the lowest. Through these states the player receives an "approval rating" which is calculated as the sum point values from NPC attitudes divided by the total possible points if all NPCs were in the adaptive state. Player approval ratings are tracked and presented for players throughout the game and ultimately contribute to the ability of the player to remain in the game. If the player's approval rating dips below a certain threshold, the player is removed from office and the game is over.

The unique combination of the four capitals approach and the attitude matrix prompts players to consider a range of values and adaptation options, rather than making their decisions purely based on finances or scientific/engineering options alone. How players engage with the community, and the adaptation options chosen, influences subsequent NPC behaviour by shifting satisfaction with current conditions and trust in decision-makers.

Each NPC contributes to the operating budget of the committee through rates (taxes). However, their contributions depend on their ability to make a living. NPCs that are not harmed by climate change can continue to contribute fully to taxes—and consequently to the committee's adaptation fund—over time. By comparison, NPCs who are somewhat harmed by climate change[32] will only be able to contribute a portion of their usual amount, while NPCs who have been fully affected by climate change[33] cease to contribute to the tax base altogether. This financial consequence is important because it means that poor or unsustainable investment choices by game players will result in losses to their tax base and impair their ability to serve the community (by protecting them from future harm). For example, an NPC with a shop located near the beach will be threatened if beach-going tourists are deterred by erosion, while they will be fully affected if the shop itself is persistently flooded.

Consulting with the Community

Another unique feature of the game is its reminder to players of the need to consult with communities on key changes. In New Zealand, the Resource Management Act (1991) sets out how councils make decisions about activities affecting the environment. It applies to major developments such as the development of new housing estates or—in the case of *Adaptative Futures*—the potential relocation of communities and/or businesses. *Adaptative Futures* requires players to spend at least one turn consulting with the community—hosting a public forum—if they seek to relocate either residents or businesses away from coastal hazards such as inundation or erosion.

When a player chooses the consultation option, a number of probabilistically selected NPCs will attend the meeting and voice their interests. Consultation has the effect of building trust with attending NPCs through a turn, which can potentially stave off a no-confidence vote. The likelihood that an NPC will attend a forum depends on both the disposition of the character and the passage of time. Consultation with certain blocs of NPCs is also necessary in order to proceed with the relocation of residents and businesses.

Winning the Game

Players win or lose a game based on their approval rating and the remaining threats to the community. So, the *Adaptive Futures* game ends when (1) the player is voted out of office (approval sinks below a critical threshold); (2) coastal residents and businesses no longer face danger from coastal inundation or erosion; or (3) the player goes through ten turns (one hundred years) without either (1) or (2) occurring. Where several participants play at once, the committee member who manages to stay in power the longest and who has the highest level of voter confidence is the overall winner. At the end of the game, each player receives a summary of the decisions that were made, the consultation history, budget, and changes in approval over time, as well as a link to a short questionnaire about the experience.

Evaluating Game Play

An early-stage evaluation of game play was undertaken at the New Zealand Coastal Society Conference in Gisborne, Aotearoa NZ in November 2018. This early prototype testing allowed researchers to obtain feedback on game development from a knowledgeable expert group. The version of the game that was tested with this group included sea-level rise trajectories, adaptation options, three characters, decision-maker popularity score, and rudimentary financial details regarding expenses and income. Human ethics approval was granted by the NIWA Human Ethics Application Process prior to field testing.

Researchers offered conference participants the opportunity to play the game on a tablet during conference breaks. Game play took approximately ten to fifteen minutes for one or two rounds of play. Prior to testing, an evaluation plan was created to assess the impact of the game on individual participants and groups.[34] The evaluation includes an introductory script and several question variants that can be tailored to suit different contexts and/or players. After being read the introductory script and agreeing to proceed with game testing, twenty volunteer players tried the game and then provided verbal feedback on their experience and suggestions for improvement. Players represented a cross-section of coastal-focused council staff, practitioners, researchers, and students.

The game was further developed based on feedback from this initial round of testing, plus several more informal testing sessions with both experts and nonexperts in coastal adaptation planning. In September 2019, a second round of formal testing was conducted with a group of approximately sixty planning students from Waikato University. These participants were more representative of our target audience (e.g., council staff, students, and Māori representatives). The version of the game that was tested with this group included sea-level rise trajectories, adaptation options, six NPCs, an approval score, the capacity for the decision-maker to be voted out of office, the capacity to consult with NPCs, financial details regarding expenses and income, and improved visuals throughout the game.

Pre-arranged groups of two to four played several rounds of the game (approximately thirty minutes play time total) on shared computers or tablets. Players discussed their responses and approaches as a group. Some of the groups completed online pre- and post-game questionnaires, but these were voluntary and not all groups chose to complete them. Players participated in a debrief discussion and several groups completed and returned a voluntary written reflection form after playing the game.

Qualitative game play results and participant observations were evaluated against a learning-and-outcomes framework based on Armitage et al.[35] (Figure 13.6). This framework assesses the enabling conditions that are required to promote two individual-scale learning effects—cognitive learning and relational learning—and then goes on to connect this learning (where possible) to either process or substantive outcomes. This typology of learning effects is adapted from previous studies that focus on cognitive, relational, and normative learning in similar collaborative governance contexts. In line with this literature, cognitive learning is defined here as changes in understanding of social and/or ecological conditions, relational learning as changes in perceptions of others in regards to efforts to learn together (e.g., levels of cooperation and trust), and normative learning as changes in participants' perceptions of the overall benefits of interactions reflected in shifts in values or a convergence of views. However, several previous studies attempting to measure normative effects have met with limited success,[36] and therefore this component was omitted from the current analysis.

In the study, the activities are game play, and the collaborative qualities are built into the game (e.g., consultation with stakeholders). However, the collaborative qualities can be further enhanced by employing facilitation techniques. Following Armitage et al.,[37] an operational definition of outcome that incorporates both process and substantive components was assigned. Outcomes are therefore measured either through perceptions of players (e.g., perceptions that game play has led to better decision-making—a process outcome) or directly (e.g., winning the game or improving game score—a substantive outcome).

Preliminary Evaluations

The response to the *Adaptive Futures* game from the initial test sample (November 2018) was overall very positive in terms of the outcomes that the game is aiming to achieve. All players felt the game would be a valuable tool to engage and teach a range of audiences about coastal adaptation strategies. Players reported liking the concept of a game and feeling that it was an appropriate learning and engagement tool. Many players became engrossed in the game themselves, describing their choices as difficult to make and agonizing over how to balance the interests of the different NPCs.

Many players reported cognitive changes resulting from the game, saying that it made them think about the consequences of climate change in more

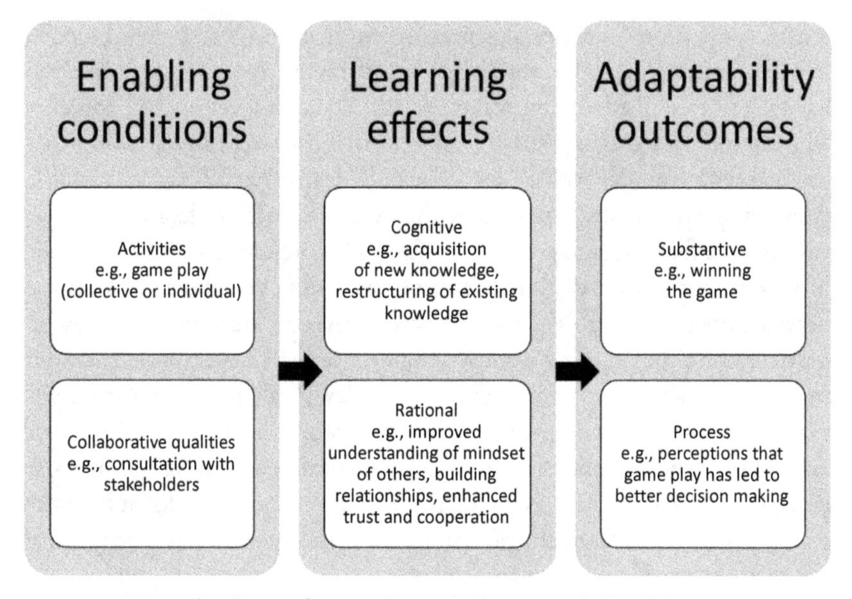

Figure 13.6. General evaluation framework to assess learning and adaptability outcomes. Adapted from Armitage et al. Used by permission.

concrete ways. For example, one player pointed out that "the moral of the story is—relocate as soon as you can." Several players also commented on early-stage relational changes that resulted from game play, describing their wish to hear more from NPCs and how they wanted more opportunities to interact with these characters. No substantive changes were reported by players, but this is not surprising given the short amount of time allocated for game play. Some players were able to make it to the very end of the game without being voted out of office, but all of these reported low popularity scores at the end of the game.

Several players reported process changes resulting from game play. For example, one player admitted that "I would make different decisions if I did it again—do nothing for a while and save money for re-nourishment and relocation." Another player commented as game play was proceeding, "This is stressful! I'm just going to install a higher wall to make the people happy." However, at the end of the game, that same player admitted that "in hindsight, I shouldn't have spent all my money at the start on a wall." In these cases, the game clearly facilitated experimentation with robust management strategies that are applicable to a range of climate change scenarios.

The response to the *Adaptive Futures* game from the second test sample (September 2019) was also positive in terms of engagement and teaching coastal adaptation strategies, although the students mostly felt that the game would be better targeted towards community members who were unlikely to have much formal education about the effects of sea-level rise. All students who completed a pregame survey reported that they were at least somewhat familiar with the effects of sea-level rise in Aotearoa New Zealand, with the majority of students reporting that they were at least moderately familiar with these effects prior to playing the game. Despite this familiarity with the topic, one area showed distinct changes in student beliefs about sea-level rise between the pre- and post-game survey responses. In the pregame survey, all participants either agreed, somewhat agreed, or were neutral when asked if sea-level rise will happen slowly enough for New Zealand coastal communities to adapt. However, in the post-game survey, all participants shifted their opinions to the other end of the spectrum, and either disagreed or somewhat disagreed with the same statement. This result indicates that, for at least the students who completed the pre- and post-game questionnaires, cognitive changes occurred because of game play.

The qualitative reflections shared by students on the written evaluation form at the end of the game play and debrief discussion, and also on the post-game survey, indicate that the majority of students also experienced relational changes as a result of game play. Several students wrote that they learned how difficult it could be to make coastal communities happy with climate change adaptation decisions.[38] Others described this social component of the game as the most challenging part of the experience.

All student players indicated that they were able to succeed at the game to some extent (by not being voted out of office for at least thirty years—although most of the groups made it to at least sixty years into the future, out of a total one hundred possible), but they also described ways that they played the game differently after their first attempt. This result indicates that the game encouraged process changes in players; players considered and experimented with a range of adaptation strategies, rather than sticking to a single approach to "win" the game.

In general, the majority of players felt the game represented plausible experiences of climate change adaptation related to decision-making at the coast, albeit with minor detail. Although there was substantial feedback generated from both testing sessions regarding specific details that could be adjusted to make the game more or less realistic, players generally felt that *Adaptive Futures* does a good job of engaging and educating players about sea-level rise adaptation options and challenges in Aotearoa New Zealand.

Discussion

Certain enabling conditions are required to promote learning effects and adaptation outcomes for participatory processes. The plausible enabling conditions provided by *Adaptive Futures* emerge primarily from the novel combination of scientific, economic, and social factors that are positioned as equally important within the structure of the game. These three evidence-based elements frame each of the climate change adaptation options, thereby encouraging players to learn about and experiment with not just one, but all three components, in their decision-making. *Adaptive Futures* encourages players to engage with these elements by providing a bounded set of collaborative social dynamics options (e.g., consultation with stakeholders), and evidence from preliminary game testing suggests that the game can promote cognitive learning about social factors in particular. This skill set is valuable for complex decision-making associated with climate change adaptation.[39]

Armitage and colleagues[40] demonstrate that more learning (both cognitive and relational together) generally leads to better outcomes, and that cognitive learning contributes more to process outcomes and relational learning more to substantive outcomes. Although the findings of this research are only preliminary at this stage, so far this study broadly supports the connection between cognitive learning effects and process outcomes, as most of the learning effects captured in the evaluation data were cognitive, and most of the changes that players described post-game were associated with experimenting with a range of adaptation strategies. The relational learning effects that were noted by participants were often associated with a mild frustration at how challenging it was to find common ground among the range of NPC perspectives and how difficult it was to "make everyone happy." It may be possible to enhance the relational learning effects of *Adaptive Futures* further through the practice of certain group facilitation techniques over others, but this link has not yet been explored in any depth.

One of the most important learning effects that the game produced was associated with climate change adaptation timeframes. Although these are only preliminary findings, there were strong indications from both the quantitative and qualitative data that study participants learned about the urgency of implementing climate change adaptation strategies now, rather than waiting another decade or more to do so. This was true even when working with experts working in the field, or students who self-identified as having some to moderate knowledge of the effects of sea-level rise.

The fact that the gaming environment forced players to consider the more social elements of climate change adaptation decision-making in order to win the game, thereby encouraging them to learn about alternative positions, may predispose participants to be open to participatory processes in reality.[41] Many participants described a new feeling of urgency regarding climate change post-game, which is an exciting and important development that we will aim to build upon in future iterations of *Adaptive Futures*.

Conclusions

Implementing adaptation strategies to address climate change implications and impacts in coastal regions is a complex and challenging problem from any angle. On their own, the scientific, economic, and social elements of these issues are rife with conflicting values and high levels of uncertainty and risk. In combination, tackling these challenges requires extensive negotiations to promote collective actions.[42] Serious games can provide a space in which to develop and practice the skills needed to navigate these challenges in the real world,[43] but to our knowledge such games have not yet taken on the social elements of climate change adaptation planning in any substantial manner, and questions regarding how to leverage these useful qualities to engage a wide range of audiences and encourage behaviour change remain unanswered. Specific challenges associated with balancing pathway diversity and social dynamics are especially important to harnessing the potential of games for application in the climate change adaptation space at large.

Adaptive Futures enables players to engage with the scientific, economic, and social elements of climate change adaptation planning in a single gaming environment. The preliminary findings indicate that through gaming, *Adaptive Futures* players learn about climate change impacts, implications, and adaptation options for coastal regions. The serious game also facilitates experimentation with robust management strategies applicable to a range of climate change scenarios.

Notes

The authors are grateful to all the test players who participated in this study, and to the Natural Hazards Research Platform for providing funding to support this work. We are also grateful for the partnerships of Marangatūhetaua/Ngati Tū, Ngāti Whakaari, Ngāi Tauira, Ngāti Kurumōkihi, Ngāi Te Ruruku, and Ngāi Tahu and Maungaharuru-Tangitū Trust (MTT) in the development of *Maraeopoly*. We are also thankful to AF, JL, and JJ for their help running game testing sessions.

1. Intergovernmental Panel on Climate Change, *Global Warming of 1.5°C*, 2018, http://www.ipcc.ch/report/sr15/.
2. Bronwyn Hayward, "'Nowhere Far from the Sea': Political Challenges of Coastal Adaptation to Climate Change in New Zealand," *Political Science* 60, no. 1 (June 2008): 47–59, https://doi.org/10.1177/003231870806000105/; Martin Manning et al., "Dealing with Changing Risks: A New Zealand Perspective on Climate Change Adaptation," *Regional Environmental Change* 15, no. 4 (April 2015): 581–94, https://doi.org/10.1007/s10113-014-0673-1/.
3. J. Barnett et al., "A Local Coastal Adaptation Pathway," *Nature Climate Change* 4, no. 12 (December 2014): 1103–8, https://doi.org/10.1038/nclimate2383/.
4. W. Neil Adger, "Place, Well-Being, and Fairness Shape Priorities for Adaptation to Climate Change," *Global Environmental Change* 38 (May 2016): A1–3, https://doi.org/10.1016/j.gloenvcha.2016.03.009/.
5. H. W. J. Rittell and M. M. Webber, "Dilemmas in a General Theory of Planning," *Policy Sciences* 4 (1973): 155–69; V. A. Brown et al., "Towards a Just and Sustainable Future," in *Tackling Wicked Problems: Through the Transdisciplinary Imagination*, ed. V. A. Brown, J. A. Harris, and J. Y. Russell (London: Earthscan, 2010).
6. P. J. Balint et al., *Wicked Environmental Problems: Managing Uncertainty and Conflict* (Washington, DC: Island Press, 2011); Brown et al., "Towards a Just and Sustainable Future."
7. Stephen Flood et al., "Adaptive and Interactive Climate Futures: Systematic Review of 'Serious Games' for Engagement and Decision-Making," *Environmental Research Letters* 13, no. 6 (June 2018): 063005, https://doi.org/10.1088/1748-9326/aac1c6/.
8. Junko Mochizuki, Piotr Magnuszewski, and Joanne Linnerooth-Bayer, "Games for Aiding Stakeholder Deliberation on Nexus Policy Issues," in *Managing Water, Soil and Waste Resources to Achieve Sustainable Development Goals: Monitoring and Implementation of Integrated Resources Management*, ed. Stephan Hülsmann and Reza Ardakanian (Cham: Springer International Publishing, 2018), 93–124, https://doi.org/10.1007/978-3-319-75163-4_5/.
9. Diana Reckien and Klaus Eisenack, "Climate Change Gaming on Board and Screen: A Review," *Simulation & Gaming* 44, nos. 2–3 (April 2013): 253–71, https://doi.org/10.1177/1046878113480867/.
10. Margarida Romero, Mireia Usart, and Michela Ott, "Can Serious Games Contribute to Developing and Sustaining 21st Century Skills?," *Games and Culture* 10, no. 2 (March 2015): 148–77, https://doi.org/10.1177/1555412014548919/.
11. Judy Lawrence and Marjolijn Haasnoot, "What It Took to Catalyse Uptake of Dynamic Adaptive Pathways Planning to Address Climate Change

Uncertainty," *Environmental Science & Policy* 68 (February 2017): 47–57, https://doi.org/10.1016/j.envsci.2016.12.003/.
12. Danya Rumore, Todd Schenk, and Lawrence Susskind, "Role-Play Simulations for Climate Change Adaptation Education and Engagement," *Nature Climate Change* 6, no. 8 (August 2016): 745–50, https://doi.org/10.1038/nclimate3084/.
13. Aleks Krotoski, "Serious Fun with Computer Games," *Nature* 466, no. 7307 (August 2010): 695, https://doi.org/10.1038/466695a/.
14. Christophe Le Page et al., "Exploring How Knowledge and Communication Influence Natural Resources Management with ReHab," *Simulation & Gaming* 47, no. 2 (April 1, 2016): 257–84, https://doi.org/10.1177/1046878116632900/; Jane McGonigal, *Reality Is Broken: Why Games Make Us Better and How They Can Change the World* (London: Penguin, 2011).
15. Phil Wilkinson, "A Brief History of Serious Games," in *Entertainment Computing and Serious Games: International GI-Dagstuhl Seminar 15283, Dagstuhl Castle, Germany, July 5–10, 2015, Revised Selected Papers*, ed. Ralf Dörner et al., Lecture Notes in Computer Science (Cham: Springer International Publishing, 2016), 17–41, https://doi.org/10.1007/978-3-319-46152-6_2/.
16. Flood et al., "Adaptive and Interactive Climate Futures."
17. Derek Armitage et al., "An Approach to Assess Learning Conditions, Effects and Outcomes in Environmental Governance," *Environmental Policy and Governance* 28, no. 1 (2018): 3–14, https://doi.org/10.1002/eet.1781/.
18. The most recent version of the game can be found here: https://www.niwa.co.nz/natural-hazards/research-projects/serious-games-for-climate-change-adaptation/.
19. Fikret Berkes, "Understanding Uncertainty and Reducing Vulnerability: Lessons from Resilience Thinking," *Natural Hazards* 41, no. 2 (May 1, 2007): 283–95, https://doi.org/10.1007/s11069-006-9036-7/.
20. See https://www.niwa.co.nz/news/fieldays-farming-for-the-future/.
21. Affiliated hapū (subtribes) include Marangatūhetaua/Ngati Tū, Ngāti Whakaari, Ngāi Tauira, Ngāti Kurumōkihi, Ngāi Te Ruruku, and Ngāi Tahu, as well as Maungaharuru-Tangitū Trust (MTT).
22. A region on the east coast of New Zealand's North Island.
23. See https://www.deepsouthchallenge.co.nz/projects/exploring-coastal-adaptation-pathways-tangoio-marae/.
24. Jackie Colliar and Paula Blackett, "Tangoio Climate Change Adaptation Decision Model: A Process for Exploring Adaptation Pathways for Tangoio Marae. Report to Maungaharuru-Tangitū Trust and Deep South National Challenge" (Auckland: National Institute of Water and Atmospheric Research, 2018).
25. Frank W. Geels, "Technological Transitions as Evolutionary Reconfiguration Processes: A Multi-Level Perspective and a Case-Study," *Research Policy* 31, no. 8 (December 1, 2002): 1257–74, https://doi.org/10.1016/S0048-7333(02)00062-8/.
26. See http://www.twinery.org/.

27. Anastasia Salter, "Learning Through Making: Notes on Teaching Interactive Narrative," *Syllabus* 4, no. 1 (January 29, 2015), http://syllabusjournal.org/syllabus/article/view/109/.
28. See Tables 10 and 11 in R. G. Bell et al., "Coastal Hazards and Climate Change: Guidance for Local Government" (Wellington: Ministry for the Environment Manatū Mō Te Taiao, 2017).
29. The ten-year time step is based on the resolution of climate change scenarios used in the game. This allows players to perceive the effects of climate change (e.g., sea-level rise, frequency of heavy storms) more acutely than they might at a more fine-grained resolution.
30. Paula Blackett et al., "How Can We Engage with Coastal Communities over Adaptation to Climate Change? A Case Study in Whitianga on the Coromandel Peninsula," in *Planning Pathways to the Future* (New Zealand Planning Institute Conference, Christchurch, 2010), 20–25; H. L. Rouse et al., "Coastal Adaptation to Climate Change in Aotearoa-New Zealand," *New Zealand Journal of Marine and Freshwater Research* 51, no. 2 (April 3, 2017): 183–222, https://doi.org/10.1080/00288330.2016.1185736/.
31. See https://treasury.govt.nz/information-and-services/nz-economy/living-standards/our-living-standards-framework/.
32. Defined as experiencing harm from negative events less than three times in a decade.
33. Defined as experiencing harm from negative events three or more times in a decade.
34. Adapted from Julia Baird et al., "Learning Effects of Interactive Decision-Making Processes for Climate Change Adaptation," *Global Environmental Change* 27 (July 2014): 51–63, https://doi.org/10.1016/j.gloenvcha.2014.04.019; Julia Baird et al., "Introducing Resilience Practice to Watershed Groups: What Are the Learning Effects?," *Society & Natural Resources* 29, no. 10 (October 2, 2016): 1214–29, https://doi.org/10.1080/08941920.2015.1107788; Armitage et al., "An Approach to Assess Learning Conditions, Effects and Outcomes in Environmental Governance."
35. Armitage et al., "An Approach to Assess Learning Conditions, Effects and Outcomes in Environmental Governance."
36. Stefania Munaretto and Dave Huitema, "Adaptive Comanagement in the Venice Lagoon? An Analysis of Current Water and Environmental Management Practices and Prospects for Change," *Ecology and Society* 17, no. 2 (June 2012), https://www.jstor.org/stable/26269041; Constanze Haug, Dave Huitema, and Ivo Wenzler, "Learning through Games? Evaluating the Learning Effect of a Policy Exercise on European Climate Policy," *Technological Forecasting and Social Change* 78, no. 6 (July 1, 2011): 968–81, https://doi.org/10.1016/j.techfore.2010.12.001; Baird et al., "Learning Effects of Interactive Decision-Making Processes for Climate Change Adaptation."
37. Armitage et al., "An Approach to Assess Learning Conditions, Effects and Outcomes in Environmental Governance."

38. See also Adger, "Place, Well-Being, and Fairness Shape Priorities for Adaptation to Climate Change."
39. Mochizuki, Magnuszewski, and Linnerooth-Bayer, "Games for Aiding Stakeholder Deliberation on Nexus Policy Issues"; Romero, Usart, and Ott, "Can Serious Games Contribute to Developing and Sustaining 21st Century Skills?"; David Rumeser and Margaret Emsley, "Can Serious Games Improve Project Management Decision Making Under Complexity?," *Project Management Journal* 50, no. 1 (February 1, 2019): 23–39, https://doi.org/10.1177/8756972818808982/.
40. Armitage et al., "An Approach to Assess Learning Conditions, Effects and Outcomes in Environmental Governance."
41. Hannah R. Parker et al., "Using a Game to Engage Stakeholders in Extreme Event Attribution Science," *International Journal of Disaster Risk Science* 7, no. 4 (December 1, 2016): 353–65, https://doi.org/10.1007/s13753-016-0105-6/.
42. Balint et al., *Wicked Environmental Problems: Managing Uncertainty and Conflict*; Brown et al., "Towards a Just and Sustainable Future."
43. Mochizuki, Magnuszewski, and Linnerooth-Bayer, "Games for Aiding Stakeholder Deliberation on Nexus Policy Issues."

CHAPTER 14

The Human Face of the Ocean

Creative Collaboration for Conservation

TIERNEY THYS

From one dimly lit corner of the stage, a lithe dancer glides forth—her form-fitting silver bodysuit glinting in the stage light. Dipping her head between outstretched arms, she stretches her streamlined, muscular body out towards the open space. Haunting violin music with a woman's wordless melancholic singing fills the air. As the dancer strains, pirouettes and twists, we see something is amiss. A thick rope is wrapped around her waist and she wrestles against it, making every one of her graceful movements strained with effort. At the far end of rope is a set of dark figures lurking in the shadows. Swaying menacingly, they alternately yank the rope taut and then relax their pull—taunting the dancer like some kind of sadistic puppeteer toying with a living marionette. With beauty and strength, the dancer pulls away from the darkened figures. She twists her glinting body against the rope and struggles to continue her dance. The rope slackens and she rushes forward with newfound freedom and palpable joy, but then the rope tightens and she is caught again. This tug of war continues—frantic bouts of struggle punctuated with exhausted slumped-over surrender. A narrator's voice announces that "Since the 1950s, we've lost 90 percent of our big fish. Our commercial fish stocks are being devastated. With millions of lines in the water, will we only stop after we've caught the very last fish?" The dancer resurrects one last time and throws herself against the pull of the rope, but the pullers of the rope are unrelenting. She collapses, goes limp and bends over—her bright silvery body fading to gray. The dark figures at the edge of the stage reel her in, toss a large black net over her tiny spent frame and the music stops. The lights go out.

—A description of "Fish on a Line" from *Okeanos: A Love Letter to the Sea*

DESPITE HAVING SEEN this dance performed many times as part of *Okeanos*,[1] I still feel a profound sense of loss each time I see it. I viscerally feel the loss of vibrant life and beauty extinguished on stage. And as I look around the theater, I'm not alone in these feelings. Many members of the audience are dabbing their eyes with tissues, visibly moved to tears. This performance has resonated in a universal way and managed to transport us all into another world—the underwater and at times brutal world of a fish fighting for its life. The kind of raw emotional reaction this dance elicits differs markedly from the standard academic response to scientific publications.

Working as the chief science consultant for *Okeanos: A Love Letter to the Sea* opened for me a window into a new way of connecting with the public and encouraging them to appreciate, understand, re-evaluate, and revalue the ocean and its inhabitants. As we watched the dancers, we were swept up into a set of diverse journeys. We, too, became crustaceans tussling in the fronds of a three-dimensional kelp forest or scuttling across a dappled seafloor wrestling for space and territory; we, too, became an octopus contorting and sliding itself into the nooks and crannies of a cave-like sculpture; we, too, became seahorses entwined in a tender mating ritual. Engaging in this kind of aesthetic experience allowed us as audience members access into the liquid realm and its inhabitants, imagined through the movement of the dancers.

This kind of embodiment can help foster a deeper empathy for otherwise unfamiliar creatures, far removed from our day-to-day existence. Engaging in this way can help build an audience's capacity to better relate to the ocean, foster empathy, and open a pathway to greater interest, learning, and ideally lasting stewardship. Dance is just one of many ways to encourage the public to rethink, reappreciate, and revalue the ocean.

After decades of working in marine conservation conducting research and outreach, I've found that effectively messaging facts and findings about marine science and environmental issues requires going far beyond presenting peer-reviewed publications. To truly have the public revalue the ocean, facts need to be woven into narrative frameworks that make issues relatable and that speak to the values and emotions of target audiences. We humans are a cerebral, irrational, superstitious, meticulous, messy, tenacious, procrastinating, compassionate, pugnacious, self-serving lot—all rolled into one complex species. For our diverse masses to revalue the ocean and work towards lasting solutions, we need to adopt a multipronged approach. Interdisciplinary collaborations with humanities, social sciences, and the arts (literary, visual, performance, digital) offer us the tools we need to broaden our cultural perspectives, confront our many biases, address counterproductive habits, bridge growing societal divides, and move ourselves forward into a more balanced relationship with the sea.

Communicating Marine Science

It's no secret that we humans are harming the ocean in myriad ways and that the challenges we present to ocean life are increasing. More than one-third of the world's population (2.4 billion people) lives within sixty miles of the coast, and our coastal footprint keeps expanding. A mere half-century ago, it would have been anathema to think of our vast ocean, covering 71 percent of Earth's surface and providing more than 98 percent of available living space, as a fragile, vulnerable environment. But in the last fifty years, we have fished out more than 70 percent of the large predators. In the last two hundred years, we've played a major role in altering ocean pH by 30 percent—faster than any known change in ocean chemistry in the last fifty million years. According to the World Meteorological Organization's State of the Global Climate 2020 report, atmospheric carbon dioxide emissions have contributed to more than a 1°C ocean warming, which has led to more intense storms, displaced communities, and the bleaching of 30 to 50 percent of the world's coral reefs. We have filled our marine waters with 500 times more pieces of plastic than there are stars in the sky.[2]

Wittingly or unwittingly, we have transformed our ocean life-support system into what William McDonough, author of *Cradle to Cradle: Remaking the Way We Make Things*, calls "the giant toilet that doesn't flush." As our global population expands to ten billion by 2050, according to a 2019 UN Population Division Report, the long-term consequences of our marine exploitations are being tallied in the loss of commodities like seafood and felt psychologically in the loss of beautiful restorative vistas, eco-tourism areas, and recreational spaces that are now frequently off limits because of our pollutants.

While the need to nurture a healthy ocean may seem obvious, changing human behavior to encourage marine stewardship on a global scale remains a tremendous challenge for numerous reasons. First, the sheer scale of marine issues is disheartening and can discourage public action. From global climate pollution and sea level rise to overfishing, erosion, piracy, plastics, and ocean noise pollution, marine problems can seem too large and intractable for any individual to solve. Second, the number of diverse stakeholders involved in any long-term solutions makes it difficult to find common ground. Third, the evidence-based scientific community speaks one language, while governmental policymakers speak another and indigenous populations speak yet another. Often, experts in one community lack sensitivity to the ways and wisdom of other communities.

Intrinsic to this communication conundrum is the fact that we humans are often irrational beings.[3] We have a long track record of ignoring sound advice and pertinent evidence and not acting in our own best interest.[4] As social scientist Mark Stern writes in *Social Science of Theory for Environmental Sustainability*, "no matter what technical solutions we develop, we [still] must navigate the complexities of human behavior to activate their utility."[5]

Lying at the heart of these complexities is our tendency to privilege our own personal set of values above all others. These personal values can be deeply entrenched, but powerful narratives, if delivered through intentional, well-informed, emotion-based storytelling, can open the mind to healthier, sustainable life choices and work to dislodge pre-existing beliefs that may be in conflict with scientific evidence.[6] Researching and compiling the robust scientific basis of the narrative is the first step.

Quantifying the Ocean's Goods and Services

Programs like the Millennium Ecosystem Assessment, TEEB (The Economics of Ecosystems and Biodiversity), and UN World Ocean Assessment use tools such as ecosystem services valuation (ESV)[7] to quantify nature's myriad contributions. Dominated by two main disciplines—the natural sciences and economics—these efforts have commodified nature's value into three broad categories: material goods (e.g., seafood, labor, medicinal resources); regulating services (e.g., climate, pH, air quality); and nonmaterial goods (e.g., inspiration, cultural identities, psychological experiences).[8] From the material and regulating categories, the ocean's seminal role is evident. It helps provide much of our atmospheric oxygen; serves as Earth's largest carbon sink, climate modifier, heat distributor, and source of rainfall; provides the transportation network for 90 percent of our global cargo and goods;[9] and supplies 140 million tons of seafood annually, which provides 17 percent of humans' global animal protein.[10]

Quantifying the ocean's *nonmaterial* goods and services, however, is much trickier. Over the last decade, a growing body of evidenced-based research has identified links between human psychological and physical health and contact with the ocean.[11] For example, a landmark Danish study, involving more than ten thousand participants, showed that people living in areas with more blue space (ocean and other water areas) had significantly fewer symptoms of ill health.[12] In Spain, children who spent more time at local beaches had better social and behavioral skills.[13] A 2015 UK study found that people engaged in greater amounts of healthy, physical activity when they spent time in blue spaces compared to other natural spaces.[14]

Blue space[15] interventions, such as immersion therapy in lakes, rivers, and the ocean, blue gyms, and nature art in health care facilities and prisons (all examples of blue care) have also demonstrated the benefits of ocean contact and imagery for speeding healing, reducing stress, and lowering violence, respectively.[16] These health effects are catching the attention of numerous organizations and leading to significant investments. For example, the US Navy recently invested $1 million USD in a research project to quantify the therapeutic value of surfing for military personnel with post-traumatic stress disorder, depression, or sleep disorders.[17] One of my collaborative research projects, led by Nalini Nadkarni at the University of Utah, found that offering digital nature imagery including images of the ocean to inmates in solitary confinement lowered violent infractions by 26 percent in an Oregon prison.[18] This work has led to more correctional institutions adopting similar practices in Maine, Utah, Kansas, Florida, and Washington State (pers comm N. Nadkarni in Washington, D. Conover in Maine, K. Lockwood in Florida, and C. Naugle in Oregon).

Identifying links between the ocean and psychological health is particularly vital given the rapid urbanization of humanity. People in developed countries spend 90 percent of their time indoors[19] which translates to an extinction of experience with the natural world[20] and less exposure to its myriad health benefits.[21] Concurrently, mental illness rates in the United States are rising, particularly among young people (ages eighteen to twenty-five), with rates of suicidal thoughts or actions increasing nearly 50 percent between 2008 to 2017.[22]

Tallying the ocean's goods and services through ESV has been highly influential in policy-making and instrumental in establishing the UN Sustainable Development Goals (https://sustainabledevelopment.un.org/), but it has also generated controversy. Criticisms have called for ESV to more fully embrace the social sciences and humanities,[23] be more inclusive of different perspectives on people's relationships with nature (including the ocean), and place the ocean's goods and services into a more comprehensive cultural context.[24] Many of these critiques highlight the need to better understand differing audience values e.g., from indigenous populations of Oceania who are tightly tied to the ocean to land-locked urban populations who experience the ocean only digitally, if at all. This diversity of lived experience calls for new forms of art and narratives to illuminate the relevance and value of the ocean to our daily lives, bring it more clearly into our consciousness, and pull it more closely into our circles of compassion. For this to happen, the humanities play an essential role.

Improving the Narratives of Quantitative Findings

The humanities can help place the ocean's contributions into historical (e.g., our seafaring traditions) and philosophical (e.g., ocean as metaphor) contexts—making these contributions more relevant and accessible to diverse audiences. At the same time, insights from social sciences can help expose our common mental traps and reveal our cognitive biases. For example, because of the Identifiable Victim Effect, we are more likely to feel the pain of watching a single fish struggling for life (as described at the start of this chapter) than we are the pain and loss of the 2.7 trillion fish caught from the wild and killed globally every year according to 2020 FAO global counts. The social sciences can also help unpack how information is received in our minds and identify different frames and word choices that may better enhance the chances of being heard. These interdisciplinary collaborations can lend valuable insight into *what* moves us and *why* we make the (sometimes irrational) decisions we do.

The arts offer vehicles into our emotional beings. As author and environmentalist Rachel Carson wrote, "It is not half so important to know, as to feel."[25] When emotional input is added to any learning experience, it makes that experience more memorable and more exciting. The brain deems emotional experiences more important and enhances our memory of them.[26] Presenting facts alone is not as likely to bring about long-term changes in either feelings or behaviors. If these facts can be tied to an engaging character, one that embarks on an exciting journey, overcomes a series of obstacles, endures hardships, fails yet perseveres, and undertakes something akin to the classic Campbellian hero's journey, then those facts can gain purchase within our fast-moving minds, connect up with similar memories, and start to influence our ingrained beliefs. Some examples of this tactic can be found in animated children's films like *Fern Gully: The Last Rainforest*

(1992) which took on deforestation; *Happy Feet* (2006) which took on overfishing; and *Wall-E* (2008) which took on consumerism and environmental destruction. Because of art's ability to harness these emotional connections, it has the potential to attract new audiences, awaken deep feelings, and facilitate dialog. Each art form—be it dance, theater, music, film, photography, painting, sculpture, video games, poetry, or virtual reality—possesses its own strengths. By melding emotions and imaginative sensory stimuli, the arts can sweep us out of our analytical selves and into more receptive mental states.

Exploring Mechanisms Underlying the Arts' Ability to Enact Behavioral Change

Ground-breaking research is unveiling various mechanisms underlying how various art forms can influence behavior. For example, Uri Hasson's neurocinematics lab at Princeton University has demonstrated how effective films are able to activate identical brain pathways in viewers and essentially create a collective "mind meld." Jeremy Bailenson, founder of the Virtual Human Interaction Lab at Stanford University, demonstrates how virtual reality (VR) creates experiences that promote empathy which can translate into pro-social behaviors. For example, "swimming breast-stroke" through a progressively degraded virtual coral reef environment causes students to care more about the fate of real reefs than just reading about the reef or passively watching a video. It turns out that moving one's body is central to the efficacy of the individual VR experience. Bailenson has found that the greater the participants' movements within the VR experience, the more their minds are moved to do something in response.[27] Well-orchestrated VR experiences can have significant conservation impacts. For example, at a 2018 Fundraising Live conference in the UK, a Greenpeace manager reported that transporting attendees into the Amazon rainforest through a VR experience doubled the number of people who signed up for donations to protect the rainforest.[28]

At a more granular level, Nik Sawe from Stanford's Environmental Decision-Making and Neuroscience Lab and I are exploring brain responses to photographic elements and conservation imagery. Our work has identified common aesthetic preferences by coding more than 40 content parameters of 890 images selected from three months of National Geographic's popular Instagram account (200 million followers). These elements include high-level features such as biome type, levels of naturalness, and presence of people, as well as algorithmically derived, low-level features such as fractal dimensionality, edge density, saturation, and entropy (calculated with the help of Marc Berman's lab at University of Chicago). We have found that images with long-range views, particularly of mountains, without people and containing water and greenery receive the highest number of likes. Using functional magnetic resonance imaging (fMRI), we have also measured and mapped the brain activity of participants while (1) viewing a representative subset of the images, and (2) answering questions as to whether they would like and/or donate to the subject matter in the photographs. From this work, we've found that the nucleus accumbens (part of the brain's reward region) is activated when a participant likes an image and the anterior insula is activated when participants don't like an image. Both the nucleus accumbens and the caudate are activated when they decide to donate.[29] As we pull together all these discoveries, we are steadily adding details and insight into how and why the arts (particularly visual arts) can and do impact our cognitive activities and life choices.

Experimenting with Narratives through Interdisciplinary Collaborations

Of course, the arts can also alienate audiences, conjure confusion, and create unwanted subconscious associations. Learning how best to wield the arts and humanities with insights from social sciences to share the most current, scientifically supported conservation messages is a challenging task, but numerous groups are making headway. A wave of collaborative momentum is building within the worlds of ocean conservation and beyond. In fact, this essay was inspired by such a collaboration at the University of Utah, where a one-day conference entitled Revaluing the Ocean gathered speakers from the fields of humanities, law, natural sciences, indigenous cultural studies, the social sciences, and more. The National Geographic Society, one of the world's largest nonprofit funders for research and exploration, now requires more social scientists be included in several of its requests for proposals. At the university level, the number of Environmental Humanities programs is growing in the United States and abroad. Additional creative collaborations include the National Academy of Science's *Science and Entertainment Exchange*. With a desire to offer easy access to accurate science in both film and TV programming, this group links entertainment industry professionals with scientists and engineers. Their principal aim is to use popular entertainment media to "deliver sometimes subtle, but nevertheless powerful, messages about science." The StoryCollider podcast also brings storytelling front and center with a goal of helping people from all walks of life tell true, personal stories about science. This podcast's objective is to humanize scientific work and show how it is an integral part of all our lives. The StoryCollider group also offers public and private storytelling workshops. Leonardo (The International Society for the Arts, Sciences and Technology ISAST) is another excellent virtual community resource for networking, resource-sharing, disseminating best practices, supporting research, and offering events in the art/science/technology sector through a range of publications, residencies, and live performances. A host of bricks-and-mortar institutions also offer creative and interactive experiences to enhance public understanding of science through the arts, e.g., the Exploratorium in San Francisco, California, and Science Gallery International. Lastly, the popularity of science festivals and live science events worldwide is also on the rise.[30] Given the viciously divisive nature of today's political scene and rising antiscience biases in the United States,[31] the importance of nurturing such collaborative undertakings has never been more urgent.

Experimenting with Narrative: A Personal Case Study

As a trained research scientist with a focus on marine conservation, working as a consultant for a dance troupe wasn't on my radar. However, several years before working on *Okeanos: A Love Letter to the Sea*, I had the opportunity to watch the Capacitor dance troupe perform a powerful show entitled *Biome*. In one of their pieces, the dancers portrayed seedlings on the floor of a Costa Rican rainforest, slithering and inching along, jockeying for space, driven upwards by their thirst for sunlight. Watching them, I could feel what it might be like to be part of that deliciously intertwined thicket of tropical life. For me, it was a visceral lesson in interconnectedness as I'd never experienced before.

Memories of that primal feeling stayed with me. As my own research became more conservation focused, I thought to myself, what if we could conjure those same powerfully emotional, empathetic feelings for ocean animals and convey their immense struggles in today's overfished ocean? Coincidently, Jodi Lomask, the director of Capacitor, had just begun what would become a two-year journey aimed at enriching her personal ocean experiences to inspire content for her troupe's next big performance piece. Our individual journeys converged and I became part of the Capacitor's monthly "lab" meetings, helping to brainstorm about ocean concepts and issues that could be both vital and intriguing to audiences. One weekend, I led Jodi and members of her troupe on a SCUBA dive into the chilly waters of the Monterey Bay kelp forest. I also helped source high resolution video materials whose imagery from the Great Barrier Reef inspired the dancers and gave presentations about the intricacies and interactions of various animals' lives. Based on our interactions and her own experiences, Jodi began creating 3D sculptures and movement scenarios in her mind and with her dancers. As science advisor, I offered research and information while Jodi and her dancers offered the movements and framing. Our shared goal was to entertain the audience and place them into the fins and tentacles of ocean life so they could experience the ocean in a novel way. Walking in another's shoes—or flippers, as the case may be—has been shown to increase empathy and positive social behaviors.[32]

As we worked together, themes began to emerge organically such as the three-dimensionality of the ocean world, the energetic jockeying for space on a crowded reef or swaying fronds of a kelp forest, the frenetic movements of some creatures juxtaposed with the slower more laconic movements of other animals, the intimate courtships rituals of seahorses and oozing movements of octopi—the interdependence of all ocean life. As the dances started to take shape, marine legend Sylvia Earle and I provided narration. Whereas I aimed to add more science and more clarity in the narration and movements, Jodi was keen to not spoon-feed the audience and rather find that sweet spot of abstraction wherein the audience would still have to work for meaning and in so doing take the content's impact more to heart. Jodi and her dancers learned much during this process about the ocean—its life and its challenges. I took away a pivotal lesson in the power of emotional messaging which subsequently inspired me to conduct research on the social and cognitive effects of nature imagery described above.

With its contortionists, live music, video screens, and climbable objects suspended from the ceiling, *Okeanos* was described as a kind of *Cirque de Soleil* meets the ocean. The show boasted multiple sold-out performances in the San Francisco Bay Area and received positive reviews. Audience members commented on its beauty and immersive qualities, stating that "it was an extraordinary event—part dance performance, part educational voyage, and part novel music acoustic immersion. At moments what was created was positively breath-taking and at other moments the experience was sheer emotion. One felt like a pioneer exploring the ocean's depths while witnessing the world of human beings and nature's wonders merging into a scene of awe." Others commented similarly: "it's incomparable to anything I've experienced, even Cirque du Soleil, as this was far more intimate. The choreography was stunning, original, delightful, unique and passionate.... it truly personified the human spirit that draws us effortlessly to our oceans." Another commented that they were relieved to see that *Okeanos* is never preachy or depressing. In fact, there is very little narration and few "downers" of any kind. Instead, "it's a really thoughtful experience filled with love and appreciation for the ocean and its creatures." Not all reviews were positive, however. Some viewers felt that parts of the performance missed the point, e.g., "We only enjoyed the 'octopus.' The rest of the pieces were either too obvious (lacking subtlety) or just acrobatics." The majority of reviews, though, did convey that the performance achieved its goals, as summarized eloquently in this audience member's comment:

> I was amazed at how Capacitor was able to make the creatures of the marine realm come to life on stage; as well as how they were able to convey the issues the ocean is currently facing—empowering people to take action.... *Okeanos* is a dance/cirque portrait of the ocean as body, environment, resource, metaphor, and force. It is a performance to inspire and educate audiences, including young people, about the ocean and connect them deeply to ocean conservation. Baba Dioum put it right: "In the end we will conserve only what we love. We will love only what we understand. We will understand only what we are taught." It's all about love (isn't everything :) and *Okeanos* helps emotionally connect people to the sea.

In the final scene of *Okeanos*, five dancers wrap themselves around a large spinning globe suspended in the darkness. Their bodies are intertwined as they cling to one another—enveloping the entire surface of the rotating globe. As they slowly turn, you feel the power of their union but also the precariousness of their position. If one dancer releases, they all run the risk of plummeting into the abyss below. Their grip, their strength, their beauty—their survival—all depend upon their ability to coexist, to make space for each other and to hold onto one another. What an apt and poignant metaphor for the sublime power of inclusion, the importance of everyone's role in conserving our finite resources and the exalted interconnectedness that makes all our lives possible. The profound power of the arts to both communicate and humanize such facts and ideas is essential if we are to ever tally and recognize the full value of a healthy ocean for humanity.

Notes

1. *Okeanos: A Love Letter to the Sea.* Performed by Capacitor dance troupe, April 2012, Fort Mason, San Francisco, https://vimeo.com/71582969/.
2. Geyer, R., Jambeck, J.R., and Lavender Law, K. (2017). Production, use, and fate of all plastics ever made, *Science Advances*, July 19, 2017, Vol. 3, no. 7, doi: 10.1126/sciadv.1700782. See also Rogers, A., and Laffoley, D. (2013). Introduction to the special issue: The global state of the ocean; interactions between stresses, impacts and some potential solutions. Synthesis papers from the International Programme on the State of the Ocean 2011 and 2012 workshops, *Marine Pollution Bulletin*, 74, 491–494.
3. Ariely, D. (2010). *Predictably Irrational*, Harper Collins, USA.
4. Stern, M. (2018). *Social Science of Theory for Environmental Sustainability: A Practical Guide*, Oxford University Press.
5. Ibid., 2.
6. National Academies of Sciences E, Medicine. Using narrative and data to communicate the value of science: Proceedings of a workshop—in brief. Washington, DC: The National Academies Press; 2017.
7. Daily, G.C., T. Söderqvist, S. Aniyar, et al. (2000). The value of nature and the nature of value. *Science* 289 (5478): 395–396. doi:10.1126/science.289.5478.395.
8. Díaz, S., U. Pascual, M. Stenseke, B. Martín-López, et al. (2018). Assessing nature's contributions to people. *Science* 359 (6373): 270–272. http://dx.doi.org/10.1126/science.aap8826/.
9. International Maritime Organization 2017, https://www.imo.org/en/About/Pages/Default.aspx/.
10. Millennium Ecosystem Assessment, TEEB (The Economics of Ecosystems and Biodiversity), and UN World Ocean Assessment, https://worldoceanobservatory.org/content/un-world-ocean-assessment#/.
11. Reviewed by Hartig, T., Mitchell, R., de Vries, S., et al. (2014). Nature and health. *Annual Review of Public Health* 35, 207–228, PMID: 24387090, 10.1146/annurev-publhealth-032013-182443; and Nichols, W.J. (2015). Blue mind: the surprising science that shows how being near, in, on, or under water can make you happier, healthier, more connected, and better at what you do. Little Brown and Company, New York, pp. 368.
12. De Vries, S., Verheij, R.A., Groenewegen, P.P., et al. (2003). Natural Environments—Healthy Environments? An Exploratory Analysis of the Relationship between Greenspace and Health. *Environment and Planning*, 35 (10): 1717–1731. doi:10.1068/a35111.
13. Amoly, E., Dadvan, P., Forns, J., et al. (2014). Green and Blue Spaces and Behavioral Development in Barcelona Schoolchildren: The BREATHE Project, *Environmental Health Perspectives*, 122 (12): 1351–1358.
14. Elliott, L.R., White, M.P., Taylor, A.H., et al. (2015). Energy expenditure on recreational visits to different natural environments, *Soc Sci Med*, 139: 53–60. doi: 10.1016/j.socscimed.2015.06.038.

15. Defined as all visible, outdoor, natural surface waters with potential for the promotion of human health and wellbeing according to Britton, E., Kindermann, K., Domegan, C., et al. (2018). Blue Care: A systematic review of blue space interventions for health and wellbeing. *Health Promotion International* 35, no. 1 (February 2020): 50–69 (definition on p. 50).
16. Britton et al., Blue Care; and Dempsey, S., Devine, M.T., Gillespie, T., Lyons, S., et al. (2018). Coastal blue space and depression in older adults. *Health & Place*, 54: 110–117. doi:10.1016/j.healthplace.2018.09.002; and Nadkarni, N., Hasbach, P., Thys, T., Gaines, E., and Schnacker, L. (2017). Impacts of nature imagery on people in severely nature-deprived environments. *Frontiers in Ecology and the Environment*, 15: 395–403, and Ulrich, R.S., Zimring, C., Zhu, X., DuBose, J., Seo, H., et al. (2008). A review of the research literature on evidence-based healthcare design. *HERD*, 1 (3): 61–125.
17. Perry, T. (2018). Riding the waves to better health, Navy studies the therapeutic value of surfing, www.washingtonpost.com/national/health-science/riding-the-waves-to-better-health-navy-studies-the-therapeutic-value-of-surfing/2018/03/09/254df9e2-06ca-11e8-94e8-e8b8600ade23_story.html?utm_term=.8e944d32cbf9, retrieved May 1, 2019.
18. Nadkarni et al., Impacts of nature imagery.
19. Klepeis, N.E., Nelson, W.C., Ott, W.R., et al. (2001). The National Human Activity Pattern Survey (NHAPS): a resource for assessing exposure to environmental pollutants. *Journal of Exposure Analysis and Environmental Epidemiology*, 11: 231–252.
20. Soga, M., and Gaston, K. (2016). Extinction of experience: the loss of human-nature interactions. *Frontiers in Ecology and the Environment*, 14: 94–101.
21. Reviewed in Nichols, W.J. (2015); and Gascón, M., Zijlema, W., Vert, C., White, M.P., et al. (2017). Outdoor blue spaces, human health and well-being: A systematic review of quantitative studies. *International Journal of Hygiene and Environmental Health*, 220: 1207–1221. doi:10.1016/j.ijheh.2017.08.004 pmid:28843736.
22. Twenge, J.M., Cooper, A.B., Joiner, T.E., et al. (2019). Age, period, and cohort trends in mood disorder indicators and suicide-related outcomes in a nationally representative dataset, 2005–2017. *Journal of Abnormal Psychology*, 128(3): 185–199. http://dx.doi.org/10.1037/abn0000410/.
23. Díaz et al., Assessing nature's contributions to people.
24. Peterson, G.D., Harmackova, Z.V., Meacham, M., Queiroz, C., et al. (2018). Welcoming different perspectives in IPBES: "Nature's contributions to people" and "Ecosystem services." *Ecology and Society* 23(1): 39. https://doi.org/10.5751/ES-10134-230139/.
25. Rachel Carson, *The Sense of Wonder* (New York: Harper 1998), 21.
26. Tyng, C.M., Amin, H.U., Saad, M.N.M., and Malik, A.S. (2017). The influences of emotion on learning and memory. *Frontiers in Psychology* 8: 1454. doi: 10.3389/fpsyg.2017.01454.

27. Bailenson, J.N. (2018). *Experience on Demand: What Virtual Reality Is, How It Works, and What It Can Do*. New York: W.W. Norton.
28. See www.civilsociety.co.uk/news/greenpeace-doubles-sign-up-rates-through-virtual-reality.html.
29. Nik Sawe and Tierney Thys, unpublished data.
30. Durant, J., Buckley, N., Comerford, D., et al. (2016). *Science Live: Surveying the Landscape of Live Public Science Events [Online]*. Boston & Cambridge, UK: ScienceLive. Available at https://livescienceevents.org/portfolio/read-the-report/.
31. See www.sciencealert.com/fauci-is-worried-about-anti-science-bias-in-the-us-as-stay-at-home-orders-are-thwarted, June 19, 2020.
32. Bailenson, *Experience on Demand*.

CHAPTER 15

Conclusion

Ocean Wildlife Photography as a Metaphor for the Anthropocene Ocean

ROBIN KUNDIS CRAIG

OCEAN WILDLIFE PHOTOGRAPHY can feel like a fool's errand. The attempt to photograph marine creatures in their native habitats comes with all the normal difficulties that attend terrestrial wildlife photography: will any animals actually show up? will they do something interesting if they do? and, of course, will the photographer be in any position to capture the action if and when it occurs?

On top of all that, however, in marine wildlife photography, *everything* is in motion. If the photographer is on a boat, the boat is moving, regardless of whether the engine is running. If the photographer is underwater, the sea itself is moving. The photographer is also moving—and, notably, not always in the same direction or at the same speed as the boat or the water at the moment the shutter clicks. Whatever creature the photographer is shooting is probably also in motion, and *its* vector may again have nothing whatsoever to do with whatever the boat, the current, the swell, and the photographer are doing. The underwater photographer also has the challenge of pursuing a decent photograph in a completely alien environment, one in which arms and legs are of limited use and the surrounding medium is not breathable.

The net result is a lot of fuzzy shots of blue (or gray, or green, or even orange with the right timing and sunlight) water. I am absolutely sure that I am not the only (amateur) marine wildlife photographer who cheered the arrival of digital underwater cameras.

Amazingly, sometimes it all works out, and the photographer can convey a new vision of a beautiful and mysterious world-in-motion, full of life. Often captivating, these pictures allow others to see indirectly a world that

Figure 15.1. School of Achilles Tangs Swim Over a Coral Reef. Kauai, Hawaii. © 2019 by Robin Kundis Craig. Used with permission of the photographer.

many will never experience directly, connecting them to a realm that might otherwise remain unfathomable and alien. Jacques Cousteau, the producers of *Flipper*, Jean-Michel Cousteau, Sylvia Earle, *National Geographic*, David Attenborough, James Cameron, *BBC Nature*, and numerous other photographers and filmmakers have made the ocean's ecosystems real places that need protection for millions, perhaps billions, of people globally.

Marine wildlife photography provides an apt metaphor for human interaction with the ocean in the Anthropocene. Everything about the Anthropocene ocean is in motion. Physically, the ocean is heating up, sea levels are rising, ice is melting, currents are changing. Chemically, the ocean is acidifying, responding to nutrient pollution, deoxygenating. Biologically, species and ecosystems are shifting ranges, responding to unnatural noise, overheating, unsuccessfully dodging propylene nets. Humanity is simultaneously the driver and the victim of these changes, with consequences still difficult to predict in detail. Capturing a focused picture can thus seem like an impossible task.

Figure 15.2. Humpback Whale and California Sea Lions. Monterey Bay, California. © 2019 by Robin Kundis Craig. Used with permission of the photographer.

Nevertheless, the essays in this volume demonstrate that in the midst of change we can also re-envision the ocean and our relationship to it. We can tell ourselves new stories of connectedness and interdependency. We can demystify the ocean's struggles, write better laws, educate the next generation. We can learn humility. We can become guardians.

We can, at the end of the day, envision a future where humans work with the ocean, shoring up its resilience and biodiversity, acknowledging our dependence on its goods and services. In so doing, we create a picture worth all the blurry images that might have accumulated along the way, a vision of the Anthropocene ocean both to keep and to share with others.

We hope that you have enjoyed our efforts here.

APPENDICES

Inspiring Ocean Voices

Editors' Introduction

THE AUTHORS IN THIS VOLUME so far have contributed new perspectives on the ocean in the twenty-first century. They are not, of course, the first voices raised about the ocean. These appendices provide additional voices, both historical and modern, as teaching and learning resources that both contextualize and broaden the essays presented so far. Specifically, these excerpts are organized to expand the reader's perspective on the ocean in three ways: a deeper historical perspective on humans' interdisciplinary interactions with the ocean; a broader global perspective on ocean issues to balance the primarily Western perspective of the original essays in Parts I, II, and III; and a snapshot of how deepening scientific understanding of the ocean over the last century has boosted momentum towards ocean protection.

Appendix A

A Deeper Historical Perspective

IN EUROPE, INCREASED VENTURING onto the Atlantic Ocean became the occasion for important legal and political developments as well as exploration and trade. In response to ongoing disputes between the Dutch and the Spanish—and, increasingly, the English and the Spanish—over rights to Southeast Asia (the "East Indies"), Hugo Grotius published *Mare Liberum—The Free Sea*—in 1609. His principle of "freedom of the seas" remains the default international rule of law even into the twenty-first century, underpinning global treaties such as the 1982 United Nations Convention on the Law of the Sea.

Exploration of the Atlantic, and then other parts of the ocean, by Europeans had, of course, many other important legal, commercial, and cultural consequences, many of which continue to influence twenty-first-century relations. In 2010, Simon Winchester ably traced these developments in his "biography" of this ocean, *Atlantic*.

Nevertheless, as the Blue Humanities continue to teach us, what the ocean "means" is a human construction that depends on context and perspective. Philip E. Steinberg explores these variations in the context of a lost shipment of Nike shoes in the introduction to his 2001 *The Social Construction of the Ocean*.

APPENDIX A.1

The Free Sea

or, A Disputation Concerning the Right Which the Hollanders Ought to Have to the Indian Merchandise for Trading

HUGO GROTIUS

To the Princes and Free States of the Christian World

It is no less ancient than a pestilent error wherewith many men (but they chiefly who abound in power and riches) persuade themselves, or (as I think more truly) go about to persuade, that right and wrong are distinguished not according to their own nature but by a certain vain opinion and custom of men. These men therefore think that both laws and show of equity were invented for this purpose: that their dissensions and tumults might be restrained who are born in the condition of obeying; but unto such as are placed in the height of fortune they say that all right is to be measured by the will and the will by profits. And it is not so great a wonder that this absurd opinion, and altogether contrary to nature, hath procured unto itself some little authority, seeing to that common disease of mankind (whereby, as vice, so we follow the defense thereof) the craft and subtlety of flatterers is added, whereunto all power is subject.

But on the contrary part, in all ages there have been some wise and religious men (not of servile condition) who would pluck this persuasion out of the minds of simple men and convince the others, being defenders thereof, of impudency. For they declared God to be the creator and governor of the world, [and] ... that great prince and householder had written certain laws of his ... in the minds and senses of everyone, where they shall offer

Excerpt from Hugo Grotius, *The Free Sea (Mare Liberum)*, trans. Richard Hakluyt (1609; Indianapolis: Liberty Fund, 2004). Spelling and ellipses in original.

themselves to be read of the unwilling and such as refuse. By these laws both high and low are bound....

* * *

If any think it hard that those things should be exacted of him which the profession of so holy a name requireth (the least whereof is to abstain from injuries) surely everyone may know what his duty is by that which he commandeth another. There is none of you who would not publicly exclaim that everyone should be moderator and arbitrator in his own matter, who would not command all citizens to use rivers and public places equally and indifferently, who would not with all his power defend the liberty of going hither and thither and trading.

If that little society which we call a commonwealth is thought not to stand without this (and indeed cannot stand without it), why shall not the self-same things be necessary to uphold the society and concord of all mankind?...

Yours only he hath excepted to himself who, though he hath reserved to himself the highest degree of punishment, slow, secret and inevitable, yet hath he assigned two judges from himself to be always present in men's affairs, whom the most happy offender cannot escape: to wit, every man's own conscience and fame, or other men's estimation of them. These seats of judgement stand always open to them to whom other tribunals are shut up; to these the weak and poor complain; in these they that master others in strength are vanquished themselves who are licentious out of measure, who esteem that at a base rate which was bought with man's blood, who defend injuries with injuries, whose manifest wickedness must needs be both condemned by the consenting judgment of the good and also not to be absolved in the opinion of their own mind.

To both these judgment places we bring a new case. Not truly of sinks or gutters or joining one rafter in another (as private men's cases are wont to be), nor yet of that kind which is usual among the people, of the right of a field bordering upon us or of the possession of a river or island, but almost of the whole sea, of the right of navigation and the liberty of traffic. These things are litigious between the Spaniards and us: whether the huge and vast sea be the addition of one kingdom (and that not the greatest); whether it be lawful for any people to forbid people that are willing neither to sell, buy nor change nor yet to come together; and whether any man could ever give that which was never his or find that which was another's before, or whether the manifest injury of long time give any right.

In this disputation we offer the counters to those who among the Spaniards are the principal doctors of the divine and humane law; and, to conclude, we desire the proper laws of Spain. If that prevail not, and covetousness forbid them to desist whom some reason convinceth, we appeal, oh ye princes, to your majesty; we appeal to your upright conscience and fidelity, oh ye nations, how many soever you be, wheresoever dispersed.

* * *

That law by whose prescript form we are to judge is not hard to be found out, being the same with all and easy to be understood, which being bred with everyone is engrafted in the minds of all. But the right which we desire is such as the king himself ought not deny unto his subjects, nor a Christian to infidels, for it hath his original from nature, which is an indifferent and equal parent to all, bountiful towards all, whose royal authority extendeth itself over those who rule the nations and is most sacred amongst them who have profited most in piety.

Understand this cause, oh yea, princes, and consider it, oh yea, people. If we demand any unjust thing, ye know of what account your authority and theirs who amongst you are nearer unto us hath always been with us: advise us, and we will obey. But if we have offended anything in this matter, we beseech you not to be offended; the hatred of mankind we pray not against. But if the matter fall out otherwise, we leave it to your religion and equity what you censure of it and what is to be done.

In times past, among the milder people it was accounted great impiety to assail them by war who would put their cause to arbitrement; on the contrary part, they who would refuse so equal a condition were repressed by the common aid not as enemies of one but of all. Therefore to that purpose we have seen truces made and judges appointed, kings themselves and puissant nations accounted nothing so glorious and honorable as to restrain others' insolency and to support others' infirmity and innocence.

Which custom, if it were in use at this day, that men thought no human thing strange unto them, surely we might have a more quiet world, for the presumption of many would wax cold and they who now neglect justice for profit's sake should learn to forget injustice with their own loss.

But as in this cause peradventure we hope for that in vain, so this we verily believe: that things being well weighed, you will all think the delays of peace are no more to be imputed unto us than the causes of war, and therefore as hitherto you have been well-willers and favorable friends unto us so you will much more befriend us hereafter, than the which nothing more desired can befall them who think it the first part of felicity to do well and the other to be well reported.

Chapter 1

By the law of nations navigation is free for any to whomsoever

Our purpose is shortly and dearly to demonstrate that it is lawful for the Hollanders, that is the subjects of the confederate states of the Low Countries, to sail to the Indians as they do and entertain traffic with them. We will lay this certain rule of the law of nations (which they call primary) as the foundation, the reason whereof is clear and immutable: that it is lawful for any nation to go to any other and to trade with it.

God himself speaketh this in nature, seeing he will not have all those things, whereof the life of man standeth in need, to be sufficiently ministered by nature in all places and also vouchsafeth some nations to excel others in arts. To what end are these things but that he would maintain human friendship by their mutual wants and plenty, lest everyone thinking themselves sufficient for themselves for this only thing should be made insociable? Now it cometh to pass that one nation should supply the want of another by the appointment of divine justice, that thereby . . . that which is brought forth anywhere might seem to be bred with all. . . .

They, therefore, that take away this, take away that most laudable society of mankind; they take away the mutual occasions of doing good and, to conclude, violate nature herself. For even that ocean wherewith God hath compassed the Earth is navigable on every side round about, and the settled or extraordinary blasts of wind, not always blowing from the same quarter, and sometimes from every quarter, do they not sufficiently signify that nature hath granted a passage from all nations unto all? . . . This right therefore equally appertaineth to all nations, which the most famous lawyers enlarge so far that they deny any commonwealth or prince to be able wholly to forbid others to come unto their subjects and trade with them. . . .

We know also that wars began for this cause, as with the Magarensians against the Athenians, and the Bononians against the Venetians, and that these also were just causes of war to the Castilians against the Americans, and more probable than the rest. Victoria also thinketh it a just cause of war if they should be forbidden to go on pilgrimage and to live with them; if they were denied from the participation of those things which by the law of nations or customs are common; if, finally, they were not admitted to traffic.

The like whereof is that which we read in the history of Moses, and Augustine thereupon: that the Israelites made just war against the Amorites because a harmless passage was denied which by the most just law of human society ought to have been open to them. And for this cause Hercules made war with the King of the Orchomenians, the Grecians under Agamemnon with the king of the Mysians, as if naturally . . . ways and passage should be free, and the Romans in Tacitus are accused of the Germans because they barred the conference and resort of the nations and shut up rivers and earth and heaven itself after a certain manner. Nor did any title against the Saracens in times past please the Christians better than that they were stopped by them from entering into the land of Jewry.

It followeth upon this opinion that the Portugals, although they had been lords of those countries whither the Hollanders go, yet they should do wrong if they stopped the passage and trade of the Hollanders.

How much more unjust is it therefore for any that are willing to be secluded from intercourse and interchange with people who are also willing, and that by their means in whose power neither these people are nor the thing itself whereby we make our way, seeing we detest not thieves and pirates more for any other cause than that they beset and molest the meetings of men among themselves?

APPENDIX A.2

They That Occupy Their Business on Great Waters

SIMON WINCHESTER

1. Laws and Order

The far north Atlantic is where parliaments began. The first lawmaking assemblies were founded there in the tenth century, and soon thereafter some kind of justice and order began to settle, not just on the lands where assemblies convened and laws were first made, but also on the seas between.

The first true parliament is reckoned by most to have assembled in Iceland and somewhat symbolically, in the curiously fashioned valley in the west of the country known as Thingvellir, where the world's American and Eurasian plates are still tugging apart from one another and new ocean floor is being created.

There is a large basalt slab protruding upward from the western wall of the valley, and it was here beneath it that more than a thousand years ago farmers and peasants and priests and merchants passing through the valley would agree to stop and camp and meet each year to hammer out in some fashion the manner in which they thought their island nation should be run. The assembly was eventually called the *Althing*, and once it had a formal structure the date generally agreed for this formation being 930 A.D. it became the sole body charged with fashioning Iceland's laws. The rock, from which the Icelandic flag flies still, day and night, is today without doubt the most revered monument in the Atlantic north: the Rock of Laws, which set the patterns for the governance of much of the rest of the world.

Soon afterward the processes and customs of the Icelandic *thing*—and yes, it is the very word that today signifies an object or a concept—were mimicked by men who made laws in the Faroe Islands nearby, and later on in Norway and Sweden and Denmark, too. It was also mimicked on that British half-possession the Isle of the Man, where the assembly was and still is known as the Tynwald. It first met in 979 A.D., and since it has gathered without interruption in all the years since (unlike the Icelandic Althing, which was suspended for many years when the country dissolved itself into anarchy), it lays claim to being the oldest continuously and regularly meeting democratic institution in the world.

There are many other competing contenders for primacy among the various parliamentary assemblies dotted around the Nordic world, and there is little value in delving into their arguments. But accepting that the idea born in Iceland did spread, rapidly and over large distances, one overarching truth appears: that in a large quadrant of the world's northern nations and all of them nations that happen to have been intimately involved with the Atlantic Ocean, there was from the tenth century onward both a popularly established means of creating codes of laws of sorts, and popularly elected or otherwise assembled bodies that were established and designed to promulgate and administer those laws.

No such institutions were created this early in Russia, say, nor in China, nor even in Greece, despite the ancient Athenian origins of a rather different kind of popular governance. Parliamentary democracy, as it is understood in today's world, was very much an Atlantic creation—a further reminder, if one were needed, that while the Mediterranean Sea was clearly central to the makings of the classical world, the North Atlantic and many of the countries bordering it were witness to the construction of many of the foundations, ties, and crossbeams of what we now know as the modern world.

2. The Rules of Trade

It would be some centuries before the ocean was fully crossed, east to west—notwithstanding the eleventh century Vikings, who visited Labrador and settled in Newfoundland—and long sea trade began. Until then, major voyages into the ocean were performed not in pursuit of trade but so men of great courage and daring could exploit the one resource with which all the world's seas, and most especially the North Atlantic, were once replete: fish.

It was the Hanseatic League that established a proper footing for commercial fishing in the North Atlantic. The popularity of the highly nutritious and economical cold-water fish prompted the Hansa merchants to order the construction of two fleets of vessels to exploit the massive shoals of fish in two quite distinct Atlantic fishing grounds: the so-called Scania waters off southern Sweden, where there were plenty of herring; and the Lofoten Islands, above the Arctic Circle in northern Norway, where there were unimaginably large stocks of *Gadus morhua*, the Atlantic cod.

The importance of this remarkable white fleshed, protein rich, almost fat free fish in the Atlantic's history can hardly be gainsaid. It dominated the trade of the Hansa; it stimulated the transoceanic adventuring of the Basques; it provided hundreds of thousands of Britons with work and tens of millions of Britons with food; and for decades it formed the central plank of the economics of all of maritime Canada and the coastal states of New England. . . .

And while the Europeans involved themselves in this voyaging, the new Americans did so as well. Whether they were sailing as settlers or as colonials or whether, as after 1776, they did so as citizens of a newly independent nation, Americans were particularly quick off the mark in exploiting all manner of transatlantic venturings.

They first got their sea legs by chasing whales. . . . Rather than heading north to mix it up with the Europeans, who were so bitterly and busily fighting among themselves, the Americans decided early on that their crews would head into virgin Atlantic territory—they would let the Danes, the Dutchmen, and the English have the rights and bowheads of the north, while they would concentrate on the largely untouched stocks of baleens—the fins, seis, minkes, grays, humpbacks, southern rights, and the gigantic and unforgettably grand blue whales—as well as the sperm whale, renowned for its superior oil, and which lived in what came to be called the Southern Whale Fishery. . . .

And so by the middle of the eighteen century the whalers, now equipped with bigger ships, thicker sails, more capacious oil barrels, stronger harpoons, more enduring rope, and more lasting ironware, swept out from the Americas east into what they called *ye deep*.

Until now their voyages lasted only a matter of days, perhaps a week or two. But the more enterprising whaling men, most of them of stout Quaker stock and not given to excitement or fear, began to sail their ships as far as Brazil, or the Guinea coast, or even the Falkland Islands or South Georgia, and were away for months; there was much terror but they also spent many hours drifting idly, good for the crafting of scrimshaw. In time the more adventurous took their vessels to the south of Isla de los Estados, and doubled Cape Horn against the vast winds and storms of those lethal latitudes known as the Roaring Forties, and with luck and good seamanship they emerged whole and passed into the whale-rich emptiness of the Pacific.

But their long sojourns in the Atlantic gave American sailors a confidence and a profound knowledge of the deep sea that was shared by few others. Whalers ranged to the farther ends of the ocean and discovered as many of its secrets as did the navigators and surveyors sent out by the maritime states: their legacy—and especially the legacy of New England whalers in the Atlantic—is profound.

APPENDIX A.3

From Davy Jones' Locker to the Foot Locker*

The Case of the Floating Nikes

PHILIP E. STEINBERG

ON MAY 27, 1990, 600 miles south of Anchorage, Alaska, the Seattle-bound container ship *Hansa Carrier* encountered heavy storms, and twenty-one containers were lost overboard. Inside five of these containers were some 80,000 Nike sneakers, hiking boots, and sandals, with a retail value of approximately $2.5 million. Four of the five Nike containers opened, and 61,280 shoes began a long journey eastward to the coast of North America. Over the next two years, more than 1,600 Nikes were recovered on the beaches of British Columbia, Washington, and Oregon.

West Coast residents soon discovered that the shoes, although somewhat stiff after a year or two in the ocean, were wearable after a thorough washing in warm water. Unfortunately for the sneaker scavengers, Nike had shipped the shoes untied, and so pairs of like models, colors, and sizes did not wash ashore together. Beachcombers responded by holding "swap meets" in West Coast beach towns throughout 1991. One particularly enterprising beachcomber, Oregon artist Steve McLeod, reported earning $568 by collecting, matching, cleaning, and selling washed-up Nikes. Meanwhile,

Excerpt from Philip E. Steinberg, introduction to *The Social Construction of the Ocean* (Cambridge: Cambridge University Press, 2001).

*Davy Jones is nautical slang, dating back to at least the mid-eighteenth century, for the spirit of the sea. Thus, Davy Jones' locker refers to "the bottom of the sea, the final resting place of sunken ships, of articles lost or thrown overboard, of men buried at sea" (Kemp 1988: 232).

two Seattle-based oceanographers, Curtis Ebbesmeyer and James Ingraham, took advantage of the spill, calibrating shoe recovery data and the release site to existing ocean-current models to gain new insights into the variability of ocean currents. . . .

This story provides a fitting beginning for a narrative about the various ways the world-ocean has been perceived, constructed, and managed under modernity. For the Nike Corporation, the sea represented the least expensive means of transporting commodities from point of production to point of sale. It enabled the company to reproduce geographical hierarchy and the global division of labor. By utilizing the sea for transporting a commodity, Nike was able to move the labor and capital embedded in each shoe from a low-wage region (Asia) to a high-wage region (North America), permitting profit realization. The sea, for Nike, was a space of distance, a space across which shoes had to travel so that their sale could generate a profit for corporate shareholders. For Leonhardt & Blumberg, the Hamburg-based shipping firm that operated the *Hansa Carrier*, the sea also was a surface to be crossed, and its distance similarly presented profit-making opportunities.

For residents of the West Coast, including beachcombers like McLeod and his customers, the sea was a provider. It was the means by which goods arrived to satisfy North American footwear needs. For these North Americans, the sea was more than just a space across which goods traveled; it was the entity that *brought* low-cost goods to the consumer. The sea provided the resource of connection. In this particular instance, the role of the sea as resource provider is especially clear, since the sea itself (not a ship) "carried" the shoes and they arrived decommodified, like a "natural" resource. Essentially, however, the sea would have performed a similar "provider" function had the containers not fallen overboard. Once the sneakers had been extracted from the ocean, McLeod performed a service similar to more conventional Nike distributors, adding labor value to the product by packaging it and bringing it to the customer.

While Nike and Leonhardt & Blumberg saw the ocean as a transportation surface and McLeod and his customers saw it as a resource provider, the oceanographers viewed the ocean as a set of discrete locations: places of ocean currents, storm centers, coastlines, islands, past message-in-bottle releases, and so on. These places, each with its own distinct nature, related to one another to make the ocean one grand physical system. Leonhardt & Blumberg's insurer similarly perceived the ocean as a space of discrete places and events, each of which might interfere with the transport-surface construction favored by Nike and Leonhardt & Blumberg.

Each actor in the Nike drama perceived the ocean as an arena of both predictability and unpredictability. Even as it constructed the ocean as a predictable, formless transport surface, Nike and Leonhardt & Blumberg were aware that the ocean was not totally "frictionless"; this is why Nike required Leonhardt & Blumberg to retain insurance on the vessel. Nike and its customers pay for this unpredictability of the sea in the price of every sneaker, a portion of which covers shipping insurance. For McLeod, the sea, while a provider, was an exceedingly capricious one. He could not know whether his time spent combing the beaches or the "swap meets" would result in his finding a matched pair of sneakers. His customers paid for this unpredictability of the sea in that the price of each pair compensated McLeod for time spent wandering the beaches and not finding any shoes. The oceanographers and the insurers, meanwhile, were attempting to conquer the unpredictability of the ocean by fine-tuning models of ocean-current variability and calculating the degree to which the ocean's features might interfere with its transport-surface properties. Indeed, if the oceanographers were to succeed in rendering the ocean totally predictable, much of the uncertainty facing the other actors would disappear: Leonhardt & Blumberg would be able to redirect its ships to safe routes and cancel its insurance policy; McLeod would know exactly where shoes would wash ashore; and Nike would benefit from lower transportation costs, which it then could pass on to the consumer by lowering the price for its sneakers.

Of course, the Nike case is exceptional. The vast majority of ocean-transported goods arrive safely. Yet the story demonstrates succinctly how the ocean is perceived and used by various social actors. Furthermore, it shows how each of these uses is characterized by that combination of predictability and unpredictability that reproduces cycles of capital speculation and investment, whether on the scale of a multinational insurance firm, the Nike Corporation, or Steve McLeod. The story of the floating Nikes demonstrates how multiple constructions of the ocean serve to maintain the concentrations and movements of wealth that characterize modern capitalism.

One could take this scenario one step further and speculate how each actor might favor certain ocean policies (which in turn would imply further perceptions and constructions of the ocean) so as to strengthen their specific interests. The insurance firm, working with the oceanographers, might seek to restrict navigation to calm regions and times. Nike, Leonhardt & Blumberg, and the consumers, on the other hand, would favor a regime that preserved for each ship the right to weigh risks on a case-by-case basis and choose the least costly and quickest route for each trip from South Korea to Seattle. McLeod too would favor an absence of regulation. As a non-traditional Nike distributor, McLeod has a direct interest in the ships embarking on high-risk journeys, but even if he were a more typical Nike retailer he could be expected to favor low-cost wholesale goods that would enable him to resell the shoes at a competitive price.

The various parties also would differ on whether or not individual entities should be permitted to claim exclusive access (or exclusive use privileges) to specific areas of the sea. McLeod would probably favor a mechanism whereby he could obtain property rights (or at least exclusive usufruct) to areas of the beach and coastal waters that he had prospected and found to be sneaker-rich. Likewise, Leonhardt & Blumberg might favor a means by which it could gain exclusive access to specific ocean routes. McLeod and Leonhardt & Blumberg then would find themselves supporting some degree of militarization of the sea in order to exclude others from their domains. By contrast, the insurer, Nike, and the consumers would probably oppose this territorial division of the sea. Militarization would increase the risks associated with each shipment, something that Nike and the insurer would seek to avoid. Additionally, these increased risks—especially when combined with the monopolization of shipping routes that would be facilitated by the territorialization of the sea—would increase cost, which would negatively impact Nike and the consumers. Ebbesmeyer and Ingraham, the oceanographers, would also oppose this territorialization of the sea, since it would probably interfere with their research.

If each actor were to pursue its strategy, the result would be a set of social institutions, attitudes, and norms that would reproduce the construction of the ocean as unclaimable transport surface, claimable resource space, a set of discrete places and events, and a field for military adventure. The balance between these often contradictory constructions would shift from time to time, as the power of the actors varied and as the need for certain ocean uses waxed and waned, but the overall competition among the various actors would serve to reproduce the ocean as a uniquely constructed space with a complex regime designed to serve a multiplicity of functions.

The narrative presented [here] resembles that of the Nike story, with the number of actors, the geographical extent, and the time frame greatly extended to encompass the whole of the modern era, roughly from 1450 to the present. Through a historical narrative, it is argued that each period of capitalism, besides having a particular spatiality on land, has had a complementary—if often contrapuntal—spatiality at sea, with specific interest

Table A.3.1. The Nike story: characteristics of key actors

Actor	Conception of sea's function	Conception of sea's physical processes	View on regulations and restrictions to limit risk	View on militarization/ possession/ territorialization
Nike	Transport surface	Featureless and placeless	Against	Against
Leonhardt & Blumberg	Transport surface	Featureless and placeless	Against	For
Insurer	Transport surface	Contains distinct features and places	For	Against
Steve McLeod	Resource provider	Contains distinct features and places	Against	For
Sneaker consumers	Resource provider	Featureless and placeless	Against	Against
Ebbesmeyer and Ingraham	Arena for physical processes	Contains distinct features and places	For	Against

groups during each period promoting specific constructions of ocean-space. As is the case with land-space, the contradictions and changes within each period's construction have been intertwined with contradictions and changes in that period's politicaleconomic structures. The ever-changing uses, regulations, and representations of ocean-space have been as much a part of each period's spatiality as have the spatial constructions of land-space.

Appendix B

A Broader Global Perspective

AS TAYLOR CUNNINGHAM suggests in "Mobilizing Vessels and Voices" (chapter 5, this volume), many cultures, but especially those cultures of the South Pacific, inhabit homelands made up of intertwined islands and ocean. Epeli Hau'ofa explores a similar change in perspective in "Our Sea of Islands" (1994), emphasizing the importance of ocean space *as* place—a vast territory that connects peoples. In Joshua Reid's powerful study of the Makah people, *The Sea Is My Country* (2015), marine space is the locus of spiritual and physical health. The Makah continue to build their cultural identity from the sea, and Reid presents contemporary Makah whale hunting as one element of their sea-based autonomy. A similar celebration of indigenous sovereignty emerges in "Praise Song for Oceania," a 2020 poem by Chamoru writer Craig Santos Perez. His view from Guam emphasizes the Pacific's many healing roles, and a broad, oceanic logic of inclusion, connection, and care. In his praise song, Perez speaks for many who perceive the ocean as a place of interconnection and hope.

Unfortunately, the islands scattered within this territory are some of the places most threatened by climate change–induced sea-level rise. Maxine Burkett takes up the idea that the industrialized West owes reparations to Pacific Islanders—and Small Island Developing States more generally—for the disappearance of the small but necessary terrestrial components of their homelands in Oceania.

APPENDIX B.1

Our Sea of Islands

EPELI HAU'OFA

THIS ESSAY RAISES some issues of great importance to our region, and offers a view of Oceania that is new and optimistic. What I say here is likely to disturb a number of men and women who have dedicated their lives to Oceania and for whom I hold the greatest respect and affection, and always will.

In our region, two levels of operation are pertinent to the purposes of this paper. The first is that of national governments and regional and international diplomacy, in which the present and future of Pacific island states and territories are planned and decided on. Discussions here are the preserve of politicians, bureaucrats, statutory body officials, diplomats and the military, and representatives of the financial and business communities, often in conjunction with donor and international lending organizations, and advised by academic and consultancy experts. Much that passes at this level concerns aid, concessions, trade, investment, defense and security, matters that have taken the Pacific further and further into dependency on powerful nations.

The other level is that of ordinary people, peasants and proletarians, who, because of the poor flow of benefits from the top, skepticism about stated

Essay appears in *The Contemporary Pacific* 6, no. 1 (Spring 1994): 147–61. First published in *A New Oceania: Rediscovering Our Sea of Islands*, edited by Vijay Naidu, Eric Waddell, and Epeli Hau'ofa (Suva: School of Social and Economic Development, The University of the South Pacific, 1993). Permission to reprint courtesy of University of Hawai'i Press.

policies and the like, tend to plan and make decisions about their lives independently, sometimes with surprising and dramatic results that go unnoticed or ignored at the top. Moreover, academic and consultancy experts tend to overlook or misinterpret grassroots activities because they do not fit with prevailing views about the nature of society and its development.

Views of the Pacific from the level of macroeconomics and micropolitics often differ markedly from those from the level of ordinary people. The vision of Oceania presented in this essay is based on my observations of behavior at the grass roots.

Having clarified my vantage point, I make a statement of the obvious—views held by those in dominant positions about their subordinates could have significant consequences for people's self-image and for the ways they cope with their situations. Such views, which are often derogatory and belittling, are integral to most relationships of dominance and subordination, wherein superiors behave in ways or say things that are accepted by their inferiors, who in turn behave in ways that serve to perpetuate the relationships.

In Oceania, derogatory and belittling views of indigenous cultures are traceable to the early years of interactions with Europeans. The wholesale condemnation by Christian missionaries of Oceanic cultures as savage, lascivious, and barbaric has had a lasting and negative effect on people's views of their histories and traditions. In a number of Pacific societies people still divide their history into two parts: the era of darkness associated with savagery and barbarism; and the era of light and civilization ushered in by Christianity.

In Papua New Guinea, European males were addressed and referred to as "masters" and workers as "boys." Even indigenous policemen were called "police boys." This use of language helped to reinforce the colonially established social stratification along ethnic divisions. A direct result of colonial practices and denigration of Melanesian peoples and cultures as even more primitive and barbaric than those of Polynesia can be seen in the attempts during the immediate postcolonial years by articulate Melanesians to rehabilitate their cultural identity by cleansing it of its colonial taint and denigration. Leaders like Walter Lini of Vanuatu and Bernard Narokobi of Papua New Guinea have spent much of their energy extolling the virtues of Melanesian values as equal to if not better than those of their erstwhile colonizers.

Europeans did not invent belittlement. In many societies it was part and parcel of indigenous cultures. In the aristocratic societies of Polynesia parallel relationships of dominance and subordination with their paraphernalia of appropriate attitudes and behavior were the order of the day. In Tonga, the term for commoners is *me'a vale* "the ignorant ones," which is a survival from an era when the aristocracy controlled all important knowledge in the society. Keeping the ordinary folk in the dark and calling them ignorant made it easier to control and subordinate them.

I would like, however, to focus on a currently prevailing notion about Islanders and their physical surroundings that, if not countered with more constructive views, could inflict lasting damage on people's images of themselves, and on their ability to act with relative autonomy in their endeavors to survive reasonably well within the international system in which they have found themselves. It is a belittling view that has been unwittingly propagated, mostly by social scientists who have sincere concern for the welfare of Pacific peoples.

According to this view, the small island states and territories of the Pacific, that is, all of Polynesia and Micronesia, are much too small, too poorly endowed with resources, and too isolated from the centers of economic growth for their inhabitants ever to be able to rise above their present condition of dependence on the largesse of wealthy nations.

Initially, I agreed wholeheartedly with this perspective, and I participated actively in its propagation. It seemed to be based on irrefutable evidence, on the reality of our existence. Events of the 1970s and 1980s confirmed the correctness of this view. The hoped-for era of autonomy following political independence did not materialize. Our national leaders were in the vanguard of a rush to secure financial aid from every quarter; our economies were stagnating or declining; our environments were deteriorating or were threatened and we could do little about it; our own people were evacuating themselves to greener pastures elsewhere. Whatever remained of our resources, including our exclusive economic zones, was being hawked for the highest bid. Some of our islands had become, in the words of one social scientist, "MIRAB societies"—pitiful microstates condemned forever to depend on migration, remittances, aid, and bureaucracy, and not on any real economic productivity. Even the better resource-endowed Melanesian countries were mired in dependency, indebtedness, and seemingly endless social fragmentation and political instability. What hope was there for us?

This bleak view of our existence was so relentlessly pushed that I began to be concerned about its implications. I tried to find a way out but could not. Then two years ago I began noticing the reactions of my students when I described and explained our situation of dependence. Their faces crumbled visibly, they asked for solutions, I could offer none. I was so bound to the notion of smallness that even if we improved our approaches to production, for example, the absolute size of our islands would still impose such severe limitations that we would be defeated in the end.

But the faces of my students continued to haunt me mercilessly. I began asking questions of myself. What kind of teaching is it to stand in front of young people from your own region, people you claim as your own, who have come to university with high hopes for the future, and you tell them that our countries are hopeless? Is this not what neocolonialism is all about? To make people believe that they have no choice but to depend?

Soon the realization dawned on me. In propagating a view of hopelessness, I was actively participating in our own belittlement. I decided to do something about it, but I thought that since any new perspective must confront some of the sharpest and most respected minds in the region, it must be well researched and thought out if it was to be taken seriously. It was a daunting task, and I hesitated.

Then came invitations for me to speak at Kana and Hilo on the Big Island of Hawai'i at the end of March 1993. The lecture at Kona, to a meeting of the Association of Social Anthropologists in Oceania, was written before I left Suva. The speech at the University of Hawai'i at Hilo was forming in my mind and was to be written when I got to Hawai'i. I had decided to try out my new perspective, although it had not been properly researched. I could hold back no longer. The drive from Kana to Hilo was my "road to Damascus." I saw such scenes of grandeur as I had not seen before: the eerie blackness of regions covered by recent volcanic eruptions; the remote majesty of Maunaloa, long and smooth, the world's largest volcano; the awesome craters of Kilauea threatening to erupt at any moment; and the lava flow on the coast not far away. Under the aegis of Pele, and before my very eyes, the Big Island was growing, rising from the depths of a mighty sea. The world of Oceania is not small; it is huge and growing bigger every day.

The idea that the countries of Polynesia[1] and Micronesia are too small, too poor, and too isolated to develop any meaningful degree of autonomy is an economistic and geographic deterministic view of a very narrow kind that overlooks culture history and the contemporary process of what may be called world enlargement that is carried out by tens of thousands of ordinary Pacific Islanders right across the ocean—from east to west and north to south, under the very noses of academic and consultancy experts, regional and international development agencies, bureaucratic planners and their advisers, and customs and immigration officials—making nonsense of all national and economic boundaries, borders that have been defined only recently, crisscrossing an ocean that had been boundless for ages before Captain Cook's apotheosis.

If this very narrow, deterministic perspective is not questioned and checked, it could contribute importantly to an eventual consignment of groups of human beings to a perpetual state of wardship wherein they and their surrounding lands and seas would be at the mercy of the manipulators of the global economy and "world orders" of one kind or another. Belittlement in whatever guise, if internalized for long, and transmitted across generations, may lead to moral paralysis, to apathy, and to the kind of fatalism that we can see among our fellow human beings who have been herded and confined to reservations or internment camps. People in some of our islands are in danger of being confined to mental reservations, if not already to physical ones. I am thinking here of people in the Marshall Islands, who have been victims of atomic and missile tests by the United States.

Do people in most of Oceania live in tiny confined spaces? The answer is yes if one believes what certain social scientists are saying. But the idea of smallness is relative; it depends on what is included and excluded in any calculation of size. When those who hail from continents, or islands adjacent to continents—and the vast majority of human beings live in these regions—when they see a Polynesian or Micronesian island they naturally pronounce it small or tiny. Their calculation is based entirely on the extent of the land surfaces they see.

But if we look at the myths, legends, and oral traditions, and the cosmologies of the peoples of Oceania, it becomes evident that they did not conceive of their world in such microscopic proportions. Their universe comprised not only land surfaces, but the surrounding ocean as far as they could traverse and exploit it, the underworld with its fire-controlling and earth-shaking denizens, and the heavens above with their hierarchies of powerful gods and named stars and constellations that people could count on to guide their ways across the seas. Their world was anything but tiny. They thought big and recounted their deeds in epic proportions. One legendary Oceanic athlete was so powerful that during a competition he threw his javelin with such force that it pierced the horizon and disappeared until that night when it was seen streaking across the sky like a meteor. Every now and then it reappears to remind people of the mighty deed. And as far as I'm concerned it is still out there, near Jupiter or somewhere. That was the first rocket ever sent into space. Islanders today still relish exaggerating things out of all proportion. Smallness is a state of mind.

There is a world of difference between viewing the Pacific as "islands in a far sea" and as "a sea of islands."[2] The first emphasizes dry surfaces in a vast ocean far from the centers of power. Focusing in this way stresses the smallness and remoteness of the islands. The second is a more holistic perspective in which things are seen in the totality of their relationships. I return to this point later. Continental men, namely Europeans, on entering the Pacific after crossing huge expanses of ocean, introduced the view of "islands in a far sea." From this perspective the islands are tiny, isolated dots in a vast ocean. Later on, continental men—Europeans and Americans—drew imaginary lines across the sea, making the colonial boundaries that confined ocean peoples to tiny spaces for the first time. These boundaries today define the island states and territories of the Pacific. I have just used the term *ocean peoples* because our ancestors, who had lived in the Pacific for over two thousand years, viewed their world as "a sea of islands" rather than as "islands in the sea." This may be seen in a common categorization of people, as exemplified in Tonga by the inhabitants of the main, capital, island, who used to refer to their compatriots from the rest of the archipelago not so much as "people from outer islands" as social scientists would say, but as *kakai mei tahi* or just *tahi* "people from the sea." This characterization reveals the underlying assumption that the sea is home to such people.

The difference between the two perspectives is reflected in the two terms used for our region: *Pacific Islands* and *Oceania*. The first term, *Pacific Islands*, is the prevailing one used everywhere; it denotes small areas of land sitting atop submerged reefs or seamounts. Hardly any anglophone economist, consultancy expert, government planner, or development banker in the region uses the term *Oceania*, perhaps because it sounds grand and somewhat romantic, and may denote something so vast that it would compel them to a drastic review of their perspectives and policies. The French and other Europeans use the term *Oceania* to an extent that English speakers, apart from the much-maligned anthropologists and a few other sea-struck scholars, have not. It may not be coincidental that Australia, New Zealand, and the United States, anglophone all, have far greater interests in the Pacific and how it is perceived than have the distant European nations.

Oceania denotes a sea of islands with their inhabitants. The world of our ancestors was a large sea full of places to explore, to make their homes in, to breed generations of seafarers like themselves. People raised in this environment were at home with the sea. They played in it as soon as they could walk steadily, they worked in it, they fought on it. They developed great skills for navigating their waters, and the spirit to traverse even the few large gaps that separated their island groups.

Theirs was a large world in which peoples and cultures moved and mingled, unhindered by boundaries of the kind erected much later by imperial powers. From one island to another they sailed to trade and to marry, thereby expanding social networks for greater flows of wealth. They traveled to visit relatives in a wide variety of natural and cultural surroundings, to quench their thirst for adventure, and even to fight and dominate.

Fiji, Samoa, Tonga, Niue, Rotuma, Tokelau, Tuvalu, Futuna, and Uvea formed a large exchange community in which wealth and people with their skills and arts circulated endlessly. From this community people ventured to the north and west, into Kiribati, the Solomon Islands, Vanuatu, and New Caledonia, which formed an outer arc of less intensive exchange. Evidence of this voyaging is provided by existing settlements within Melanesia of descendants of these seafarers. [Only blind landlubbers would say that settlements like these, as well as those in New Zealand and Hawai'i, were made through accidental voyages by people who got blown off course—presumably while they were out fishing with their wives, children, pigs, dogs, and food-plant seedlings—during a hurricane.] The Cook Islands and French Polynesia formed a community similar to that of their cousins to the west; hardy spirits from this community ventured southward and founded settlements in Aotearoa, while others went in the opposite direction to discover and inhabit the islands of Hawai'i. Also north of the equator is the community that was centered on Yap.

Melanesia is supposedly the most fragmented world of all: tiny communities isolated by terrain and at least one thousand languages. The truth is that large regions of Melanesia were integrated by trading and cultural exchange systems that were even more complex than those of Polynesia and Micronesia. Lingua francas and the fact that most Melanesians were and are multilingual (which is more than one can say about most Pacific rim countries), make utter nonsense of the notion that they were and still are babblers of Babel. It was in the interest of imperialism and is in the interest of neocolonialism, to promote this blatant misconception of Melanesia.[3]

Evidence of the conglomerations of islands with their economies and cultures is readily available in the oral traditions of the islands, and in blood ties that are retained today. The highest chiefs of Fiji, Samoa, and Tonga, for example, still maintain kin connections that were forged centuries before Europeans entered the Pacific, to the days when boundaries were not imaginary lines in the ocean, but rather points of entry that were constantly negotiated and even contested. The sea was open to anyone who could navigate a way through.

This was the kind of world that bred men and women with skills and courage that took them into the unknown, to discover and populate all the habitable islands east of the 130th meridian. The great fame that they have earned posthumously may have been romanticized, but it is solidly based on real feats that could have been performed only by those born and raised with an open sea as their home.

Nineteenth-century imperialism erected boundaries that led to the contraction of Oceania, transforming a once boundless world into the Pacific Island states and territories that we know today. People were confined to their tiny spaces, isolated from each other. No longer could they travel freely to do what they had done for centuries. They were cut off from their relatives abroad, from their far-flung sources of wealth and cultural enrichment. This is the historical basis of the view that our countries are small, poor, and isolated. It is true only insofar as people are still fenced in and quarantined.

This assumption is no longer tenable as far as the countries of central and western Polynesia are concerned, and may be untenable also of Micronesia. The rapid expansion of the world economy in the years since World War II may have intensified third world dependency, as has been noted from certain vantage points at high-level academia, but it also had a liberating effect on the lives of ordinary people in Oceania, as it did in the Caribbean islands. The new economic reality made nonsense of artificial boundaries, enabling the people to shake off their confinement. They have since moved, by the tens of thousands, doing what their ancestors did in earlier times: enlarging their world as they go, on a scale not possible before. Everywhere they go, to Australia, New Zealand, Hawai'i, the mainland United States, Canada, Europe, and elsewhere, they strike roots in new resource areas, securing employment and overseas family property, expanding kinship networks through which they circulate themselves, their relatives, their material goods, and their stories all across their ocean, and the ocean is theirs because it has always been their home. Social scientists may write of Oceania as a Spanish Lake, a British Lake, an American Lake, and even a Japanese Lake. But we all know that only those who make the ocean their home and love it, can really claim it as their own. Conquerors come, conquerors go, the ocean remains, mother only to her children. This mother has a big heart though; she adopts anyone who loves her.

The resources of Samoans, Cook Islanders, Niueans, Tokelauans, Tuvaluans, I-Kiribati, Fijians, Indo-Fijians, and Tongans, are no longer confined to their national boundaries. They are located wherever these people are living, permanently or otherwise, as they were before the age of western imperialism. One can see this any day at seaports and airports throughout the central Pacific, where consignments of goods from homes abroad are unloaded as those of the homelands are loaded. Construction materials, agricultural machinery, motor vehicles, other heavy goods, and myriad other things are sent from relatives abroad, while handcrafts, tropical fruits and root crops, dried marine creatures, kava, and other delectables are dispatched from the homelands. Although this flow of goods is generally not included in official statistics, much of the welfare of ordinary people of Oceania depends on an informal movement along ancient routes drawn in bloodlines invisible to the enforcers of the laws of confinement and regulated mobility.

The world of Oceania is neither tiny nor deficient in resources. It was so only as a condition of the colonial confinement that lasted less than a century in a history of millennia. Human nature demands space for free movement, and the larger the space the better it is for people. Islanders have broken out of their confinement, are moving around and away from their homelands, not so much because their countries are poor, but because they were unnaturally confined and severed from many of their traditional sources of wealth, and because it is in their blood to be mobile. They are once again enlarging their world, establishing new resource bases and expanded networks for circulation. Alliances are already being forged by an increasing number of Islanders with the *tangata whenua* of Aotearoa and will inevitably be forged with the native Hawaiians. It is not inconceivable that if Polynesians ever get together, their two largest homelands will be reclaimed in one form or another. They have already made their presence felt in these homelands, and have stamped indelible imprints on the cultural landscapes.

We cannot see the processes outlined here clearly if we confine our attention to things within national boundaries and to events at the upper levels of political economies and regional and international diplomacy. Only when we focus on what ordinary people are actually doing, rather than on what they should be doing, can we see the broader picture of reality.

The world of Oceania may no longer include the heavens and the underworld, but it certainly encompasses the great cities of Australia, New Zealand, the United States, and Canada. It is within this expanded world that the extent of the people's resources must be measured.

In general, the living standards of Oceania are higher than those of most third world societies. To attribute this merely to aid and remittances—misconstrued deliberately or otherwise as a form of dependence on rich countries' economies—is an unfortunate misreading of contemporary reality. Ordinary Pacific people depend for their daily existence much, much more on themselves and their kin, wherever they may be, than on anyone's largesse, which they believe is largely pocketed by the elite classes. The funds and goods that homes-abroad people send their homeland relatives belong to no one but themselves. They earn every cent through hard physical toil in the new locations that need and pay for their labor. They also participate in the manufacture of many of the goods they send home; they keep the streets and buildings of Auckland clean, and its transportation system running smoothly; they keep the suburbs of the western United States (including Hawai'i) trimmed, neat, green, and beautiful; and they have contributed much, much more than has been acknowledged.

On the other hand Islanders in their homelands are not the parasites on their relatives abroad that misinterpreters of "remittances" would have us believe. Economists do not take account of the social centrality of the ancient practice of reciprocity, the core of all oceanic cultures. They overlook the fact that for everything homeland relatives receive, they reciprocate with goods they themselves produce, by maintaining ancestral roots and lands for everyone, homes with warmed hearths for travelers to return to permanently or to strengthen their bonds, their souls, and their identities before they move on again. This is not dependence but interdependence, which is purportedly the essence of the global system. To say that it is something else and less is not only erroneous, but denies people their dignity.

What I have stated so far should already have provided sufficient response to the assertion that the islands are isolated. They are clearly not. Through developments in high technology, communications and transportation systems are a vast improvement on what they were twenty years ago. These may be very costly by any standard, but they are available and used. Telecommunications companies are making fortunes out of lengthy conversations between breathless relatives thousands of miles apart.

But the islands are not connected only with regions of the Pacific rim. Within Oceania itself people are once again circulating in increasing numbers and frequency. Regional organizations—intergovernmental, educational, religious, sporting, and cultural—are responsible for much of this mobility. The University of the South Pacific, with its highly mobile staff and student bodies comprising men, women, and youth from the twelve island countries that own it and from outside the Pacific, is an excellent example. Increasingly the older movers and shakers of the islands are being replaced by younger ones; and when they meet each other in Suva, Honiara, Apia, Vila, or any other capital city of the Pacific, they meet as friends, as people who have gone through the same place of learning, who have worked and played and prayed together.

The importance of our ocean for the stability of the global environment, for meeting a significant proportion of the world's protein requirements, for the production of certain marine resources in waters that are relatively clear of pollution, for the global reserves of mineral resources, among others, has been increasingly recognized, and puts paid to the notion that Oceania is the hole in the doughnut. Together with our exclusive economic zones, the areas of the earth's surface that most of our countries occupy can no longer be called small. In this regard, Kiribati, the Federated States of Micronesia, and French Polynesia, for example, are among the largest countries in the world. The emergence of organizations such as SPACHEE (South Pacific Action Committee for Human Environment and Ecology), SPREP (South Pacific Regional Environment Programme), the Forum Fisheries Agency, and SOPAC (South Pacific Applied Geosciences Commission); of movements for a nuclear-free Pacific, the prevention of toxic waste disposal, and the ban on the wall-of-death fishing methods, with linkages to similar organizations and movements elsewhere; and the establishment at the University of the South Pacific of the Marine Science and Ocean Resources Management programs, with linkages to fisheries and ocean resources agencies throughout the Pacific and beyond; all indicate that we could play a pivotal role in the protection and sustainable development of our ocean. There are no people on earth more suited to be guardians of the world's largest ocean than those for whom it has been home for generations. Although this is a different issue from the ones I have focused on for most of this paper, it is relevant to the concern for a far better future for us than has been prescribed and predicted. Our role in the protection and development of our ocean is no mean task; it is no less than a major contribution to the well-being of humanity. Because it could give us a sense of doing something very worthwhile and noble, we should seize the moment with dispatch.

The perpetrators of the smallness view of Oceania have pointed out quite correctly the need for each island state or territory to enter into appropriate forms of specialized production for the world market, to improve their management and marketing techniques, and so forth. But they have so focused on bounded national economies at the macrolevel that they have overlooked or understated the significance of the other processes I have outlined here, and have thereby swept aside the whole universe of Oceanic mores and just about all our potentials for autonomy. The explanation seems clear: one way or another, they or nearly all of them are involved directly or indirectly in the fields of aided development and Pacific rim geopolitics, for whose purposes it is necessary to portray our huge world in tiny, needy bits. To acknowledge the larger reality would be to undermine the prevailing view and to frustrate certain agendas and goals of powerful interests. These perpetrators are therefore participants, as I was, in the belittlement of Oceania, and in the perpetuation of the neocolonial relationships of dependency that have been and are being played out in the rarefied circles of national politicians, bureaucrats, diplomats, and assorted experts and academics, while far beneath them exists that other order, of ordinary people, who are busily and independently redefining their world in accordance with their perceptions of their own interests and of where the future lies for their children and their children's children. Those who maintain that the people of Oceania live from day to day, not really caring for the long-term benefits, are unaware of the elementary truth known by most native Islanders: that they plan for generations, for the continuity and improvement of their families and kin groups.

As I watched the Big Island of Hawai'i expanding into and rising from the depths, I saw in it the future for Oceania, our sea of islands. That future lies in the hands of our own people, not of those who would prescribe for us, get us forever dependent and indebted because they can see no way out.

At the Honolulu Airport, while waiting for my flight back to Fiji, I met an old friend, a Tongan who is twice my size and lives in Berkeley, California. He is not an educated man. He works on people's yards, trimming hedges and trees, and laying driveways and footpaths. But every three months or so he flies to Fiji, buys eight-to-ten-thousand dollars' worth of kava, takes it on the plane flying him back to California, and sells it from his home. He has never heard of dependency, and if he were told of it, it would hold no real meaning for him. He told me in Honolulu that he was bringing a cooler full of T shirts, some for the students at the university with whom he often stays when he comes to Suva, and the rest for his relatives in Tonga, where he goes for a week or so while his kava is gathered, pounded, and bagged in Fiji. He later fills the cooler with seafoods to take back home to California, where he has two sons he wants to put through college. On one of his trips he helped me renovate a house that I had just bought. We like him because he is a good storyteller and is generous with his money and time, but mostly because he is one of us.

There are thousands like him, who are flying back and forth across national boundaries, the international dateline, and the equator, far above and completely undaunted by the deadly serious discourses below on the nature of the Pacific Century, the Asia–Pacific coprosperity sphere, and the dispositions of the post–cold war Pacific rim, cultivating their ever-growing universe in their own ways, which is as it should be, for therein lies their independence. No one else would give it to them—or to us.

Oceania is vast, Oceania is expanding, Oceania is hospitable and generous, Oceania is humanity rising from the depths of brine and regions of fire deeper still, Oceania is us. We are the sea, we are the ocean, we must wake up to this ancient truth and together use it to overturn all hegemonic views that aim ultimately to confine us again, physically and psychologically, in the tiny spaces that we have resisted accepting as our sole appointed places, and from which we have recently liberated ourselves. We must not allow anyone to belittle us again, and take away our freedom.

Notes

I would like to thank Marshall Sahlins for convincing me in the end that not all is lost, and that the world of Oceania is quite bright despite appearances. This paper is based on lectures delivered at the University of Hawai'i at Hilo, and the East-West Center, Honolulu, March/April 1993. Vijay Naidu and Eric Waddell read a draft of this paper and made very helpful comments. I am profoundly grateful to them for their support.

1. For geographic and cultural reasons I include Fiji in Polynesia. Fiji, however, is much bigger and better endowed with natural resources than all tropical Polynesian entities.
2. I owe much to Eric Waddell for these terms (pers comm).
3. I use the terms *Melanesia*, *Polynesia*, and *Micronesia* because they are already part of the cultural consciousness of the peoples of Oceania. Before the nineteenth century there was only a vast sea in which people mingled in ways that, despite the European-imposed threefold division, the boundaries today are still blurred. This important issue is, however, beyond the purview of this paper.

APPENDIX B.2

Just Where *Does* One Get a License to Kill Indians?

JOSHUA L. REID

ON THE MORNING of May 17, 1999, eight Makah men paddled the *K̓ʷiti·k̓ʷitš* ("kwih-tee-kwihtsh," *Hummingbird)* up to the three-year-old gray whale. Ignoring the drizzling rain, buzz of news copters above, and watchful eyes of a National Marine Fisheries biologist, Theron Parker thrust the harpoon. Unlike his misses on the prior two days of hunting, this throw sank into the thirty-ton leviathan and stuck. From a nearby support craft, a modified .577 caliber rifle roared three times. Fired by an experienced game hunter and decorated Vietnam War combat veteran, the third shot lanced through the water and into the whale's brain, killing it within seconds. As the female whale died off Washington State's Pacific coast, harpooner Theron led the crew in prayer, thanking her for offering herself to the Makahs. Surrounded by a small fleet of canoes from neighboring American Indian nations, the *Hummingbird* brought the whale ashore at Neah Bay about twelve hours later. Hundreds of men pulled on two heavy chains, hauling her onto the beach where generations of whalers had beached them before. Theron sprinkled eagle feathers on the whale's head while the community welcomed her—the first in seventy years—to the Makah nation.[1]

A coalition of indigenous peoples and non-Natives throughout the world supported the hunt at several critical stages. When the United States removed the gray whale from the list of endangered wildlife in 1994, the

Excerpt from the introduction to Joshua L. Reid, *The Sea Is My Country: The Maritime World of the Makahs* (New Haven: Yale University Press, 2015), 1-12. Reprinted with permission from Yale University Press.

tribal nation expressed interest in resuming customary whale hunts. With the support of the National Oceanic and Atmospheric Administration, in 1997, the Makah petitioned the International Whaling Commission (IWC) for approval of annual subsistence hunts. After negotiating with other North Pacific indigenous groups, the IWC granted the tribal nation a yearly quota of five whales, one for each ancestral Makah village near Cape Flattery, the northwesternmost tip of the contiguous United States. As part of this deal, Alaska Innuits traded twenty bowhead whales to the Chuktchis, a Siberian people, for five gray whales. Chuktchis have an annual quota of two hundred grays, and this reallocation undercut the potential criticism that Makah hunts would add stress on the species.[2]

Indigenous peoples continued to provide encouragement during the 1999 hunt and afterward. A few years earlier, several Vancouver Island Nuu-chah-nulth carvers, relatives of Makahs, helped to make the *Hummingbird*, and American Indian communities and Canadian First Nations from across North America offered prayers for the hunters' success. After landing the whale at Neah Bay, members of the community stripped blubber and meat from the carcass and hosted a large feast, reminiscent of potlatches from earlier centuries. Makah elder Dale Johnson remembers that getting the whale "brought a gathering of people; tribes from all over came in" as the whalers shared the catch. In addition to American Indians from Alaska, the Great Plains, and the West Coast, indigenous peoples from across North America, the Pacific, and Africa were honored guests at the celebratory feast. Billy Frank Jr., a Nisqually elder, president of the Northwest Indian Fisheries Commission, and veteran of the Indian fishing wars of the 1960s and 1970s, spoke passionately about the importance of exercising treaty rights. A Maasai warrior from Kenya expressed the need to preserve unique cultural characteristics in the world today. Many non-Natives also supported the hunt because whaling is a treaty right that Makahs reserved for themselves in the 1855 Treaty of Neah Bay, which the tribal nation signed with the United States. The 1999 hunt has inspired self-determination struggles by others. Currently negotiating treaties with the Canadian government, Nuu-chah-nulths point to the Makah whale hunt to support their efforts to resume a customary lifestyle based on whaling.[3]

Coming at the close of the twentieth century, however, the whalers' actions drew passionate opposition. Longtime foes of American Indian treaty rights, such as Congressman Jack Metcalf (R-Washington), railed against the federal government's bias of "giving Indians special rights," a statement exhibiting his ignorance that Native negotiators, not the government, reserved rights for their tribal nations in the treaties.[4] A vocal minority argued that whaling is barbaric and out of step with today's more "enlightened" views. Racism peppered much of their rhetoric, echoing criticisms levied against Washington's American Indians during the fishing rights conflict of earlier decades. Makah commercial fisher Dan Greene observed that "the same people who are racist against the Indian tribes are still there" since the treaty fishing wars of the 1960s and 1970s, "and that this [the 1999 whale hunt] just brought it to the surface again." After the May 17 hunt, "the floodgates of hate were opened."[5] Calling Makahs "'Red' necks with rifles" and accusing them of "playing Indian," non-Natives revealed strong anti-Indian sentiments still present in mainstream society. Phillips Wylly, an accomplished film and television producer, wrote to the editor of the *Seattle Times*, asking, "I am anxious to know where I may apply for a license to kill Indians. My forefathers helped settle the west and it was their tradition to kill every Redskin they saw. 'The only good Indian was a dead Indian,' they believed. I also want to keep with the faith of my ancestors." Protestors blocked the road to Neah Bay, issued bomb threats, and harassed gray whales in local waters in order to chase them away from Makah hunters. Anti-Indian sentiment became so vociferous that the governor deployed the National Guard to the reservation to protect Makah lives. The US Coast Guard defended the whalers at sea from activist organizations such as the Sea Shepherd Society.[6] These critics were racist, and they were also wrong. This kind of criticism of the Makahs' continuing whaling efforts reveals a deep lack of understanding about the issue. They have overlooked—and continue to ignore—the historical and cultural connections the Makahs have to the ocean.

Calling themselves the Qʷidiččaʔa·tx̌ ("kwi-dihch-chuh-aht"), meaning "the People of the Cape," Makahs shaped marine space in and around the Strait of Juan de Fuca, rather than terrestrial spaces, as the primary locus of their identity. They placed marine space at the center of their culture. Strategic exploitation of customary waters enabled the People of the Cape to participate in global networks of exchange, to resist assimilation, and to retain greater autonomy until the early twentieth century, later than many other land-based reservation communities in North America. When explorers and maritime fur traders entered the Pacific Northwest at the end of the eighteenth century, Makah chiefs protected their control over customary waters and resources. During talks for the 1855 Treaty of Neah Bay, Makah negotiators forced Territorial Governor Isaac Stevens to alter the treaty language to fit the tribal nation's maritime needs. Networks of exchange, kinship, and conflict made the waters around the Strait of Juan de Fuca, today an international border that separates the state of Washington from the Canadian province of British Columbia, into a space of connections. Indigenous whalers, sealers, and fishers combined customary practices with modern opportunities and technology to maintain Makah identity amid the cultural and environmental changes of the nineteenth and twentieth centuries. Challenges included overfished marine species, environmental degradation, rising state power, and assimilation and conservation efforts. By understanding the contours of the Makahs' connection to the ocean, we can see more easily why reviving the active practice of whaling is critical to the People of the Cape today.

The Environmental and Cultural Context of the Northwest Coast

As a marine people, Makahs developed a deep understanding of and relationship with the waters around them. Oceanographic features created what some scholars call a bioregion, acknowledging that powerful forces of the natural environment are more important than divisions created by borders. As in other places such as the Medicine Line country of the Alberta-Montana borderlands, environment and geography—rather than gender, age, ethnicity, or nationality—dictated community bonds in this region. Unpredictable and difficult forces of nature exerted a "constant push and pull" on the lives of those living in the Medicine Line country. Similarly, natural forces such as the ebb and flow of tides, winds and currents, powerful storms, spawning salmon, breaching whales, and gamboling seals and sea otters exerted a constant push and pull on the lives of the People of the Cape and other indigenous inhabitants of the Northwest Coast. Environmental features made these marine waters a complex, rich space that drew many peoples to this region and encouraged a spectrum of exchanges from violence to trade.[7]

Surface and submarine geological features bound this marine space in three dimensions. At the seascape's edges, the watersheds of mountainous Vancouver Island and the glaciated ramparts of the Olympic and Cascade Ranges sent mineral-laden freshwater into inshore coastal waters. Separating Vancouver Island from the Olympic Peninsula, the Strait of Juan de Fuca funneled cold seawater into the warmer waters of Puget Sound.[8] Underwater west of the Olympic Peninsula and Vancouver Island, the continental shelf kept the seafloor shallow farther offshore and worked with seasonal currents to prevent winter waters from becoming too frigid, thereby improving conditions for marine organisms during the cold months. Numerous reefs, rocky outcrops, and other localized submarine features attracted many types of fish.

Possessing a detailed indigenous knowledge of the local seascape, Makahs understood that distinct water masses in the region circulated in a complex yet regular manner as currents interacted with seasonally changing winds and geological features.[9] Within a large geographic context, permanent ocean currents moved water masses into and out of this region. Flowing east across the Pacific Ocean, the North Pacific Current split when it drove into the coast of North America. One branch, the Alaska Current System, headed north along the west coast of Vancouver Island. Pushing warmer water northward, it moderated conditions in the southeast Bering Sea. A second branch, the California Current System, headed south along the Washington coast to the Baja Peninsula. These currents deposited flotsam on beaches in front of Makah villages, bringing giant redwood logs from northern California in the winter and carrying bamboo and other tropical flora from Asia in the summer. More important, these water masses affected the marine biology of local waters, making species available at regular times. When the California Current System swung closer to the coast each spring, it brought fur seals, migrating to their breeding grounds on the Pribilof Islands in the Bering Sea, within safe reach of Makah men hunting from canoes. Observing these natural cycles helped coastal indigenous peoples understand that these larger currents affected local marine space and the resources within it.[10]

The effects of regional movements of water also benefited the People of the Cape. Rivers discharged nutrient-rich water into local marine space. Freshwater from the numerous smaller watersheds of Vancouver Island and the Olympic Peninsula carried minerals into this environment. Large outlets such as the Strait of Juan de Fuca and the mouth of the Columbia River funneled plumes seaward. Draining an approximate area of 258,200 square miles, the Columbia River discharged nutrients that moved north along the coast as the Davidson Current picked them up. In the Strait of Juan de Fuca, estuarine runoff from the Fraser River mixed with deep, nutrient-rich seawater before being discharged into the ocean. This mixture contributed substantially to the nutrient supply and the resultant production of phytoplankton—microscopic, single-cell plants around Cape Flattery. The wide continental shelf off the west coast of Vancouver Island created an extensive foundation for spreading this nutrient-rich water mass into ocean waters farther offshore. Experienced Makah mariners used the strong tidal current around Cape Flattery to get to and from fishing grounds off the cape and to travel among villages. Oral histories highlight Makah knowledge of local currents and tides and relate how indigenous mariners used them to their advantage.[11]

Because they could literally see it through the resultant high concentration of sea life, the People of the Cape also knew about one of the most important regional oceanographic processes, upwelling.[12] This raised cold, nutrient-rich water into the photic zone where sunlight fueled photosynthesis, which created a highly productive biomass. Promontories and capes such as Cape Flattery and the related Juan de Fuca Eddy created "upwelling centers" in the surrounding waters. The deep California Undercurrent and the outflow from the strait interacted with the submarine canyon system off the cape to facilitate upwelling from an extreme depth. The Juan de Fuca Eddy made this marine space into a dense biomass of phytoplankton (microscopic single-celled plants), which, in turn, enhanced the quantity of larger marine species. During its peak, the eddy propelled dissolved inorganic nutrients about sixty miles from the mouth of the Strait of Juan de Fuca and marked the water with a dark stain as it literally "boil[ed] with bait." It also facilitated the mixing of fresh- and salt water in the straits and flushed the inlets of the west coast of Vancouver Island.[13]

The physical features of this complex marine environment supported a rich food web, which radiated out from the phytoplankton. These microorganisms got energy from sunlight and dissolved inorganic nutrients. In the spring, the Juan de Fuca Eddy facilitated an explosive bloom of these phytoplankton, which fed the large amount of krill at the coastal-oceanic interface along the Washington coast. A type of zooplankton, krill was a keystone species, the primary food source for most other marine life. This included fish, seabirds, and marine mammals, especially whales, all significant to Makahs and Nuu-chah-nulth peoples of Vancouver Island. Found throughout the Strait of Juan de Fuca and along the coast, kelp beds formed dense marine forests that supported fish, invertebrates, marine mammals, and seabirds.[14]

Keeping temperatures warmer in the winter and cooler in the summer, the ocean also shaped the weather patterns experienced by the Makahs. Driven by the Aleutian Low, an atmospheric pressure cell that dropped south from the Bering Sea during the winter, wet storms often swept in from the ocean, annually depositing from 90 to 110 inches of precipitation, most in the form of rain due to the temperate climate. Calmer and warmer, summer was rarely hot. Cloudy weather, foggy conditions, and variable winds bedeviled many a non-Native mariner and at times presented challenges for Makahs when on the water. The latitude, various ocean currents, and climate kept the waters off Cape Flattery cold, yet warm enough to support the rich marine life of the region.[15] Together, winds, geological features, the circulation of water masses, and marine biology composed the oceanography of marine space in which the People of the Cape and neighboring indigenous peoples lived.

Located on the shores of the Pacific Northwest, the People of the Cape are part of the Northwest Coast culture. This cultural area extends along the Pacific coast from the Copper River delta on the Gulf of Alaska to the Oregon-California border and inland to the Coast Mountains of British Columbia and the Cascade Mountains of Washington and Oregon. Speaking forty-five separate languages that scholars have organized into thirteen language families, the many distinct peoples of the Northwest Coast composed the second most diverse (after California) linguistic area of indigenous North America in the fifteenth century. Despite this diversity, these peoples shared several commonalities differentiating them from the interior peoples east of the Cascades, far northern peoples of the Pacific, and those in California.[16]

Northwest Coast peoples exploited a range of natural resources, but most relied on cedar and salmon. Natives used cedar, an evergreen tree found throughout the region and valued for its durability and decay resistance, for so many necessities that some referred to the spirit of the tree as "Long Life Maker" or "Rich Woman Maker." They spun, wove, and plaited its fibrous bark into clothing, baskets, and rope. Northwest Coast peoples split cedar into planks for longhouses, steamed boards into bentwood boxes, carved totem poles and masks, and fashioned canoes from logs. They fished for five species of salmon, an anadromous fish that hatches in freshwater streams, swims to the ocean to spend most of its life fattening up, and returns to natal streams to spawn and die. Using reef nets, seines, weirs, lines, and spears, families took from three hundred to a thousand pounds of salmon per person each year. Because the separate species return to spawn at different seasons and only for limited durations, Native women developed a range of methods for preserving and storing salmon. The differential availability of cedar and particular species of salmon—and other food and natural resources—fueled a vast trade network encompassing this culture area and beyond. Makahs had unique access to enormous amounts of halibut, whales, and seals, which they traded to other Northwest Coast peoples for salmon and cedar, two resources that they did not have in abundance in comparison to peoples living on major river systems.[17]

Indigenous communities formed around villages situated to exploit marine resources. Makahs resided at five primary villages—Biʔidʔa ("bih-ih-duh"), Di·ya ("dee-yuh," Neah Bay), Waʔač̓ ("wuh-uhch"), Č̓u·yas ("tsoo-yuhs"), and ʔuse·ʔiɫ ("oo-sa-ihlth," Ozette)—located near Cape Flattery. During the spring and summer, families dispersed to residences located to access seasonal resources more easily. During the late eighteenth century, many Makahs annually relocated to the summer village on Tatoosh Island, about half a mile northwest of Cape Flattery, to fish for halibut and hunt whales. Others lived in smaller communities such as Q̓idi·q̓abit ("kih-dee-kuh-biht")—called "Warm Houses" in English because of the many smoke-houses for drying fish—and fishing camps along the Hoko River, where they caught and preserved various fish. During the stormy winter months, they relocated back to sheltered, more permanent villages. Families moved from one area to another because they had ownership and usufruct rights (the right to use something owned by someone else) based on kinship to specific resources. These rights shifted as social connections changed due to marriages and divorces, births and deaths, and the waxing and waning power of particular individuals and communities. Because of these changes, and periodic fluctuations in the availability of particular species, families did not always follow the same particular pattern of seasonal movements.[18]

Before the appearance of non-Natives in the eighteenth century, Northwest Coast peoples recognized the status of individuals within a three-tiered system of social stratification, which included chiefs, commoners, and slaves. A range of ranked leaders occupied the highest social stratum in each village. Claiming ancestral ties to supernatural ancestors, chiefs owned the rights and titles to leadership positions within communities. Titleholders owned lucrative fishing and hunting grounds and other resource areas. "Outside resources"—called such because they were in marine spaces outside bays, inlets, and rivers—were the most important property rights, and only the highest-ranking chiefs owned them. Among the People of the Cape, the most powerful chiefs were whalers. In addition to tangible property such as cranberry bogs, freshwater streams, driftwood, game, timber, and wild plants, chiefs owned intangible items, like names, dances, songs, and stories. Sometimes sold but usually given as gifts, these propertied items passed within kinship networks and crossed gender lines, which meant that women occasionally held positions of authority.

Except for personal items such as canoes, clothing, and fishing gear, commoners did not own any culturally significant property like resource areas and titles. However, they did have access to hunting and fishing grounds because chiefs extended these rights to people who respected their authority. In return, commoners often gave a portion of what they hunted, harvested, and caught to the high-status owner. Slaves, seized during conflicts between communities and traded throughout the Northwest Coast, occupied the lowest social stratum and made up 20 to 40 percent of a village's population. Slave status was hereditary—the children of slaves inherited this condition. Because they did a range of labor and augmented an owner's wealth, slaves were both valuable property and status items. High-status Makahs kept many slaves, and the People of the Cape were key traders in the regional slave trade.[19]

Ownership of important titles and rights provided Northwest Coast chiefs with their foundation for leadership within villages and sometimes across larger sociopolitical spaces. Chiefs exercised authority through influence rather than coercion. Although titles, rights, and privileges passed from one generation to the next through kinship networks, an individual had to maintain his—and sometimes her—noble status by providing for the people. Within a Makah village, whaling chiefs who harvested several thirty- to forty-ton whales each year provided a wealth of subsistence and trade goods. This system of leadership made authority competitive both within villages and across larger areas. The wealthiest chiefs—those who could provide the most for their people—had more influence than others with less. These leaders could muster hundreds of warriors from several villages. They protected their people from raiders or led war expeditions against neighboring villages to seize slaves and lucrative resources such as salmon streams or marine fishing and hunting grounds. Chiefs strengthened their social status by marrying members of other noble families throughout the region; for similar reasons, they married their sons and daughters into other high-status families. Marriages often brought more property items, including fishing rights, slaves, titles, and other tangible goods, although these privileges ended on divorce.[20]

Chiefs displayed and distributed wealth through rituals and feasts, which scholars often lump together under the name "potlatch." These events ranged from simple affairs, during which a chief shared a substantial catch of salmon, to more elaborate celebrations such as a marriage between two noble families. The largest events involved hundreds of guests, whom the host would feast for several consecutive weeks. Potlatches incorporated speeches, dances, songs, and other ritual activities. These feasts featured the giving of gifts ranging from food to valuable commodity trade items. Within the framework of leadership rivalries in the Northwest Coast, potlatching became competitive when chiefs expanded their influence over people by demonstrating the ability to provide increasing amounts of wealth for supporters. The People of the Cape were renowned throughout the Northwest Coast for their feasts. S'Klallams, a separate people living to the east of Cape Flattery, called them *Makahs*, a name that means "generous with food," probably because of the abundance they harvested from rich marine resources, which allowed them to host lavish events.[21]

Northwest Coast characteristics of marine resources, hierarchy, competitive power, and potlatches worked together to allow influential Makah chiefs to produce tribal space. As human geographers have demonstrated, spaces are products of societies. Makah society produced marine and terrestrial spaces—made them Makah spaces—through the ways they thought about, organized, and lived in these spaces. Many factors shaped the historical process of producing space in the Northwest Coast, including the environment, competing Native and non-Native societies and polities, the actions of individuals and groups, differing attitudes on how to best make a living, and connections to other spaces. Makah marine space became a social reality through the actions of distinct borderlands peoples and through the networks of kinship, trade, and violence drawing them together.[22] The availability of whales, seals, and halibut made the People of the Cape different from others, and Makahs funneled these unique resources into Native and non-Native trade networks. Influential chiefs exploited these resources to expand and maintain their power, which allowed them to interrupt imperial processes such as trade and colonization. Much like an interruption in a conversation pauses the flow of words, registers a counterpoint, or inserts a missed perspective, these interruptions of imperial processes questioned, suspended, or redirected non-Native plans for a time, served as bold reminders that newcomers were in Native spaces layered with specific protocols, and on occasion shaped the unfolding trajectory of empire by interjecting indigenous priorities in ways that could not be ignored. Makah leaders recognized the foundation of their power and identity, an understanding captured in the words of C̓aqa·wiƛ ("tsuh-kah-wihtl"), a Makah chief, during the negotiations for the 1855 Treaty of Neah Bay. He told the Euro-American treaty negotiators, "I want the sea. That is my country."[23]

Notes

1. Bowechop, "Contemporary Makah Whaling," 415–19; Sullivan, *Whale Hunt*, 238–65; Sepez, "Political and Social Ecology," 112–97; Miller, "Exercising Cultural Self-Determination"; Miller, "Tribal Cultural Self-Determination"; interviews by Bowechop and Pascua, NOAA Interviews, MCRC; Makah Tribal Council and Makah Whaling Commission, *Makah Nation*; Arnold, interview.
2. Claplanhoo, interview; Greene, interview; US Department of the Interior, Fish and Wildlife Service, and US Department of Commerce, NOAA, "Endangered and Threatened Wildlife and Plants." For a discussion of the IWC negotiations in 1996 and 1997, see Martello, "Negotiating Global Nature and Local Culture," 267–69.
3. Johnson, interview; Mapes, "Celebrating the Whale"; Bowechop, "Contemporary Makah Whaling," 418; Blow, "Great American Whale Hunt"; Gutthi-udaschmitt, "Makah Whale Hunt"; Adrienne Bowechop, Janine Bowechop, Micah McCarty, and Crystal Thompson, NOAA Interviews.
4. Blow, "Great American Whale Hunt"; Martello, "Negotiating Global Nature and Local Culture," 267.
5. Ellingson, *Myth of the Noble Savage*, 370; Dan Greene, NOAA Interviews.
6. Wylly quotation from Tizon, "E-Mails, Phone Messages." See also Barton, "'Red Waters,'" 200–18; Marker, "After the Makah Whale Hunt." For more examples, see Mapes, "Celebrating the Whale"; Bowechop, "Contemporary Makah Whaling," 418–19; and Sullivan, *Whale Hunt*.
7. LaDow, *Medicine Line*, 109; Evans, *Borderlands*, 354; Binnema, "Case for Cross-National and Comparative History," 18–19. For a more terrestrial-oriented perspective on the environment of the Northwest Coast, see Suttles, "Environment."
8. Edwards and MacCready, "Strait of Juan de Fuca," 2.
9. For an introduction to indigenous knowledge, see Ellen, Parkes, and Bicker, *Indigenous Environmental Knowledge*; and Gordon and Krech, "Introduction."
10. Favorite and Northwest Fisheries Center, *Ocean Environment*, 28; Roden, "Subarctic-Subtropical Transition Zone."
11. Johnson, interview; Greene, interview; McCarty, interview.
12. Makah fisher Dave Sones speaks of being able to see the change in the ocean when one arrives at the upwelling. See Sones, interview. Today's Makah fishers and whalers simply refer to it as the "Big Eddy."
13. Beak Consultants and Patricia Bay Institute of Ocean Sciences, *Examination*; Purdy, *Summary*, 14; MacFadyen, Hickey, and Cochlan, "Influences of the Juan de Fuca Eddy." Through informal conversations, Makah fishers provided details on how the eddy appears.
14. Favorite and Northwest Fisheries Center, *Ocean Environment*, 48; Berry, Sewell, and Wagenen, "Temporal Trends."

15. Suttles, "Environment," 17–18; Renker and Gunther, "Makah," 422.
16. Suttles, "Introduction"; Thompson and Kinkade, "Languages." For other culture areas, see Walker, *Plateau*; Helm, *Subarctic*; and Heizer, *California*. Throughout the book, I use modern place-names so that we can more easily understand the setting. Where appropriate, I introduce us to indigenous place-names and reference earlier terms used by non-Natives.
17. Lane, "Political and Economic Aspects," 1–13; Stewart, *Indian Fishing*; Stewart, *Cedar*; Taylor, *Making Salmon*, 13–38; Montgomery, *King of Fish*, 39–58; Hewes, "Indian Fisheries Productivity," 136.
18. The complex networks of kinship are discussed in detail in chapters 1 and 2. For an excellent analysis of this topic, see Harmon, *Indians in the Making*. For more on seasonal movements, see Kirk, *Tradition and Change*, 105–38; and Ames and Maschner, *Peoples of the Northwest Coast*, 120–21.
19. Kirk, *Tradition and Change*, 36–56; Donald, *Aboriginal Slavery*; Drucker, *The Northern and Central Nootkan Tribes*, 243–73; Renker and Gunther, "Makah"; Coté, *Spirits of Our Whaling Ancestors*, 22–23.
20. The literature cited earlier discusses the power of chiefs. For a different model on authority within the Northwest Coast, see Miller and Boxberger, "Creating Chiefdoms."
21. For the meaning of the name *Makah*, see Coté, *Spirits of Our Whaling Ancestors*, 18. As with other Northwest Coast characteristics, the literature on potlatches is extensive. For concise overviews, see Drucker, *Northern and Central Nootkan Tribes*, 370–86; Drucker, *Indians of the Northwest Coast*, 123–33; Arima, *West Coast People*, 68–82; Kirk, *Tradition and Change*, 57–69; and Lutz, *Makúk*, 58–61. More than fifty years old, the classic text still remains Codere, "Fighting with Property." For a newer interpretation, see Suttles, "Streams of Property, Armor of Wealth." For a critique of Codere's argument that potlatching replaced warfare, see Lovisek, "Aboriginal Warfare on the Northwest Coast."
22. Lefebvre, *Production of Space*; Tuan, *Space and Place*.
23. "Ratified Treaty No. 286: Documents Relating to the Negotiation of the Treaty of January 31, 1855, with the Makah Indians," p. 2, Documents Relating to the Negotiation of Ratified and Unratified Treaties, NARA-PNR.

APPENDIX B.3

Praise Song for Oceania

CRAIG SANTOS PEREZ

praise your capacity for birth
fluid currents and trenchant darkness
praise your briny beginning
source of every breath

~

praise your capacity for renewal
ascent into clouds and descent into rain
praise your underground aquifers
rivers and lakes, ice sheets and glaciers
praise your watersheds and hydrologic cycles

~

praise your capacity to endure
the violation of those who map you aqua nullius
who claim dominion over you
who pillage and divide your body
into latitudes and longitudes
who scar your middle passages

~

praise your capacity to survive
our trawling boats breaching
your open wounds and taking
from your collapsing
depths

~

praise your capacity to dilute
our heavy metals and greenhouse gases
sewage and radioactive waste
pollutants and plastics

~

praise your capacity to bury
our shipwrecks and ruined cities
praise your watery grave
human reef of bones

~

praise your capacity to remember
your library of drowned stories
museum of lost treasures
your vast archive of desire

~

praise your tidalectics
your migrant routes
and submarine roots

~

praise your capacity to smother
whales and fish and wash them ashore
to save them from our cruelty
to show us what we're no longer allowed to take

to starve us like your corals starved and bleached
liquid lungs choked of oxygen

~

praise your capacity to forgive

please forgive our territorial hands and acidic breath
please forgive our nuclear arms and naval bodies
please forgive our concrete dams and cabling veins

please forgive our deafening sonar and lustful tourisms
please forgive our invasive drilling and deep sea mining
please forgive our extractions and trespasses

~

praise your capacity for mercy

please let my grandpa catch just one more fish

please make it stop raining soon
please make it rain soon

please spare our fragile farms and fruit trees
please spare our low-lying islands and atolls
please spare our coastal villages and cities
please let us cross safely to a land without war

~

praise your capacity for healing
praise your cleansing rituals
praise your holy baptisms

~

please protect our daughter
when she swims in your currents

~

praise your halcyon nests
praise your pacific stillness
praise your breathless calm

~

praise your capacity for hope

praise your rainbow warrior and peace boat
praise your hokule'a and sea shepherd
praise your arctic sunrise and freedom flotillas

praise your nuclear free and independent pacific movement
praise your marine stewardship councils and sustainable fisheries

praise your radical seafarers and native navigators

praise your sacred water walkers
praise your activist kayaks and traditional canoes
praise your ocean conservancies and surfrider foundations

praise your aquanauts and hydrolabs
praise your Ocean Cleanup and Google Oceans
praise your whale hunting and shark finning bans

praise your sanctuaries and no take zones
praise your pharmacopeia of new antibiotics
praise your #oceanoptimism and Ocean Elders
praise your wave and tidal energy
praise your blue humanities

~

praise your capacity for echolocation
praise our names for you that translate
into creation stories and song maps
tasi : kai : tai : moana nui : vasa :
tahi : lik : wai tui : wonsolwara

~

praise your capacity for communion

praise our common heritage
praise our pathway and promise to each other
praise our most powerful metaphor
praise your vision of belonging

praise our endless saga
praise your blue planet
one world ocean

praise our trans-oceanic
past present future flowing
through our blood

APPENDIX B.4

Rehabilitation*

A Proposal for a Climate Compensation Mechanism for Small Island States

MAXINE BURKETT

Introduction

For more than two decades, vulnerable small island states have sought a means to preserve their lives and livelihoods under threat of the impacts of climate change. Formally organized as the Alliance of Small Island States (AOSIS), small islands have been the source of many novel approaches to climate governance within the United Nations Framework Convention on Climate Change (hereinafter "Framework Convention" or UNFCCC).[1] Perhaps because of their unique vulnerability to its impacts,[2] AOSIS has led a steady drumbeat for urgent and ambitious methods for arresting, and if not, adapting to climate change to survive some of the worst forecast climate phenomena. With the specter of the most threatening and unavoidable impacts becoming more certain in light of current emissions rates, the calls for mechanisms to address climate-related loss and damage become more compelling. This article seeks to respond to those calls by developing a comprehensive mechanism to compensate small island states that will suffer

Excerpt from *Santa Clara Journal of International Law* 13 (2015): 81–124.

*Publisher's Note: Notes are reprinted as they appear in the original text. Since this reprint is an excerpt, note numbers are not sequential. For the complete article, see http://digitalcommons.law.scu.edu/scujil/vol13/iss1/5.

from devastating slow-onset events, such as encroaching seas, unrelenting heat, and acidifying oceans.

From the earliest days of international negotiation—indeed, prior to the drafting of the Framework Convention—the Small Island Developing States (SIDS) anticipated the need to address the full panoply of climate challenges. In addition to mitigation and adaptation, a method for rehabilitating communities in response to, for example, the atmospheric hazards of increasing heat waves is necessary and especially important for poor and vulnerable countries.[3] It would require, as introduced and explicated by AOSIS over the years, two main approaches. First, the international community would need to reduce the risks of future loss and damage through risk management—that is, by avoiding or minimizing climate impacts—in addition to climate change adaptation and mitigation.[4] Second, states would need to address loss and damage at the time of its occurrence, now and into the future. AOSIS's proposal, referenced in the Framework Convention and widely supported by developing countries with growing developed countries' support,[5] conceives of and constructs an international mechanism, building on initial calls for an insurance mechanism for vulnerable island states, with which they can access funds immediately after a disaster. That initial insurance proposal has evolved into a multi-pronged mechanism that also includes provisions for disaster risk management and—for climate impacts that are unavoidable and irreversible[6]—compensation and rehabilitation. Now more than ever, AOSIS seeks integration of this proposal into the ongoing UNFCCC negotiations.[7]

To date, the compensation and rehabilitation portion of the proposal has not received sufficient explication.[8] Indeed, to the extent member states and researchers have considered it, the entire discussion of compensation has been hobbled by concerns regarding lack of political will and donor fatigue, at best, and explicit rejection of any measure that might vaguely resemble climate-related reparations, at worst.[9] Indeed, some developed states have clearly expressed their distaste for any proposal that "may have the potential to create open-ended financial and legal liabilities."[10] Nevertheless, anticipating the increasing need to seek aid or recompense for climate impacts, scholars and researchers have considered the possibility of a compensation mechanism and its legal and theoretical underpinnings.[11] Even without a viable legal hook,[12] however, there are numerous reasons for developed countries to agree to establish a compensation mechanism.[13] With a sound rationale for a non-retributive compensation mechanism . . . there is also significant precedent in international practice, namely in the United Nations

Compensation Commission (UNCC), that provides a useful blueprint for constructing a funding mechanism with the scale and sophistication to meet the SIDS' needs.

To advance the proposal for compensation and rehabilitation under the UNFCCC, this article explores in detail the feasibility and structure of an ambitious mechanism to compensate small island states at the advent of slow-onset events. Nested within the Framework Convention governance infrastructure and administered through the UNFCCC Secretariat, the Small Islands Compensation and Rehabilitation Commission (hereinafter CRC) would disburse monies from a global pool to aid in rehabilitating individuals, communities, and countries affected by sea-level rise, ocean acidification, and other devastating slow-onset events. . . . [T]he CRC would operate consistent with clearly laid out claims categories and coordinate payouts triggered by the crossing of agreed on thresholds. Although AOSIS has not limited its proposal to benefit small islands alone, this article structures the CRC around the needs and vulnerabilities—as well as the possibilities for rehabilitation—of small islands. It demonstrates the feasibility and grave need to introduce and implement it. Further, it argues that the CRC might break a logjam between the highest emitters and the most vulnerable states. Instituting a formal compensation mechanism would allow member countries to condition establishment of the CRC on the waiver of legal liability pursuant to terms agreed on when crafting and implementing the commission. This may well quell the liability fears of certain developed countries and allow reasoned and collaborative discussion about funding for slow-onset events to proceed.

* * *

I. Climate Change and AOSIS

A. Climate Present and Future for AOSIS Member Countries

1. AOSIS Country Vulnerability

A brief introduction to the AOSIS member states quickly reveals their heightened interest in the health and rigor of the Framework Convention. The Alliance is a "coalition of small islands and low-lying coastal countries that share similar development challenges and concerns about the environment, especially their vulnerability to the adverse effects of global climate change."[17] Within the UN system, AOSIS functions as an ad hoc lobbying and negotiating voice for SIDS,[18] advocating on behalf of roughly 59 million citizens of 44 States and observer countries from across the globe, including the Pacific, Indian, and Atlantic oceans as well as the Mediterranean, Caribbean, and South China Seas.[19] These low-lying coastal states share similar challenges to sustainable economic development, including: geographic isolation, limited resources, dependence on international trade, and pre-existing vulnerability to natural disasters.[20] Further, although their GDPs vary wildly in some cases,[21] most of these countries are middle- or low-income countries, and five also rank among the least developed in the world.[22]

AOSIS countries also have notable vulnerability to climate extremes.[23] The Pacific islands are situated in one of the most natural disaster prone regions and are highly susceptible to—among other disasters—floods, droughts, and tropical cyclones.[24] All SIDS are especially vulnerable to sea-level rise. In the Caribbean, for example, about 70 percent of the population lives on the coast, and regional experts expect that many will have to relocate away from the coasts.[25] In addition, predicted sea-level rise of roughly 3.3 feet by 2100 would "wreak havoc on the region's tourist areas,"[26] flood airports, destroy resorts, and—as is already happening—deepen the damage of saltwater intrusion on vital crops.[27] Atoll nations, such as the Marshall Islands, are already experiencing high tides, or "king tides," that surge over sea walls, repeatedly flooding its capital.[28] In the summer of 2013, in an unfortunate coincidence, the tides exacerbated the crisis situation in the northern atolls resulting from devastating drought, which damaged or destroyed local food crops, depleted water tanks, and rendered groundwater unsuitable for human consumption because of high salinity.[29]

SIDS' climate vulnerability is not only the result of the unique exposure to climate extremes but also of the severe impacts that these natural hazards mete on local and national economies. Like many countries, a great proportion of SIDS' economic activity occurs at the coastline, which is particularly true for tourist-dependent island nations. For example, 90 percent of Jamaica's gross domestic product is generated within the coastal zone.[30] Accordingly, any risks to coastal activity will have significant consequence. Not surprisingly, extensive data and a general consensus demonstrate that "developing countries are more economically vulnerable to climate extremes."[31] There are several reasons for this increased vulnerability, including: "less resilient economies" that depend more on natural capital and "climate sensitive activities" such as cropping and fishing; poor preparation for physical hazards; and an "adaptation deficit resulting from the low level of economic development and a lack of ability to transfer costs through insurance and fiscal mechanisms."[32] Accordingly, industrialized countries possess the highest income and account for most of the total economic and insured disaster losses, while fatality rates as well as losses as a proportion of GDP are greater in the developing world.[33] For SIDS, a single disaster has immense ripple effects, severely stressing public financial resources,[34] if not dwarfing annual GDP.[35] As evidenced by Fiji in 2010,[36] consecutive natural disasters in a short period of time multiplies the magnitude of losses and recovery demands and decreases overall socio-economic development.[37]

2. Climate Forecasts, Novel Slow-onset Events, and Loss and Damage

The climate forecast for SIDS is equally troubling. Changes in the "frequency, intensity, spatial extent, duration, and timing of climate extremes" can result in "unprecedented" events.[38] Further, "the crossing of poorly understood climate thresholds cannot be excluded, given the transient and complex nature of the climate system."[39] In other words, scientists anticipate that the known climate-related impacts are set to shift in extreme and novel ways. And, the biggest surprises, though not fully understood, may well come to pass. The IPCC's Special Report on Managing the Risks of Extreme Events and Disasters to Advance Climate Change Adaptation (SREX) links anthropogenic activities to known climate extremes, including sea-level rise. Further, it has identified effects like sea-level rise as inevitable.[40] Whereas some climate impacts are avoidable, either through mitigation or adaptation, many are no longer.[41] These unavoidable impacts are ones that cause significant damage regardless of future measures taken to adapt.[42] In addition to land lost to sea-level rise,[43] for example, agricultural land lost to persistent drought,[44] human health impacts of increasing heat events,[45] and the entire collapse of the fishing industry due to increased ocean heat and ocean acidification[46] all constitute unavoidable damage with which SIDS will have to contend. Not only are their territories and economic engines predicted to face significant challenges, SIDS are home to some of the most vulnerable populations—indigenous peoples,[47] internally displaced peoples, and climate-induced migrants.[48] In sum, there are unprecedented climate impacts that efforts to mitigate have failed to address and measures to adapt will fail to prevent or alleviate. And, there are countries and regions in the crosshairs of these impacts.

3. AOSIS Impacts

AOSIS's early calls for assistance identified the need to consider an expanded spectrum of necessary responses to climate change, anticipating heightened vulnerabilities over time. Beyond efforts to mitigate and adapt, efforts to insure against disaster risks and compensate for unavoidable impacts are also necessary. The latter two approaches seek to respond to the current and future loss and damage resulting from climate change–related events, for which mitigation and adaptation by definition are unable to address. Together, loss and damage describe "the actual and/or potential manifestation of impacts associated with climate change in developing countries that negatively affect human and natural systems."[49] Independently, loss refers to impacts for which restoration is not possible. For example, the total destruction of coastal infrastructure due to sea-level rise or the total collapse of a fishery due to lower ocean pH would constitute a loss. Damage refers to negative impacts for which restoration is possible. Damage to a coastal mangrove forest due to storm surge would fall under this category,[50] and presumably appropriate adaptation efforts or disaster risk management could mitigate or avoid impacts suffered as a result. Both loss and damage interact with human systems,[51] exacerbating their pre-existing socio-economic vulnerability. Both can halt or reverse development and "reinforce cycles of poverty,"[52] with particularly dire consequences for the least developed.

Whereas small islands are experiencing early and devastating climate impacts today, the prospect of future increasing loss and damage is especially concerning,[53] particularly as the world pushes beyond the worst assumptions regarding emissions and exposure variables. Nevertheless, these impacts fall through the gaps in the climate governance regime. The loss and damage discussion seeks to resurrect these critical concerns.

* * *

In small island states, the civil engineering plans typical of adaptation projects funded by the undercapitalized Adaptation and Green Climate Funds will not suffice.[55] At some point, the seawalls of the Maldives and Tuvalu will fail so consistently and completely that communities and countries will need compensation for rehabilitation from losses incurred. Further, for SIDS, the cascading effect of disrupted customary institutions and subsistence lifestyles, which have aided resilience to climate variability in the past, could mean the loss of whole cultures.[56]

For the present and forecasted climate impacts, the response regime is deeply flawed and will soon become wholly inadequate. Presently, SIDS rely on ad hoc requests for disaster aid when a devastating event occurs.[57] These ad hoc measures are often slow to arrive. They also increase the likelihood of funders and communities introducing maladaptive measures in the wake of disaster recovery. As explained by the IPCC, "[a]n emphasis on rapidly rebuilding houses, reconstructing infrastructure, and rehabilitating livelihoods often leads to recovering in ways that recreate or even increase existing vulnerabilities, and that preclude longer-term planning and policy changes for enhancing resilience and sustainable development."[58] For all of these reasons, AOSIS has concluded that the absence of a comprehensive loss and damage mechanism for the most vulnerable is a "gaping hole" in the Framework Convention process.[59]

* * *

II. Completing the Multi-Window Framework: Taking Compensation and Rehabilitation Seriously

The compensation component of the AOSIS proposal may be indispensable. Progressive, negative climate impacts can result in, among other things, the permanent or extended loss of useful land, irreparable damage to coral reefs, damage to water tables, loss of fisheries, and loss of territory with attendant human displacement.[81] These uninsurable, residual risks will yield devastating effects on SIDS and other highly vulnerable nations.[82] The effects might also reverberate well beyond the most vulnerable.

* * *

III. The Small Islands Compensation and Rehabilitation Commission

To blunt the disruption and devastation of unavoidable climate impacts, the United Nations should establish the Small Islands Compensation and Rehabilitation Commission (CRC). The CRC would process claims and pay compensation for loss, damage, and injuries to or displacement of individuals and governments that suffer as a direct result of catastrophic slow-onset events.

* * *

The CRC would perform three primary functions. It would (i) determine the total amount of damages to award for a given event, such as the collapse of a reef system due to low ocean pH, (ii) allocate payment to relevant claimants, and (iii) develop and manage an innovative system of post-award auditing of larger claims to ensure adequate environmental remediation and inform more sophisticated approaches for later rehabilitation.[115]

A. The Theoretical Basis for the CRC: Guiding Principles and Antecedent Questions

The reasons for crafting and implementing the CRC are legion. Notwithstanding copious arguments for reparations based on liability under international law,[116] there are weighty reasons to establish a compensation commission based on the Framework Convention. There are also sound reasons based on the maintenance of global stability, which is also of great interest to the largest emitters.[117] More than 190 countries established the UNFCCC to meet the many significant demands of slow-onset phenomena. The Framework Convention, at least in theory, sanctions a number of ambitious and viable actions to slow climate change and ease the impacts of unavoidable climate change. It would be naive to suggest, however, that the Framework Convention can do all it states to do, at least without significant political opposition. That said, its failure, if even partial, portends grave consequences for the global community, including the most intransigent of the largest carbon emitters.

* * *

Though the well-being of small islands has been a stated concern since the UNFCCC's inception, attention to loss and damage at the UNFCCC began in earnest in 2007 with the Bali Action Plan.[118] The Plan called for enhanced adaptation efforts, including strategies and means to address loss and damage in developing countries, particularly for those most vulnerable.[119] In 2010, the Cancun Adaptation Framework noted that approaches to address loss and damage should consider impacts, including sea-level rise, increasing temperatures, and ocean acidification.[120] It further recognized the "need to strengthen international cooperation and expertise in order to understand and reduce loss and damage associated with the adverse effects of climate change, including impacts related to extreme weather events and slow-onset events."[121] Most recently, Decision 3/CP.18 of the COP 18 meetings in Doha recognizes the importance of the work on loss and damage, including the need to build "comprehensive climate risk management approaches."[122] It also calls for advanced understanding of noneconomic loss and damage, patterns of migration and displacement, and identification and development of approaches to rehabilitate from climate-related loss and damage.[123] Further, the Doha Gateway mandated the formation of "institutional arrangements, such as an international mechanism, including its functions and modalities" for the next COP.[124] Given the staunch opposition the loss and damage proposal has received in the past, it took many by surprise when it assumed a central role in the Doha climate negotiations.[125] Although the proposal has suffered setbacks in the interim, it remains firmly on the agenda.

* * *

Given the geopolitical risks, a claims resolution facility that serves disaster relief and social welfare goals will aid the most vulnerable as well as the largest emitters.[129] The CRC would provide a social safety net that attempts to meet the fundamental needs of each claimant, thus stunting overall regional and global instability.

1. Guiding Principles

There are several guiding principles that might govern the CRC from conception to implementation. . . . [A]n expansive and flexible view of compensable harm vis-a-vis environment-based damages is necessary. Further, due to the progressive nature of climate change and our scientific understanding of its complexity, the ultimate impacts and the ability to link them to climate inputs will necessarily shift. For these reasons, it will be especially important that the CRC has clear guiding principles that will allow for consistency across claims over time.

* * *

International solidarity and equity is a theoretical hallmark of reparative efforts.[130] In this context, international solidarity captures the overarching notion that the CRC is the product of a collaborative effort of each nation to facilitate the best outcomes for all other nations and their citizens. Demonstrations of solidarity will differ across nations. For the developed world, emerging economies, and the largest present and historical emitters, the principle of common but differentiated responsibilities (CBDR)[131] will still operate.

* * *

Predictability and transparency in the determination and distribution of damages will ensure horizontal equity and maintain the legitimacy of the CRC.[135] Similarly, with the advent of triggering events likely occurring over a substantial period of time—perhaps even a century or more—this facility will need to operate for quite some time while ensuring equity of treatment over generations, thus ensuring vertical equity. As much as possible the standards of compensation should not change sharply or without clear scientific reasoning.[136] Of course, claims will shift over time. As with any claims resolution facility, flexibility to adapt to changed circumstances is essential.[137] To ensure legitimacy in implementation, internal and external quality control audits governed by the third goal of the CRC[138] will also be critical.

Although the CRC can strive toward an optimal and consistent distribution of funds for rehabilitation, the size of potential claimants and the nature of climate change—marked by unpredictable, stochastic, and accelerating shifts—means that the CRC will fail to adequately compensate all claimants. In fact, short of aggressive mitigation commenced several decades ago, there exist unavoidable and irreversible impacts that make restoration of the *status quo ante* in some cases impossible. Further, assuming full and enthusiastic participation, it is unlikely that countries could sufficiently capitalize the CRC. For these and related reasons, notions of rough justice are relevant. Although the CRC cannot preclude all negative outcomes for small island states, it can, given the circumstances, facilitate better outcomes.

* * *

To come close to an adequate and appropriate method for claims resolution, funding for scientific and technological rigor over time will also be essential. This would need to be a core part of a highly disciplined while adaptive organizational structure.[142] In other words, to make any legitimate attempt at identifying triggering events, linking them to the harm experienced by a claimant, and assessing adequate remedy, the CRC administration will require sufficient funding for its technical operations in addition to the ability to compensate satisfactorily.

* * *

2. Addressing Antecedent Questions

As contemplated in this article, the Small Islands Compensation and Rehabilitation Commission would follow the precedent of other claims resolutions facilities that have operated under the inquisitorial model of decision-making.[149] Distinct from adversarial procedures that are the hallmark of American litigation and other common law courts, the CRC would emphasize "truth-seeking and participation."[150] To that end, language and culturally appropriate notice to claimants, low access costs through standardized forms[151] or regional offices for claims administration, and direct access to the CRC rather than through counsel will be key features.

Coherent parameters to determine the "who, what, and how much" of the commission will aid the inquisitorial process. Below, I provide preliminary answers to the following antecedent questions: What events will give rise to claims? Who is and is not eligible for compensation? Who will fund the CRC at the outset and throughout the life of the commission? And, because most claims will not ripen until the advent of some climate-related event, when will payouts occur?

Eligible Claims. Generally speaking, loss and damage resulting from slow-onset events for which insurance is inadequate will be eligible for compensation.

* * *

To the extent that the CRC is limited to SIDS, impacts within terrestrial and marine geographic regions should be relevant to the claimant country or its citizens. . . . This would aid SIDS as the passing of certain thresholds for sea-level rise or ocean pH in certain regions to facilitate rapid claims resolution and funds distribution. It would be particularly helpful for the efficient resolution of individual claims. If a coastal region is no longer habitable and is within a designated compensable area, all of the inhabitants can file for compensation, with the attendant evidentiary burdens eased. This might also help ensure that future compensable damage is at least geographically bounded, thus avoiding claims for loss and damage due to remote environmental disruptions with more tenuous causal links to the climate-related slow-onset event.[161]

It is also plausible that the CRC could consider and compensate for intangible, noneconomic losses. . . . As a symbolic gesture, the CRC can determine a means to compensate and rehabilitate for losses related to loss of culture and territory. Whereas the latter might also include actual monetary compensation for the costs of displacement and lost access to natural resources, the Commission could establish a protocol for addressing the perhaps weightier loss of culture some islanders will likely experience. . . . This would be especially relevant for Pacific island countries, in which the majority is indigenous with cultures that have retained "profound cultural and spiritual connection" to nature and place.[164]

In sum, the CRC's jurisdiction would extend only to loss and damage suffered that is more likely than not due to slow-onset events.[165] The CRC's technical arm would define these events, consistent with changing scientific knowledge over time. To facilitate this task, the CRC should also coordinate initial funds to support the early establishment of baselines to aid in determining the extent of damage incurred relevant to a fixed point in time.[166]

Eligible Claimants. There are potentially billions of claimants if the eligibility question is not answered clearly and coherently. . . . The SIDS alone total almost 60 million people. All may not have claims; however, one can imagine expanding eligible claimants in time if the mission of the CRC also grows to cover other vulnerable communities. This would be consonant with the goals and spirit of the commission contemplated here. What is clear, however, is that the CRC would exclude countries like China that also have massive coastlines that sea-level rise threatens. They, and others similarly situated, would not be eligible by virtue of their current emissions rate coupled with an enormous economy.

Individual citizens of small island nations would be eligible to bring claims identified above. To avoid the piecemeal presentation of individual claims, and the attendant administrative burdens and possibilities for delay, island governments will file consolidated claims on their citizens' behalf.[167]

* * *

In sum, citizens and governments of AOSIS countries are eligible to bring claims. Governments will represent the claims of their nationals as well as claims regarding environmental damage brought on the government's own behalf.

The Global Pool. . . . The sums required to fund operation of the CRC, as well as adequate compensation over time, will be substantial. One cannot overstate this. . . . Nonetheless, decisions made regarding relative contribution to the fund can set the stage for a viable and solvent fund.

* * *

There are, of course, myriad ways in which the CRC could secure funding, at least initially.[186] . . . The forms of compensation may also vary and have been diverse in other claims resolution facilities.[190] Monetary compensation is the norm, but in-kind payments are also plausible. This might include property repair, outreach assistance, and perhaps unique to the SIDS circumstances, host country support of climate-induced migrants.

In sum, all UNFCCC parties will pay into a global pool from which countries will receive a payout or in-kind compensation based on their losses.

Triggering Events. Compensation is appropriate at two primary points—now and at the time of a triggering event.

* * *

Perhaps the more challenging technical questions for the CRC will be to determine when a slow-onset event has caused damage sufficient to trigger the payment of claims. For this determination, . . . an individual or country's claims would ripen once the impact of an event exceeds an agreed on level beyond the recorded baseline for that country.[193]

* * *

The valuation of damages will be exceedingly difficult. Creating a standing work group committed to providing appropriate definitions and methodologies for valuing harm from economic and noneconomic loss and damage over time will be an indispensable, early task for the CRC. This group may be housed under the technical arm of the CRC.

In sum, the need for financial assistance to establish and record baselines is immediate. To the extent it is not completed sufficiently under a disaster risk component, claims based on setting baselines and monitoring and prevention will ripen at the founding of the CRC. Parametric indicators will govern the ripening of claims due to slow-onset events.

* * *

Conclusion

Critics from varied perspectives might declare the Small Island Compensation and Rehabilitation Commission unviable. The hurdles are enormous. Political will may be absent for the foreseeable future. General skepticism about the efficacy of UNFCCC institutions abounds, with funding and management challenges already dogging existing efforts. The cost and scale of the CRC, based on the framework sketched here, may seem prohibitive. Notwithstanding these potential deficiencies, however, AOSIS has good reason to press its proposal. For SIDS, this approach in concept and execution is necessarily suboptimal—all, I assume, would prefer that climate change did not so fundamentally compromise their futures. The CRC, however, may be a welcome alternative to the more formidable obstacles present in pursuing international litigation for climate-related loss and damage or tackling unabated climate change without some assistance. Further, the benefits to the global community are clear and copious when compared to the costs of delayed attempts to anticipate and manage slow-onset climatic processes.

Of course, comparisons between the costs and benefits of action, on the one hand, and far costlier inaction, on the other, have not persuaded the largest emitters to act with prevention and precaution in mind. There are a few significant exceptions noted above. These countries provide hope that the AOSIS proposal and elaboration of component parts will have traction in the near term. If not, it seems clear that the proposal will serve as evidence of yet another missed opportunity to avoid worst-case scenarios.

Notes

1. See *generally* Maxine Burkett, *Climate Reparations*, 10 MELB. J. INT'L L. 509 (Fall 2009) (discussing the history and special status of AOSIS). "Indeed, the particular vulnerabilities SIDS face resulted in special recognition within the *UNFCCC* as well as advanced speaking rights, and a special linkage to climate-related financial assistance." *Id.* at 537. *See also* W. Jackson Davis, *The Alliance of Small Island States (AOSIS): The International Conscience*, ASIA-PACIFIC MAG. (May 1996), *available* at https:// web.archive .org/web/20070610034008/http://coombs.anu.edu.au/SpecialProj/APM /TXT/davis-j-02-96.html/ (stating that since its establishment during the 1990 Second World Conference in Geneva, AOSIS has played a central role in shaping international policy on climate change).

2. Articles 4.8 and 4.9 of the United Nations Framework Convention on Climate Change identify least developed countries and small island states as being the most vulnerable to the adverse effects of climate change. United Nations Framework Convention on Climate Change, May 9, 1992, S. Treaty Doc No. 102-38, 1771 U.N.T.S. 107.
3. Malia Talakai, *Climate Conversations—Small island states need action on climate loss and damage*, THOMSON REUTERS FOUNDATION (Aug. 30, 2012, 11:40 GMT), http://www.trust.org/item/?map=small-island-states-need-action-on-climate-loss-and-damage/.
4. UNFCCC, Subsidiary Body for Implementation, A literature review on the topics in the context of thematic area 2 of the work programme on loss and damage: a range of approaches to address loss and damage associated with the adverse effects of climate change, Note by the Secretariat, at 4–5, 37th Sess., Nov. 26,–Dec. 1, 2012, U.N. Doc. FCCC/SBI/2012/INF.14 (Nov. 15, 2012) [hereinafter SBI Review].
5. *See* Kim Chipman & Alex Morales, *Islands Seek Funds for Climate Damage at UN Discussions*, BLOOMBERG (Dec. 4, 2012, 3:29 AM), http://www.bloomberg.com/news/2012-12-03/islands-seek-funds-for-climate-damage-at-un-talks.html/ (citing EU Climate Commissioner Connie Hedegaard's statement that the 27-member bloc has been supportive of the concept, though there are some reservations on how to proceed). Hedegaard explained, "We think that it's not really mature enough yet to say this is exactly how we do it. We need some more work on that, but we have signaled very clearly to them that we are open to find a solution on loss and damage." *Id.* Given staunch opposition to the proposal in the past, the fact that Loss and Damage took a central role in the Doha climate negotiations took many by surprise. *Loss and Damage Reflects New Era of the Climate Talks*, ALLIANCE OF SMALL ISLAND STATES (Apr. 15, 2013), http://aosis.org/loss-and-damage-reflects-new-era-of-the-climate-talks/. The Doha decision called on the UNFCCC secretariat to carry out a number of activities that it must complete before the 39th session of the Subsidiary Body for Implementation, including an expert meeting in the fall of 2013 and the provision of technical papers on both the noneconomic losses related to climate change and the gaps in existing institutional arrangements relative to Loss and Damage. UNFCCC Conference of the Parties, 18th Sess., Decision 1/CP.18, U.N. Doc. FCCC /CP/2010/7/Add.1, (Feb. 28, 2013). For a discussion of the relevance of the Doha decision to the CRC and other Framework Convention initiatives on loss and damage, see *infra* Part III.A.
6. Five Pacific Island states are listed as both small island developing states and least developed countries. They are Kiribati, Samoa, Solomon Islands, Tuvalu, and Vanuatu. Erika J. Techera, *Climate change, legal governance and the Pacific Islands: an overview, in* CLIMATE CHANGE AND INDIGENOUS

PEOPLES: THE SEARCH FOR LEGAL REMEDIES 339, 344 (Randall S. Abate & Elizabeth Ann Kronk eds., 2013).

7. See discussion *infra* Part I.B.
8. It also falls beyond the purview of other, related UNFCCC institutions. *See* Alliance of Small Island States, Montego Bay, Jamaica, Mar. 10-12, 2013, *Informal Dialogue on Loss and Damage*, 6 [hereinafter AOSIS Dialogue] (citing existing UNFCCC institutions which were seen as having responsibilities that were relevant to loss and damage including, *inter alia*, the Conference of Parties, the Adaptation Committee, and the National Adaptation Planning Process). As explained *infra* Part III, because the compensation mechanism would respond to phenomenon to which communities cannot adapt, the Green Climate Fund, would also not be an appropriate venue for a compensation mechanism. *See* GREEN CLIMATE FUND, www.gcfund.net (last visited July 23, 2013) (the "Fund will promote the paradigm shift towards low-emission and climate-resilient development pathways by providing support to developing countries to limit or reduce their greenhouse gas emissions and to adapt to the impacts of climate change, taking into account the needs of those developing countries particularly vulnerable to the adverse effects of climate change.").
9. Kim Chipman, *Global Warming 'Damages' Spur Rift at UN Climate Treaty Talks*, BLOOMBERG (Dec. 7, 2012), http://www.bloomberg.com/news/2012-12-07/global-warming-damages-spur-rift-at- un-climate-treaty-talks/.html (quoting U.S. negotiator, Jonathan Pershing, stating that the U.S. doesn't approve of any "liability-based structure"); Chipman & Morales, *supra* note 5 (quoting Saleemul Huq, a Bangladeshi scientist based in London's International Institute for Environment and Development, who explained that U.S. State Department Envoy Todd Stern dodged a question on the matter after arriving at COP 18 in Doha, saying there are "some issues that are of concern there, but I don't want to weigh into that without being certain").
10. Chipman & Morales, *supra* note 5. Huq explains, "Developed countries hear that phrase, 'loss and damage,' and they think of an international fund for compensation and liability—taboo subjects for them. There's strong push back. The U.S. has said there is no way they are going to do it." *Id.* The U.S. approach contrasts with other more balanced responses from developed countries. U.K. Energy Secretary Ed Davey stated, "We should be cautious about saying we are strictly liable for some particular event or some particular change. That does not mean we should not work with others to help some of the very poorest adapt to the impacts of climate change." Id.
11. See *e.g.*, Ilona Millar et al., *Making Good the Loss: An Assessment of the Loss and Damage mechanism under the UNFCCC Process*, in THREATENED ISLAND NATIONS: LEGAL IMPLICATIONS OF RISING SEAS AND A CHANGING CLIMATE 433 (Michael B. Gerrard & Gregory Wannier eds., 2013) (exploring the legal basis for a loss and damage mechanism and

providing an overview of the component elements that might respond to unavoidable impacts, including migration); Daniel Farber, *The UNCC as a Model for Climate Compensation, in* GULF WAR REPARATIONS AND THE UN COMPENSATION COMMISSION 242 (Cymie R. Payne & Peter H. Sand eds., Oxford 2011) (citing the asymmetry between carbon emitters and "climate victims" and stating that compensation claims are inevitable, and therefore, discussions of compensation are not "merely academic"). Professor Farber sketches out the loose contours of an international compensation mechanism in this chapter, *id.* at 242-57, but does not provide more than a preliminary thought-piece on the topic. *See also* Melissa Farris, *Compensating Climate Change Victims: The Climate Compensation Fund as an Alternative to Tort Litigation*, 2 SEA GRANT L. & POL'Y J. 49 (Winter 2009/2010) (introducing a no-fault compensation fund as an alternative to tort litigation to compensate the victims of climate change).

12. See *generally* Maxine Burkett, *A Justice Paradox: On Climate Change, Small Island Developing States, and the Quest for Effective Legal Remedy*, 35 U. HAW. L. REV. 633 (2013) [hereinafter Burkett, *A Justice Paradox*].
13. The need to persuade the developed world to agree to rehabilitation is the result of long-standing global North-South dynamics that consistently and systemically debilitate the global South. For a trenchant discussion of this dynamic in the context of human rights and environmental justice, see Carmen G. Gonzalez, *Human Rights and Environmental Justice: the North-South Dimension*, 13 SANTA CLARA J. INT'L L. 151 (2015).

* * *

17. *About AOSIS*, ALLIANCE OF SMALL ISLAND STATES, http://aosis.org/about-aosis/ (last visited Jan. 5, 2015).
18. *Id.*
19. AOSIS member countries include: Antigua and Barbuda, Bahamas, Barbados, Belize, Cape Verde, Comoros, Cook Islands, Cuba, Dominica, Dominican Republic, Fiji, Federated States of Micronesia, Grenada, Guinea-Bissau, Guyana, Haiti, Jamaica, Kiribati, Maldives, Marshall Islands, Mauritius, Nauru, Niue, Palau, Papua New Guinea, Samoa, Singapore, Seychelles, Sao Tome and Principe, Solomon Islands, St. Kitts and Nevis, St. Lucia, St. Vincent and the Grenadines, Suriname, Timor-Leste, Tonga, Trinidad and Tobago, Tuvalu, and Vanuatu. Observers include: American Samoa, Netherlands Antilles, Guam, U.S. Virgin Islands, and Puerto Rico. *See Members*, ALLIANCE OF SMALL ISLAND STATES, http://aosis.org/members/ (last visited July 10, 2013). Thirty-seven are active members of the UN, constitute approximately 28 percent of developing countries, and 20 percent of the UN's total membership. *About AOSIS, supra* note 17. Together, SIDS communities constitute roughly five percent of the global population. *Id.*

20. *About SIDS*, SIDSNET, http://www.sidsnet.org/about-sids/ (last visited July 14, 2013).
21. For instance, Singapore's GDP per capita is the highest at $61,400, whereas Guinea Bissau's is $1,200. *The World Factbook,* CENT. INTELLIGENCE AGENCY, https://www.cia.gov/library/ publications/the-world-factbook /rankorder/2004rank.html?ga=1.188844395.154334249.141239 4817/ (last visited Oct. 9, 2014).
22. Techera, *supra* note 6, at 344.
23. Intergovernmental Panel on Climate Change, Special Rep., *Managing the Risks of Extreme Events and Disasters to Advance Climate Change Adaptation*, 1-19 (2012) [hereinafter IPCC], *available at* https://www.ipcc.ch/pdf/special -reports/srex/SREX_Full_Report.pdf/.
24. *Pacific Islands: Disaster Risk Reduction and Financing in the Pacific*, THE WORLD BANK, http://web.worldbank.org/WBSITE/EXTERNAL /NEWS/0,,contentMDK:23176294~menuPK:1413 10~pagePK:34370~ piPK:34424~theSitePK:4607,00.html/ (last visited July 17, 2013).
25. David McFadden, *Coastal living in Caribbean likely doomed due to sea-level rise*, E&E PUBLISHING CLIMATEWIRE (May 9, 2013), http://www .eenews.net/climatewire/stories/1059980837/search?keyword=coastal+ living+in+caribbean+likely+doomed+due+to+sea+level+rise/.
26. *Id.*
27. *Id.*
28. Paul Brown, *Simultaneous Disasters Batter Pacific Islands*, CLIMATE NEWS NETWORK (July 5, 2013), http://www.climatecentral.org/news /simultaneous-disasters-batter-pacific-islands-16171/.
29. *Id.* (explaining that normally, the scant fresh water supplies are topped up from frequent evening rains, however, a devastating drought, which the locals blame on climate change, has reduced a desperate population to rationing water supplies to a liter a day).
30. U.N. Framework Convention on Climate Change, Jamaica's Initial Climate Change Technology Needs Assessment, 18, http://unfccc.int/ttclear /misc_/StaticFiles/gnwoerk_static/TNR_CRE/ e9067c6e3b97459989b-2196f12155ad5/4ccc885176b94f949b83ed4b294061dc.pdf/ (last visited Jan. 3, 2015). In fact, SIDS are the most at risk when losses, that is, economic damage from natural disasters linked to rising global temperatures over coming decades, are compared against GDP. Nathanial Gronewold, *US and China Most Exposed to Costs of Climate-Related Disasters*, E&E PUBLISHING CLIMATEWIRE, (Mar. 13 2009), http://www.eenews.net/climatewire /2009/03/13/ stories/75480/.
31. IPCC, *supra* note 23, at 265.
32. *Id.*; *see also* Millar et al., *supra* note 11 (stating that losses due to natural disasters in developing countries are 20 times greater as a percentage of GDP than in industrialized countries owing to the lack of risk transfer and risk sharing

mechanisms, and consequently, the reliance on financial assistance from donor countries to respond to extreme events).

33. *See* IPCC, *supra* note 23, at 5. Regarding disaster related deaths, from 1970 to 2008 over 95% of deaths from natural disasters occurred in developing countries. *Id.* at 9. "The relative economic burden in terms of direct loss expressed as a percentage of GDP has been substantially higher for developing states." *Id.* at 270.
34. *Id.* at 266.
35. *See id.* at 270 ("In small exposed countries, particularly small island developing states, these wealth losses expressed as a percentage of GDP and averaged over both disaster and non-disaster years can be considerably higher, exceeding 1% in many cases and 8% in the most extreme cases over the period from 1970 to 2010 (World Bank and UN, 2010), and individual events may consume more than the annual GDP (McKenzie et al., 2005)."). *See also* Techera, *supra* note 6, at 351.
36. *See, e.g.*, Map from U.N. Office for the Coordination for Human Affairs, Natural Disasters and Other Events Being Monitored by the OCHA Regional Office for the Asia Pacific (Mar. 17–23, 2010), *available at* http://reliefweb.int/map/fiji/asia-pacific-region-natural-disasters-and-other - events-being-monitored-ocha-regional-offi-0/.
37. IPCC, *supra* note 23, at 266.
38. *Id.* at 7.
39. *Id.* at 8 ("Assigning 'low confidence' for projections of a specific extreme neither implies nor excludes the possibility of changes in this extreme.").
40. *See, e.g.*, *id.* (general discussion on the rise in sea levels); *see also* UNFCCC, Subsidiary Body of Implementation, Views and Information on Elements to be Included in the Recommendations on Loss and Damage in Accordance With Decision 1/CP.16, at 28, U.N. Doc. FCCC/SBI/2012/ MISC.14 (Sept. 28, 2012), *available at* http://unfccc.int/resource/docs/2012/sbi/eng /misc14.pdf/ [hereinafter SBI Views & Information].
41. Roda Verheven and Peter Roderick identify three types of climate change damage: avoided, not avoided, and unavoidable. Avoided damages are those for which mitigation of emissions or timely implementation of adaptation measures were successful. Not avoided damages are damages that occur because of insufficient mitigation efforts and delays in accessing adequate adaptation funding or technology or challenges in institutional capacity. RODA VERHEYEN & PETER RODERICK, WWF-UK CLIMATE CHANGE PROGRAMME, BEYOND ADAPTATION: THE LEGAL DUTY TO PAY COMPENSATION FOR CLIMATE CHANGE DAMAGE 11 (Nov. 2008), *available at* http://assets.wwf. org.uk/downloads /beyond_adaptation_lowres.pdf/.
42. *Id. See also* SBI Review, *supra* note 4, at 24 ("The countries with the highest levels of residual risk are those that will be the least able to manage loss and

damage in the future. They are also the countries that may be in need of the greatest support to manage loss and damage.") (citation omitted).

43. IPCC, *supra* note 23, at 15. In addition to observed changes in extreme coastal high water related to increases in sea level, it is "*very likely* that mean sea level rise will contribute to upward trends in extreme coastal high water levels in the future The *very likely* contribution of mean sea level rise to increased extreme coastal high water levels, coupled with the *likely* increase in tropical cyclone maximum wind speed, is a specific issue for tropical small island states." *Id.* (emphasis in original). *See also* Mary-Elena Carr et al., *Sea Level Rise in a Changing Climate: What Do We Know?, in* THREATENED ISLAND NATIONS: LEGAL IMPLICATIONS OF RISING SEAS AND A CHANGING CLIMATE 15 (Michael B. Gerrard & Gregory Wannier eds., 2013).
44. *See* Mary-Elena Carr et al., *supra* note 43, at 44 (explaining that "[l]onger and more severe periods of drought are of particular concern for SIDS because such periods will reduce both the amount of rainwater collected directly and in the recharge of the freshwater lens").
45. *See, e.g.,* IPCC, *supra* note 23, at 25 (explaining that "[m]odels project substantial warming in temperature extremes by the end of the 21st century. It is *virtually certain* that increases in the frequency and magnitude of warm daily temperature extremes and decreases in cold extremes will occur in the 21st century at the global scale. It is *very likely* that the length, frequency, and/or intensity of warm spells or heat waves will increase over most land areas.") (emphasis in original).
46. *See, e.g., Acidic oceans of the future show extinction*, THE DAILY CLIMATE, July 9, 2013, *available at* http://www.dailyclimate.org/tdc-newsroom/2013/07/future-acidic-oceans/ (explaining that ocean acidification may create an impact similar to extinction on marine ecosystems; according to a study published in the Proceedings of the National Academy of Sciences,"[t]oday the ocean's pH is lower than anything seen in the historical record in the past 800,000 years, scientists say. As the acidity increases, organisms such as corals, oysters, snails and urchins have trouble pulling minerals from the seawater to create protective shells . . . [the study] buttresses ecologists' fears that such changes could ripple through entire ecosystems—and that ocean acidification could prove as consequential and catastrophic for the globe as any changes in air temperature associated with climate change.").
47. *See, e.g.*, Techera, *supra* note 6, at 346 (explaining that "[s]ea level rise at the very least disrupts Pacific Islanders, who are largely comprised of indigenous peoples living traditional lifestyles at the coastal zone, and at worst will result in the displacement of whole communities").
48. IPCC, *supra* note 23, at 8.

    Disasters associated with climate extremes influence population mobility and relocation, affecting host and origin communities (*medium agreement, medium evidence).* If disasters occur more frequently and/or with greater magnitude, some local areas will become increasingly marginal as places to live or in which to maintain livelihoods. In such cases, migration and displacement could become permanent and could introduce new pressures in areas of relocation. For locations such as atolls, in some cases it is possible that many residents will have to relocate.

Id. at 16; *see also* AOSIS, *Proposal to the AWG-LCA: Multi-Window Mechanism to Address Loss and Damage from Climate Change Impacts* (2008), *available at* http://unfccc.int/files/ kyoto_protocol/application/pdf /aosisinsurance061208.pdf/ [hereinafter AOSIS Proposal] (arguing that to manage climate change effectively, the adaptive frameworks must address all categories, including unavoidable impacts such as migration).

49. SBI Review, *supra* note 4, at 4 ("Loss and damage includes the effects of the full range of climate change related impacts, from increasing (in number and intensity) extreme weather events to slow onset events and combinations of the two.").
50. *Id.* at 3, 23.
51. *Id.* at 5 ("[F]or example, sea level rise and glacial melt result from climate change stimuli, and these shifts in natural systems in turn result in loss and damage in human systems, such as loss of habitable land or freshwater.").
52. *Id.*
53. *Id.* ("Future loss and damage is likely to increase, especially considering non-economic factors and the interlinkages of phenomena leading to cascading, transnational effects.").

* * *

55. *See generally* Burkett, *Climate Reparations*, *supra* note 1.
56. *See* Techera, *supra* note 6, at 347 (citing the IPCC's acknowledgement of cultural impacts having deeper effects than first appears through the damaging of culturally-informed flexibility and resilience).
57. *See* Millar et al., *supra* note 11, at 438 (explaining that owing to the lack of risk transfer and sharing, developing countries are reliant on financial assistance from donor countries to respond to extreme events); *see also* Brown, *supra* note 28.
58. IPCC, *supra* note 23, at 10.
59. SBI Views & Information, *supra* note 40, at 4. This hole has existed in spite of consistent (w.c.) attempts to fill it. For a full chronology of attempts to advance AOSIS's loss and damage proposal at the UNFCCC, see Alliance of Small Island States, Loss and Damage in the UNFCCC Process—Workshop on the AOSIS Proposal, New York, New York, May 16–18, 2013 (on file with author) [hereinafter AOSIS Expert Meeting]; AOSIS Dialogue, *supra* note 8; AOSIS, *Loss and Damage Briefing*, 2012 [hereinafter AOSIS Briefing];

AOSIS Proposal, *supra* note 48; UNFCCC, Conference of the Parties, 18th Sess., Decision 3/CP.18, U.N. Doc. FCCC/CP/2012/8/Add.1 (Feb. 28, 2013) [hereinafter COP 18, Decision 3].

* * *

81. AOSIS Proposal, *supra* note 48, at 6. Coral reefs, for example, hold significant importance for the tourism industry as well as local communities. In St. Lucia, coral reefs contribute to roughly 20% of the nation's GDP through their benefits for tourism, fisheries, and for protecting shorelines. Hutchinson-Jafar, *supra* note 71.
82. AOSIS Proposal, *supra* note 48, at 6 (citing SBI Review, *supra* note 4).

* * *

115. Technological facilities would support all three functions of the CRC. For further discussion of this indispensable piece of the CRC and the overall multi-window mechanism, see discussion *infra* Part III.B. David Caron identifies this as one of the key lessons from the UNCC, explaining,

    [G]iven the previous lessons and the assumed presence of a significant disaster, it should likewise be assumed that the claims process extends far into the future. This is particularly the case in terms of any follow-up program where the claims institution monitors the performance of the agent; that is, tracks the execution of an agent's stated plan to address environmental concerns.

    Caron, *supra* note 87, at 274.
116. *See e.g.*, Millar et al., *supra* note 11, at 440-41; Tol & Verheyen, *supra* note 54, at 1109-11 (citing the no harm rule and law of state responsibility under customary international law); Burkett, *Climate Reparations*, *supra* note 1; AOSIS Proposal, *supra* note 48 (citing the principle of state responsibility, which requires the cessation of wrongful activity and reparation for damage caused by the wrongful act); UNEP, *supra* note 111, at 119; *see also* ILC, *supra* note 66, at arts. 30-39.
117. UNFCCC, *supra* note 2, at arts. 4.1(e), 4.4, 4.8. ("[T]he Parties shall give full consideration to what actions are necessary under the Convention, including actions related to funding, insurance and the transfer of technology, to meet the specific needs and concerns of developing country Parties arising from the adverse effects of climate change and/or impact of the implementation of response measures, especially on ... [s]mall island countries "); *see* Kyoto Protocol to the United Nations Framework Convention on Climate Change art. 3(14), Dec. 10, 1997, UN Doc. FCCC/CP/1997/7/Add.2, *reprinted in* 37 ILM 22 (1998); *see also* Millar et al., *supra* note 11, at 438.
118. UNFCCC, Conference of the Parties, 13th Sess., Decision 1/CP.13, U.N. Doc. FCCC/CP/2007/6/Add.1 (Mar. 14, 2008).

119. *Id.* at ¶ 1(c)(iii). *See also id.* at ¶¶ 1(c)(i), 1(c)(ii), 1(c)(iii) and 1(c)(v) on adaptation, as well as ¶¶ 1(e)(i), 1(e)(ii), 1(e)(iii), 1(e)(iv), 1(e)(v), and 1(e)(vi) on finance and investment.
120. UNFCCC Conference of the Parties, 16th Sess., Decision 1/CP.16, U.N. Doc. FCCC/CP/2010/7/Add.1, (Mar. 15, 2011); *see also* SBI Review, *supra* note 4, at 23.
121. *Id.* at ¶ 25.
122. UNFCCC Conference of the Parties, 18th Sess., Decision 1/CP.18, U.N. Doc. FCCC/CP/2010/7/Add.1, (Feb. 28, 2013).
123. *Id.*
124. *Id.* Although this mandate suffered setbacks in the most recent international meetings in Bonn, a loss and damage mechanism remains a key deliverable. *See* Laurie Goering, *Africa: Vulnerable States Decry Slow Progress at Bonn Climate Talks*, ALLAFRICA (June 17, 2013), http://allafrica.com/stories/201306181411.html/.
125. *Loss and Damage Reflects New Era of the Climate Talks*, *supra* note 5.

* * *

129. McGovern, *The What and Why of Claims Resolution Facilities*, *supra* note 86, at 1365-66 (arguing that a "metaphor or paradigm" for the facility is critical for the public perception of claims resolution). McGovern notes that the "welfare paradigm is that of a social safety net to insure that no fundamental needs are unmet. There is no implication of wrongdoing or malfeasance." *Id.* at 1366.
130. AOSIS Proposal, *supra* note 48, at 3.
131. UNFCCC, *supra* note 2, at art 3.1 ("The Parties should protect the climate system for the benefit of present and future generations of humankind, on the basis of equity and in accordance with their common but differentiated responsibilities and respective capabilities.").

* * *

135. *See* McGovern, *The What and Why of Claims Resolution Facilities*, *supra* note 86, at 1378.
136. *See* Van Houtte et al., *supra* note 106, at 370.
137. *See* McGovern, *The What and Why of Claims Resolution Facilities*, *supra* note 86, at 1378.
138. See discussion *infra* Part III.B.

* * *

142. On the importance of considering the "longitudinal form" of an organization design, see McGovern, *The What and Why of Claims Resolution Facilities*,

supra note 86, at 1369-70 ("If the claims resolution facility is to allocate a certain sum to claimants over any extended period of time, there will be a disproportionate ratio of administrative cost to damage payments as they facility winds down. Thus, what way be an efficient organization in the early stages of the life of a facility may not be economical as it ages.").

* * *

149. *See generally id.* at 1368.
150. *Id.* ("The organization is more reminiscent of the inquisitorial model of the courts of equity rather than the adversarial mode of the common law courts.").
151. Again, the UNCC provides precedent for this. *See* Van Houtte et al., *supra* note 106, at 343 ("The standard claim forms designed and distributed by the Secretariat to capture data in a consistent and uniform manner.").

* * *

161. A similar limitation was proposed in the context of wartime environmental damage in McManus, *supra* note 113, at 442.

* * *

164. *See* Techera, *supra* note 6, at 342-44.
165. *See* Farber, *Basic Compensation for Victims of Climate Change*, *supra* note 103, at 1635-39 (discussing various models to predict the possibility of damages for purposes of compensating climate change victims). Full explication of this piece is beyond the scope of this article.
166. IPCC, *supra* note 23, at 10 ("This would be critical, particularly in light of the dearth of data on "disasters and disaster risk reduction . . . at the local level, which can constrain improvements in local vulnerability reduction."); Sand, *supra* note 96, at 431 (stating that faced with a similar lack of reliable data, "the UNCC started by awarding funds upfront for monitoring and assessment of the damage. Total funding for this purpose amounted to $243.6 million.").
167. *See* Bederman, *supra* note 91, at 30.

* * *

186. *See, e.g.*, Millar et al., *supra* note 11, at 444-45 (discussing funding for an international insurance pool).

* * *

190. *See* McGovern, *The What and Why of Claims Resolution Facilities*, *supra* note 86, at 1373.

* * *

193. AOSIS suggest that "[f]or payouts using parametric approaches, different metrics might be applied in different countries." AOSIS Insurance, *supra* note 146, at 8.

Appendix C

A Snapshot of the Last Century of Scientific Calls to Arms

SCIENTISTS HAVE ALWAYS been the vanguard of ocean understanding, and that is particularly true in the Anthropocene, when the ocean and its systems are changing rapidly in response to climate change and ocean acidification. Like Jeremy Jackson, moreover, a few of these scientists publish for a wider audience than just their colleagues, explaining their new discoveries and new understandings to the general public and to politicians.

Before she wrote about the problems that pesticides cause in *Silent Spring* (1962), marine biologist Rachel Carson described the ocean and its many interactions with humans to an entranced general public. Her 1951 *The Sea Around Us* topped *The New York Times*' best seller list, remained on the list for thirty-one weeks, and won the 1952 National Book Award for nonfiction. Remarkably, over seventy years later, it remains eminently insightful.

The multinational team of scientists who author the Intergovernmental Panel on Climate Change's (IPCC's) many reports analyze cutting-edge science to convey to the world—notably, with special Summaries for Policymakers—the state-of-the-art consensus on what we know about how climate change is affecting Planet Ocean. In 2019, a team of these scientists published the IPCC's *Special Report on the Ocean and Cryosphere in a Changing Climate*. The report details humanity's many dependencies on the ocean and the risks that we increasingly all face as it changes, but it also recognizes human capacity to both cope with and avert the worst of those impacts.

APPENDIX C.1

The Encircling Sea

RACHEL CARSON

A sea from which birds travel not within a year, so vast it is and fearful.
—Homer

TO THE ANCIENT GREEKS the ocean was an endless stream that flowed forever around the border of the world, ceaselessly turning upon itself like a wheel, the end of earth, the beginning of heaven. This ocean was boundless; it was infinite. If a person were to venture far out upon it—were such a course thinkable—he would pass through gathering darkness and obscuring fog and would come at last to a dreadful and chaotic blending of sea and sky, a place where whirlpools and yawning abysses waited to draw the traveler down into a dark world from which there was no return.

These ideas are found, in varying form, in much of the literature of the ten centuries before the Christian era, and in later years they keep recurring even through the greater part of the Middle Ages. To the Greeks the familiar Mediterranean was The Sea. Outside, bathing the periphery of the land world, was Oceanus. Perhaps somewhere in its uttermost expanse was the home of the gods and of departed spirits, the Elysian fields. So we meet the ideas of unattainable continents or of beautiful islands in the distant ocean, confusedly mingled with references to a bottomless gulf at the edge of the world—but always around the disc of the habitable world was the vast ocean, encircling all.

Perhaps some word-of-mouth tales of the mysterious northern world, filtering down by way of the early trade routes for amber and tin, colored the conceptions of the early legends, so that the boundary of the land world came to be pictured as a place of fog and storms and darkness. Homer's *Odyssey* described the Cimmerians as dwelling in a distant realm of mist and darkness on the shores of Oceanus, and they told of the shepherds who lived in the land of the long day, where the paths of day and night were close. And again perhaps the early poets and historians derived some of their ideas of the ocean from the Phoenicians, whose craft roamed the shores of Europe, Asia, and Africa in search of gold, silver, gems, spices, and wood for their commerce with kings and emperors. It may well be that these sailor-merchants were the first ever to cross an ocean, but history does not record the fact. For at least 2000 years before Christ—probably longer—the flourishing trade of the Phoenicians was plied along the shores of the Red Sea to Syria, to Somaliland, to Arabia, even to India and perhaps to China. Herodotus wrote that they circumnavigated Africa from east to west about 600 B.C., reaching Egypt via the Straits of the Pillars and the Mediterranean. But the Phoenicians themselves said and wrote little or nothing of their voyagings, keeping their trade routes and the sources of their precious cargoes secret. So there are only the vaguest rumors, sketchily supported by archaeological findings, that the Phoenicians may have launched out into the open Pacific.

Nor are there anything but rumors and highly plausible suppositions that the Phoenicians, on their coastwise journeys along western Europe, may have sailed as far north as the Scandinavian peninsula and the Baltic, source of the precious amber. There are no definite traces of any such visits by them, and of course the Phoenicians have left no written record of any. Of one of their European voyages, however, there is a secondhand account. This was the expedition under Himlico of Carthage, which sailed northward along the European coast about the year 500 B.C. Himlico apparently wrote an account of this voyage, although his manuscript was not preserved. But his descriptions are quoted by the Roman Avienus, writing nearly a thousand years later. According to Avienus, Himlico painted a discouraging picture of the coastwise seas of Europe:

> These seas can scarcely be sailed through in four months ... no breeze drives the ship forward, so dead is the sluggish wind of this idle sea ... There is much seaweed among the waves ... the surface of the earth is barely covered by a little water ... The monsters of the sea move

> continually hither and thither, and the wild beasts swim among the sluggish and slowly creeping ships.

Perhaps the "wild beasts" are the whales of the Bay of Biscay, later to become a famous whaling ground; the shallow water areas that so impressed Himlico may have been the flats alternately exposed and covered by the ebb and flow of the great tides of the French coast—a strange phenomenon to one from the almost tideless Mediterranean. But Himlico also had ideas of the open ocean to the west, if the account of Avienus is to be trusted: "Farther to the west from these Pillars there is boundless sea . . . None has sailed ships over these waters, because propelling winds are lacking on these deeps . . . likewise because darkness screens the light of day with a sort of clothing, and because a fog always conceals the sea." Whether these descriptive details are touches of Phoenician canniness or merely the old ideas reasserting themselves it is hard to say, but much the same conceptions appear again and again in later accounts, echoing down the centuries to the very threshold of modern times.

* * *

Of the methods of those secretive master mariners, the Phoenicians, we cannot even guess. We have more basis for conjecture about the Polynesians, for we can study their descendants today, and those who have done so find hints of the methods that led the ancient colonizers of the Pacific on their course from island to island. Certainly they seem to have followed the stars, which burned brightly in the heavens over those calm Pacific regions, which are so unlike the stormy and fog-bound northern seas. The Polynesians considered the stars as moving bands of light that passed across the inverted pit of the sky, and they sailed toward the stars which they knew passed over the islands of their destination. All the language of the sea was understood by them: the varying color of the water, the haze of surf breaking on rocks yet below the horizon, and the cloud patches that hang over every islet of the tropic seas and sometimes seem even to reflect the color of a lagoon within a coral atoll.

Students of primitive navigation believe that the migrations of birds had meaning for the Polynesians, and that they learned much from watching the flocks that gathered each year in the spring and fall, launched out over the ocean, and returned later out of the emptiness into which they had vanished. Harold Gatty believes the Hawaiians may have found their islands by following the spring migration of the golden plover from Tahiti to the Hawaiian chain, as the birds returned to the North American mainland. He has also suggested that the migratory path of the shining cuckoo may have guided other colonists from the Solomons to New Zealand.

* * *

So here and there, in a few out-of-the-way places, the darkness of antiquity still lingers over the surface of the waters. But it is rapidly being dispelled and most of the length and breadth of the ocean is known; it is only in thinking of its third dimension that we can still apply the concept of the Sea of Darkness. It took centuries to chart the surface of the sea; our progress in delineating the unseen world beneath it seems by comparison phenomenally rapid. But even with all our modern instruments for probing and sampling the deep ocean, no one now can say that we shall ever resolve the last, the ultimate mysteries of the sea.

In its broader meaning, that other concept of the ancients remains. For the sea lies all about us. The commerce of all lands must cross it.

* * *

The very winds that move over the lands have been cradled on its broad expanse and seek ever to return to it. The continents themselves dissolve and pass to the sea, in grain after grain of eroded land. So the rains that rose from it return again in rivers. In its mysterious past it encompasses all the dim origins of life and receives in the end, after, it may be, many transmutations, the dead husks of that same life. For all at last return to the sea—to Oceanus, the ocean river, like the ever-flowing stream of time, the beginning and the end.

APPENDIX C.2

Summary for Policymakers

INTERGOVERNMENTAL PANEL ON CLIMATE CHANGE

Startup Box | The Importance of the Ocean and Cryosphere for People

All people on Earth depend directly or indirectly on the ocean and cryosphere. The global ocean covers 71% of the Earth surface and contains about 97% of the Earth's water. The cryosphere refers to frozen components of the Earth system. Around 10% of Earth's land area is covered by glaciers or ice sheets. The ocean and cryosphere support unique habitats, and are interconnected with other components of the climate system through global exchange of water, energy and carbon. The projected responses of the ocean and cryosphere to past and current human-induced greenhouse gas emissions and ongoing global warming include climate feedbacks, changes over decades to millennia that cannot be avoided, thresholds of abrupt change, and irreversibility.

Human communities in close connection with coastal environments, small islands (including Small Island Developing States, SIDS), polar areas and high mountains are particularly exposed to ocean and cryosphere change, such as sea level rise, extreme sea level and shrinking cryosphere.

Excerpt from *IPCC Special Report on the Ocean and Cryosphere in a Changing Climate*, ed. H.-O. Pörtner, D.C. Roberts, V. Masson-Delmotte, P. Zhai, M. Tignor, E. Poloczanska, K. Mintenbeck, A. Alegría, M. Nicolai, A. Okem, J. Petzold, B. Rama, and N. M. Weyer (IPCC, 2019), https://www.ipcc.ch/srocc/chapter/summary-for-policymakers/.

Other communities further from the coast are also exposed to changes in the ocean, such as through extreme weather events. Today, around 4 million people live permanently in the Arctic region, of whom 10% are Indigenous. The low-lying coastal zone is currently home to around 680 million people (nearly 10% of the 2010 global population), projected to reach more than one billion by 2050. SIDS are home to 65 million people. Around 670 million people (nearly 10% of the 2010 global population), including Indigenous peoples, live in high mountain regions in all continents except Antarctica. In high mountain regions, population is projected to reach between 740 and 840 million by 2050 (about 8.4–8.7% of the projected global population).

* * *

In addition to their role within the climate system, such as the uptake and redistribution of natural and anthropogenic carbon dioxide (CO_2) and heat, as well as ecosystem support, services provided to people by the ocean and/or cryosphere include food and water supply, renewable energy, and benefits for health and well-being, cultural values, tourism, trade, and transport. The state of the ocean and cryosphere interacts with each aspect of sustainability reflected in the United Nations Sustainable Development Goals (SDGs).

* * *

A. Observed Changes and Impacts

A.2 It is *virtually certain* that the global ocean has warmed unabated since 1970 and has taken up more than 90% of the excess heat in the climate system (*high confidence*). Since 1993, the rate of ocean warming has more than doubled (*likely*). Marine heatwaves have very likely doubled in frequency since 1982 and are increasing in intensity (*very high confidence*). By absorbing more CO_2, the ocean has undergone increasing surface acidification (*virtually certain*). A loss of oxygen has occurred from the surface to 1000 m[eters] (*medium confidence*).

* * *

A.3 Global mean sea level (GMSL) is rising, with acceleration in recent decades due to increasing rates of ice loss from the Greenland and Antarctic ice sheets (*very high confidence*), as well as continued glacier mass loss and ocean thermal expansion. Increases in tropical cyclone winds and rainfall, and increases in extreme waves, combined with relative sea level rise, exacerbate extreme sea level events and coastal hazards (*high confidence*).

* * *

A.5 Since about 1950 many marine species across various groups have undergone shifts in geographical range and seasonal activities in response to ocean warming, sea ice change and biogeochemical changes, such as oxygen loss, to their habitats (*high confidence*). This has resulted in shifts in species composition, abundance and biomass production of ecosystems, from the equator to the poles. Altered interactions between species have caused cascading impacts on ecosystem structure and functioning (*medium confidence*). In some marine ecosystems species are impacted by both the effects of fishing and climate changes (*medium confidence*).

* * *

A.6 Coastal ecosystems are affected by ocean warming, including intensified marine heatwaves, acidification, loss of oxygen, salinity intrusion and sea level rise, in combination with adverse effects from human activities on ocean and land (*high confidence*). Impacts are already observed on habitat area and biodiversity, as well as ecosystem functioning and services (*high confidence*).

* * *

A.8 Changes in the ocean have impacted marine ecosystems and ecosystem services with regionally diverse outcomes, challenging their governance (*high confidence*). Both positive and negative impacts result for food security through fisheries (*medium confidence*), local cultures and livelihoods (*medium confidence*), and tourism and recreation (*medium confidence*). The impacts on ecosystem services have negative consequences for health and well-being (*medium confidence*), and for Indigenous peoples and local communities dependent on fisheries (*high confidence*).

* * *

A.9 Coastal communities are exposed to multiple climate-related hazards, including tropical cyclones, extreme sea levels and flooding, marine heatwaves, sea ice loss, and permafrost thaw (*high confidence*). A diversity of responses has been implemented worldwide, mostly after extreme events, but also some in anticipation of future sea level rise, e.g., in the case of large infrastructure.

* * *

B. Projected Changes and Risks

* * *

B.2 Over the 21st century, the ocean is projected to transition to unprecedented conditions with increased temperatures (*virtually certain*), greater upper ocean stratification (*very likely*), further acidification (*virtually certain*), oxygen decline (*medium confidence*), and altered net primary production (*low confidence*). Marine heatwaves (*very high confidence*) and extreme El Niño and La Niña events (*medium confidence*) are projected to become more frequent. The Atlantic Meridional Overturning Circulation (AMOC) is projected to weaken (*very likely*). The rates and magnitudes of these changes will be smaller under scenarios with low greenhouse gas emissions (*very likely*).

* * *

B.3 Sea level continues to rise at an increasing rate. Extreme sea level events that are historically rare (once per century in the recent past) are projected to occur frequently (at least once per year) at many locations by 2050 in all RCP scenarios, especially in tropical regions (*high confidence*). The increasing frequency of high water levels can have severe impacts in many locations depending on exposure (*high confidence*). Sea level rise is projected to continue beyond 2100 in all RCP scenarios. For a high emissions scenario (RCP8.5), projections of global sea level rise by 2100 are greater than in AR5 due to a larger contribution from the Antarctic Ice Sheet (*medium confidence*). In coming centuries under RCP8.5, sea level rise is projected to exceed rates of several centimetres per year resulting in multi-metre rise (*medium confidence*), while for RCP2.6 sea level rise is projected to be limited to around 1 m[eter] in 2300 (*low confidence*). Extreme sea levels and

Extreme sea level events

Due to projected global mean sea level (GMSL) rise, local sea levels that historically occurred once per century (historical centennial events, HCEs) are projected to become at least annual events at most locations during the 21st century. The height of a HCE varies widely, and depending on the level of exposure can already cause severe impacts. Impacts can continue to increase with rising frequency of HCEs.

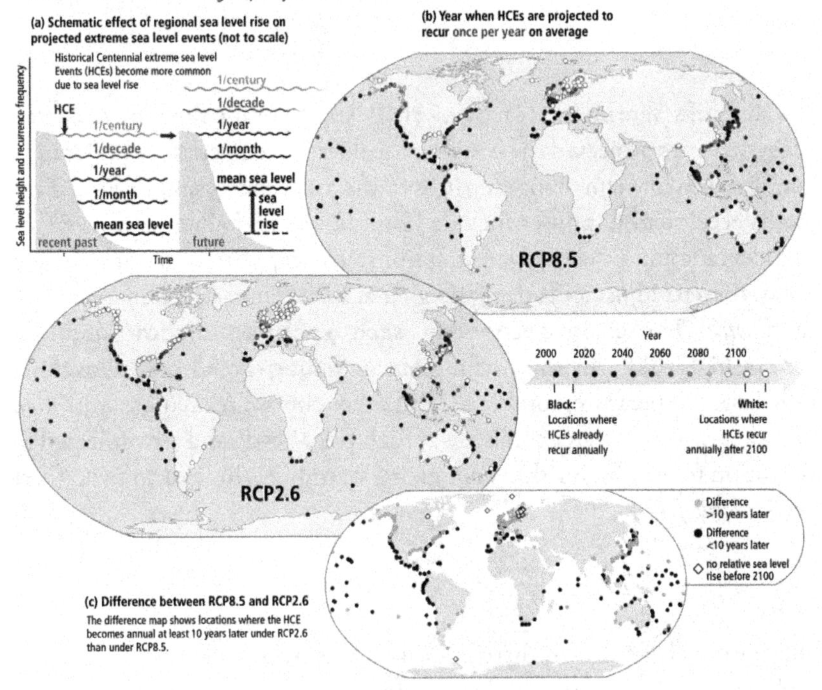

coastal hazards will be exacerbated by projected increases in tropical cyclone intensity and precipitation (*high confidence*). Projected changes in waves and tides vary locally in whether they amplify or ameliorate these hazards (*medium confidence*).

* * *

B.5 A decrease in global biomass of marine animal communities, their production, and fisheries catch potential, and a shift in species composition are projected over the 21st century in ocean ecosystems from the surface to the deep seafloor under all emission scenarios (*medium confidence*). The rate and magnitude of decline are projected to be highest in the tropics (*high confidence*), whereas impacts remain diverse in polar regions (*medium confidence*) and increase for high emissions scenarios. Ocean acidification (*medium confidence*), oxygen loss (*medium confidence*) and reduced sea ice extent (*medium confidence*) as well as non-climatic human activities (*medium confidence*) have the potential to exacerbate these warming-induced ecosystem impacts.

* * *

B.6 Risks of severe impacts on biodiversity, structure and function of coastal ecosystems are projected to be higher for elevated temperatures under high compared to low emissions scenarios in the 21st century and beyond. Projected ecosystem responses include losses of species habitat and diversity, and degradation of ecosystem functions. The capacity of organisms and ecosystems to adjust and adapt is higher at lower emissions scenarios (*high confidence*). For sensitive ecosystems such as seagrass meadows and kelp forests, high risks are projected if global warming exceeds 2°C above pre-industrial temperature, combined with other climate-related hazards (*high confidence*). Warm-water corals are at high risk already and are projected to transition to very high risk even if global warming is limited to 1.5°C (*very high confidence*).

* * *

B.8 Future shifts in fish distribution and decreases in their abundance and fisheries catch potential due to climate change are projected to affect income, livelihoods, and food security of marine resource–dependent communities (*medium confidence*). Long-term loss and degradation of marine ecosystems compromises the ocean's role in cultural, recreational, and intrinsic values important for human identity and well-being (*medium confidence*).

* * *

B.9 Increased mean and extreme sea level, alongside ocean warming and acidification, are projected to exacerbate risks for human communities in low-lying coastal areas (*high confidence*). In Arctic human communities without rapid land uplift, and in urban atoll islands, risks are projected to be moderate to high even under a low emissions scenario (RCP2.6) (*medium confidence*), including reaching adaptation limits (*high confidence*). Under a high emissions scenario (RCP8.5), delta regions and resource rich coastal cities are projected to experience moderate to high risk levels after 2050 under current adaptation (*medium confidence*). Ambitious adaptation including transformative governance is expected to reduce risk (*high confidence*), but with context-specific benefits.

* * *

C.Implementing Responses to Ocean and Cryosphere Change

C.1 Impacts of climate-related changes in the ocean and cryosphere increasingly challenge current governance efforts to develop and implement adaptation responses from local to global scales, and in some cases pushing them to their limits. People with the highest exposure and vulnerability are often those with lowest capacity to respond (*high confidence*).

* * *

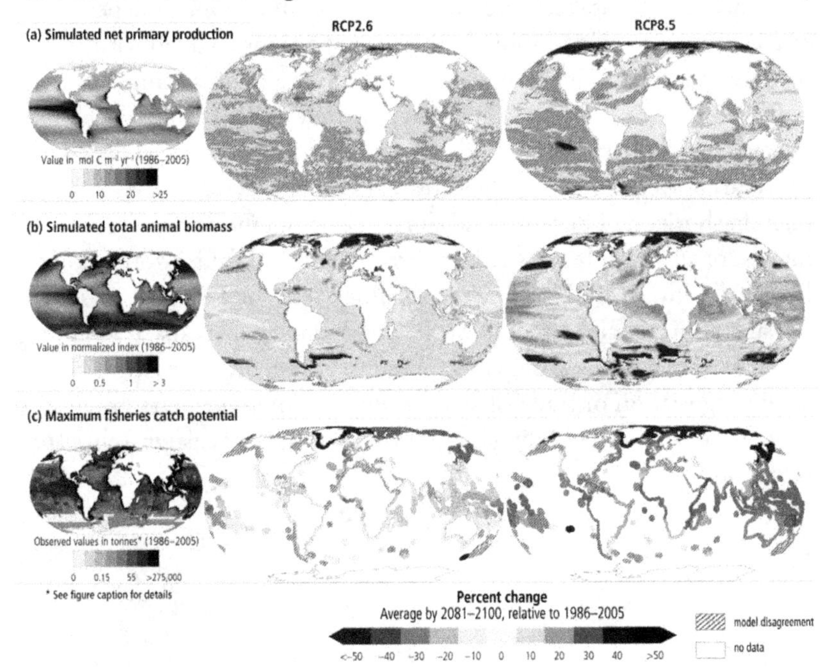

C.2 The far-reaching services and options provided by ocean and cryosphere-related ecosystems can be supported by protection, restoration, precautionary ecosystem-based management of renewable resource use, and the reduction of pollution and other stressors (*high confidence*). Integrated water management (*medium confidence*) and ecosystem-based adaptation (*high confidence*) approaches lower climate risks locally and provide multiple societal benefits. However, ecological, financial, institutional and governance constraints for such actions exist (*high confidence*), and in many contexts ecosystem-based adaptation will only be effective under the lowest levels of warming (*high confidence*).

* * *

C.3 Coastal communities face challenging choices in crafting context-specific and integrated responses to sea level rise that balance costs, benefits and trade-offs of available options and that can be adjusted over time (*high confidence*). All types of options, including protection, accommodation, ecosystem-based adaptation, coastal advance and retreat, wherever possible, can play important roles in such integrated responses (*high confidence*).

C.4 Enabling climate resilience and sustainable development depends critically on urgent and ambitious emissions reductions coupled with coordinated sustained and increasingly ambitious adaptation actions (*very high confidence*). Key enablers for implementing effective responses to climate-related changes in the ocean and cryosphere include intensifying cooperation and coordination among governing authorities across spatial scales and planning horizons. Education and climate literacy, monitoring and forecasting, use of all available knowledge sources, sharing of data, information and knowledge, finance, addressing social vulnerability and equity, and institutional support are also essential. Such investments enable capacity-building, social learning, and participation in context-specific adaptation, as well as the negotiation of trade-offs and realisation of co-benefits in reducing short-term risks and building long-term resilience and sustainability (*high confidence*).

Contributors

Shaul Bassi is professor of English and postcolonial literature at Ca' Foscari University of Venice, Italy, where he directs the Master's Degree in Environmental Humanities. His work has focused on Shakespeare, postcolonial literature, Jewish studies and the environmental and social issues of Venice. His books include *Shakespeare's Italy and Italy's Shakespeare: Place, "Race," Politics* (2016); *The Merchant* in *Venice. Shakespeare in the Ghetto* (edited with Carol Chillington Rutter, 2021); *Turbo Road. Il Kenya, i suoi scrittori, un bambino* (2022).

Abigail Benesh is a licensed attorney currently working for the Animal Legal Defense Fund. Abbey earned her J.D. from the University of Utah's S.J. Quinney College of Law in 2020 with certificates in environmental and natural resources law as well as public interest law. Before joining the Animal Legal Defense Fund, Abbey worked for the Humane Society of Utah, the Humane Society of the United States, and Southern Utah Wilderness Alliance.

Paula Blackett is an Environmental Social Scientist at the National Institute of Water and Atmospheric Research (NIWA) in Aotearoa New Zealand. Her research addresses the human dimensions of environmental change, but she has focused much of her work on the social impacts of climate change. Some of her recent work has explored how serious games can be developed to enhance the uptake of climate change adaptation strategies.

Brenda B. Bowen is a professor of Geology and Geophysics and director of the Global Change and Sustainability Center at the University of Utah in Salt Lake City. Bowen is an interdisciplinary geoscientist whose work focuses on how changing environmental conditions influence landscape evolution. In addition to geologic research and teaching, Bowen works to facilitate interdisciplinary environmental research and education that addresses critical issues related to understanding global change and creating sustainable solutions.

Nathaniel E. Broadhurst is a judicial clerk at the U.S. District Court for the District of Utah. He earned his J.D. degree from the University of Utah S.J. Quinney College of Law with a specialty in environmental law. Before clerking in the federal district court, Broadhurst served as a judicial clerk at the Utah Court of Appeals. Prior to his clerkships, Broadhurst completed an externship at the Environment and Natural Resources Division of the Department of Justice and worked on water rights litigation at the firm of Clyde Snow & Sessions and in the Office of the City Attorney for Salt Lake City.

Nicholas Cradock-Henry is a Senior Scientist at Manaaki Whenua Landcare Research, a Crown Research Institute based in Aotearoa New Zealand. His research focuses on enhancing resilience to global change processes in coupled human-natural systems. As part of this work, he collaborates closely with stakeholders and end-users on targeted and strategic research to enable adaptation to climate change, reduce vulnerability from low-frequency, high-impact events, and enhance science-policy outcomes.

Robin Kundis Craig is the Robert C. Packard Trustee Chair in Law at the University of Southern California Gould School of Law in Los Angeles, where she teaches environmental law subjects, including Ocean & Coastal Law. Craig is the author of *Comparative Ocean Governance: Place-Based Protection in an Era of Climate Change* (Edward Elgar, 2012) and coauthor (with Melinda Harm Benson) of *The End of Sustainability: Resilience and the Future of Environmental Governance in the Anthropocene* (University of Kansas Press, 2017), among other volumes. She has also authored or coauthored twenty-five book chapters and over one hundred articles in both legal and scientific publications, about one-third of which deal directly with ocean issues.

Taylor Cunningham is a graduate assistant at the University of Utah Press. She graduated from the University of Utah in 2019 with a B.S. in marine science/biology, then completed a master's degree in Utah's Environmental Humanities Program working on *Ecologies of Thirst* in Arizona. She has helped conduct research on sharks at One Ocean Diving in Hawaii.

Benjamin Davies is a postdoctoral fellow in data science at Yale University. His research is primarily oriented around questions of population dynamics and mobility of human groups in the past, and how patterning forms in archaeological landscapes. He regularly uses quantitative and computational modeling methods to address these questions in ongoing projects in Africa, Australia, and the Pacific Islands. He is also interested in social simulations more broadly, and how they may be applied to enhance resilience and human adaptation to climate change.

Kathryn K. Davies is a lecturer in the Department of Urban and Environmental Policy and Planning at Tufts University where she focuses on navigating equitable transformations to sustainability using a range of collaborative and participatory methods. Recent projects include developing serious games to address air quality challenges in the Salt Lake Valley and using scenario-planning techniques to improve cumulative effects management in Aotearoa New Zealand's coastal and marine areas. Prior to joining Tufts, Davies worked as a social scientist for the National Institute of Water and Atmospheric Research (NIWA) in Aotearoa New Zealand, as an urban planner in Salt Lake City, and as a research assistant professor in geography at the University of Utah.

Christopher Finlayson is a barrister and arbitrator with Bankside Chambers in Auckland, New Zealand, where he specializes in the Treaty of Waitangi and Public Law. From 2005 to 2019 he served as a Minister of Parliament in New Zealand. In 2008, Finlayson became Attorney-General and Minister for Treaty of Waitangi Negotiations and served in those capacities until 2017, helping to negotiate many settlements with the Maori. In addition, in 2013 he represented New Zealand in the International Court of Justice in a case where Australia sued Japan seeking to stop commercial whaling in the Southern Oceans. He recently published *He Kupu Taurangi: Treaty Settlements in New Zealand* (2021).

Paula Holland is an Environmental Economist at the National Institute of Water and Atmospheric Research (NIWA) in Aotearoa New Zealand. She is a development practitioner with specialist skills and knowledge in natural resource economics for sustainable development and risk management (climate change adaptation and disaster risk management). In economic analysis, her experience covers issues such as cost benefit analysis (e.g., of risk management interventions), economic valuation and costing of policies (e.g., assessment of national disaster risk management action plans). She also has considerable experience in building the capacity of Pacific Island government staff in natural resource economics.

Jeremy B. C. Jackson holds emeritus positions at the Smithsonian Tropical Research Institute and Scripps Institution of Oceanography, where he led the Center for Marine Biodiversity and Conservation. He is also a research associate at the American Museum of Natural History. Jackson studies threats and solutions to human impacts on the environment and the ecology and evolution of tropical seas. He is a member of the United States National Academy of Sciences and the American Academy of Arts and Sciences and has won numerous international prizes and awards. He is the author of more than 170 scientific publications and eleven books, most recently *Breakpoint: Reckoning with America's Environmental Crises* and *Shifting Baselines in Fisheries: Using the Past to Manage the Future* (2018).

Jeffrey Mathes McCarthy is a professor in the Honors College at the University of Utah and director of the Environmental Humanities graduate program. His interdisciplinary scholarship examines the cultural function of nature in precarious social moments. One of his research specialties is artistic representations of the ocean and their significance to climate change. McCarthy's books include *Green Modernism: Nature and the English Novel* (Palgrave-MacMillan, 2015), *Contact: Mountain Climbing and Environmental Thinking* (Nevada, 2008), and *Conrad & Nature* (edited with Lissa Schneider-Rebozo and John Peters) (Routledge, 2019).

Steve Mentz is a professor of English at St. John's University in Queens, NY, where he teaches Shakespeare, literary theory, and the blue humanities with a focus on environmental questions. Widely credited with coining the term "Blue Humanities," his recent books include *Ocean* (2020); *Break Up the Anthropocene* (2019); *Shipwreck Modernity: Ecologies of Globalization, 1550–1719* (2015); and *At the Bottom of Shakespeare's Ocean* (2009). He is also editor or coeditor of six collections, including *The Cultural History of the Sea in the Early Modern Age* (2021); *The Routledge Companion to Marine* *and Maritime Worlds, 1400–1800* (2020); *The Sea in Nineteenth-Century Anglophone Literary Culture* (2017); and *Oceanic New York* (2015).

Thomas Michael Swensen is an assistant professor of ethnic studies at the University of Utah. His research focuses on Native American history, law, the environment, finance, and punk studies. He was born and raised in the Kodiak Archipelago. An original shareholder in the Alaska Native Claims Settlement corporations Koniag and Leisnoi Village, he is also enrolled in the federally recognized Tangirnaq Native Village, aka the Woody Island tribe. With pride, he serves the Alutiiq people on the board of directors of the Koniag education foundation, an organization that promotes the educational goals and economy of the Koniag Alutiiq and their descendants. Swensen holds undergraduate degrees in English, art, and urban planning, Master's degrees in English and ethnic studies, and a PhD in ethnic studies from the University of California, Berkeley.

Tierney M. Thys is a biologist, filmmaker and Research Associate at California Academy of Sciences. Tierney was named a National Geographic Emerging Explorer in 2004. As past Director of Research for Sea Studios Foundation, she helped produce award-winning PBS documentaries *Strange Days on Planet Earth* and *Shape of Life*. As an independent filmmaker, she frequently contributes to TEDed including The Secret Life of Plankton,which earned recognition as a Wildscreen Panda Winner. Thys is cofounder of Around the World in 80 Fabrics—an educational non-profit elevating the voices of maker community working to solve the fast fashion pollution crisis. She holds an AB in Biology from Brown University, a PhD in Zoology from Duke University and serves on the advisory boards of Think Beyond Plastic, the Plastic Pollution Coalition, and the Journal of Animal Sentience.

Index